STUDIES IN MODERN THERMODYNAMICS 2

PRINCIPLES OF THERMODYNAMICS

STUDIES IN MODERN THERMODYNAMICS

1. Biochemical Thermodynamics, edited by M.N. Jones

STUDIES IN MODERN THERMODYNAMICS 2

PRINCIPLES OF THERMODYNAMICS

JAMES A. BEATTIE and IRWIN OPPENHEIM
Department of Chemistry, Massachusetts Institute of Technology, Cambridge, M.A. 02139, U.S.A.

ELSEVIER SCIENTIFIC PUBLISHING COMPANY
Amsterdam — Oxford — New York   1979

ELSEVIER SCIENTIFIC PUBLISHING COMPANY
335 Jan van Galenstraat
P.O. Box 211, 1000 AE Amsterdam, The Netherlands

Distributors for the United States and Canada:

ELSEVIER/NORTH-HOLLAND INC.
52, Vanderbilt Avenue
New York, N.Y. 10017

Library of Congress Cataloging in Publication Data
Beattie, James A   1895-
  Principles of thermodynamics.

  (Studies in modern thermodynamics ; 2)
  Includes index.
  1. Thermodynamics. I. Oppenheim, Irwin,
joint author. II. Title. III. Series.
QC311.B352   536'.7   79-23755
ISBN 0-444-41806-7

ISBN 0-444-41806-7 (Vol. 2)
ISBN 0-444-41762-1 (Series)

© Elsevier Scientific Publishing Company, 1979
All rights reserved. No part of this publication may be reproduced, stored in a retrieval system or transmitted in any form or by any means, electronic, mechanical, photocopying, recording or otherwise, without the prior written permission of the publisher,
Elsevier Scientific Publishing Company, P.O. Box 330, 1000 AH Amsterdam, The Netherlands

Printed in The Netherlands

PREFACE

The main purpose of this book is to present a rigorous and logical discussion of the fundamentals of thermodynamics. We have been particularly careful to specify and discuss the postulates and laws which form the bases for the science of thermodynamics. We have attempted to give a clear and full account of the first, second and third laws of thermodynamics. Some applications of these laws are treated in detail; others are not considered here.

Partial molar properties are treated fully and emphasis has been placed on the importance of writing changes in state for thermodynamic calculations. The Gibbs criteria for equilibrium and stability are discussed in great detail.

The inspiration for this work comes from the classic work of Gibbs and from discussions with our teachers and colleagues.

In his lectures on thermodynamics, Professor Henry B. Phillips has given the clearest and most complete treatment of the second law which we have found anywhere. With his consent we have used some of these results in our Chapter 8.

Professor Louis J. Gillespie was always ready to give his time and knowledge to solve problems, especially those arising from the *Thermodynamics* of J. Willard Gibbs.

Professor Clark C. Stephenson has helped us to understand certain features of the third law of thermodynamics.

Professor John G. Kirkwood introduced one of us (I.O.) to the beauty and logical structure of thermodynamics.

This book would not have appeared without the aid and superb typing abilities of Doris L. Currier, Frances M. Doherty, Virginia Prescott and Vera M. Spanos who have participated in its preparation at various stages in its development.

James A. Beattie
Irwin Oppenheim

Cambridge, Massachusetts
February 1979

Dedicated to...

Doris B. Beattie and Bernice R. Buresh

CONTENTS

PREFACE V

Chapter 1

DEFINITIONS 1

1.1 System, boundary, surroundings, medium 1
1.2 Closed and open systems 2
1.3 Isolated system 2
1.4 Thermodynamic properties of a system 2
1.5 State of thermodynamic equilibrium 2
1.6 State variables, extensive and intensive properties, independent and dependent variables 5
1.7 Specification of the state of a simple system 5
1.8 Change in state, chemical reaction 6
1.9 Path, process, cycle 8
1.10 Quasistatic process 8
1.11 Reversible and irreversible processes, cycles 9
1.12 Phase 9
1.13 Homogeneous and heterogeneous systems 10
1.14 States of aggregation 10
1.15 Pure substances and mixtures 12
1.16 Elements and compounds 13
1.17 Extra-thermodynamic laws 13
1.18 Chemical units of mass 14
1.19 Species and components 14
1.20 Composition of a system expressed in terms of its species and its components ... 17
References 23

Chapter 2

PRESSURE AND TEMPERATURE 24

2.1 Pressure 24
2.2 Temperature 26
2.3 Thermal contact and thermal equilibrium 26
2.4 Zeroth law of thermodynamics 27
2.5 Definition of temperature 28
2.6 Thermometry 28
2.7 Temperature scales 30
2.8 Thermometers 35
2.9 Perfect gas temperature scales 41
2.10 Equation of state of a perfect gas 42
2.11 Virial equations of state of a real gas 44
2.12 Thermostats 46
References 46

Chapter 3

WORK AND HEAT 48

- 3.1 Work 49
- 3.2 Measure and units of work 50
- 3.3 Kinds of work 51
- 3.4 Expansion work 52
- 3.5 A reversible, isothermal process 54
- 3.6 A two-step, reversible process 56
- 3.7 A reversible cycle 57
- 3.8 Two fast, irreversible processes 57
- 3.9 A quasistatic, irreversible process 59
- 3.10 Work and location of the boundary in reversible and irreversible processes .. 61
- 3.11 Equations for the expansion work of several processes 63
  - 3.11A A constant-volume process 63
  - 3.11B A quasistatic process under a constant applied pressure .. 63
  - 3.11C A reversible, isothermal expansion of a gas 64
  - 3.11D A reversible, polytropic expansion of a gas 65
  - 3.11E A reversible, isothermal expansion of a condensed phase .. 65
  - 3.11F A steady-flow process 66
- 3.12 Heat 67
- 3.13 Measure and units of heat 68
- 3.14 Definition of heat capacity 70
- 3.15 Constant-pressure and constant-volume heat capacities 72
- 3.16 Saturation heat capacities and latent heat of phase change ... 75
- 3.17 Five thermal experiments 77
- 3.18 Calorimetry 79
- 3.19 Bomb calorimeter 80
- 3.20 Open calorimeter 83
- 3.21 Low temperature heat capacity calorimeter 83
- 3.22 High temperature heat capacity calorimeter 85
- 3.23 Calorimeters for measurement of saturation thermal properties . 87
- 3.24 Flow calorimeter 87
- 3.25 Heat, work, and the location of the boundary 89
- 3.26 Review of the properties of work and heat 92
- 3.27 Quantities that determine the behavior of a simple system 93
- 3.28 Calorimetric coefficients 94
- 3.29 Thermometric coefficients 95
- 3.30 Elastic coefficients 96

References 97

Chapter 4

THE FIRST LAW OF THERMODYNAMICS 99

- 4.1 Historical introduction 99
- 4.2 Mechanical equivalent of heat 103
- 4.3 First law of thermodynamics 105
- 4.4 Definition of the energy of a closed system 105
- 4.5 Law of the conservation of energy 107
- 4.6 Measure and units of energy 108
- 4.7 Review of the properties of energy 108
- 4.8 Definition of the enthalpy of a closed system 109
- 4.9 Some first-law equations for closed systems 110
- 4.10 Tabulation of some first-law equations for closed systems 112

References 114

Chapter 5

APPLICATIONS OF THE FIRST LAW TO PHYSICAL CHANGES 115

5.1 Effect of volume or pressure and temperature on energy and enthalpy . . . 115
5.2 Heat capacities 117
5.3 Latent heats 118
5.4 Free expansion experiment, Joule effect 119
5.5 Porous-plug experiment, Joule—Thomson effect 122
5.6 Relation of the Joule coefficient to the Joule—Thomson coefficient 126
5.7 Isothermal Joule—Thomson experiment 127
5.8 Perfect gases 128
References 130

Chapter 6

THERMOCHEMISTRY 131

6.1 Enthalpy, energy, and heats of reaction 131
6.2 Standard states 133
    6.2A Standard states of pure substances 133
    6.2B Standard states of substances in solution 134
6.3 Calorimetric determination of enthalpies of reaction 134
6.4 Standard enthalpy of formation 136
6.5 Calculation of enthalpies of reaction from tables 137
6.6 Effect of pressure and temperature on enthalpies of reaction 138
    6.6A When no reactant or product of a change in state undergoes a change in aggregation state in the pressure and temperature interval under consideration 138
    6.6B When one or more of the reactants and products of a change in state undergo a change in aggregation state in the temperature and pressure range under consideration 139
6.7 Maximum flame and explosion temperatures 141
6.8 Enthalpies of reaction in solution 142
References 142

Chapter 7

PARTIAL MOLAR PROPERTIES 143

7.1 Definition of a partial molar property 143
7.2 Relations among partial molar properties 144
7.3 Effect of pressure and temperature on partial molar properties 145
7.4 Relations for binary solutions 145
7.5 Determination of partial molar properties 147
    7.5A From the definition 147
    7.5B Method of intercepts 148
7.6 Integral enthalpy of solution 149
7.7 Integral enthalpy of dilution 150
7.8 Differential enthalpy of dilution 150
7.9 Differential enthalpy of solution 150
7.10 Differential enthalpy of transfer of solvent and solute 151
7.11 Effect of pressure and temperature on partial molar enthalpies 152
7.12 Calculation of the partial molar enthalpy of one component from that of the other 152
7.13 The quantities $\bar{h}_1$, $\bar{h}_2$, and ϕ_{h_2} 152
7.14 Standard states and standard molar enthalpies 153

	7.15	Enthalpies of reaction in solution	155
References			156

Chapter 8

THE SECOND LAW OF THERMODYNAMICS ... 157

	8.1	Historical introduction	157
	8.2	Heat engine	159
	8.3	Heat reservoir	159
	8.4	Second law of thermodynamics	159
	8.5	The second-law equation	160
	8.6	The second-law equation for reversible cycles	160
	8.7	The second-law equation for irreversible cycles	161
	8.8	The Carnot cycle	162
	8.9	Efficiency of the Carnot cycle	164
	8.10	Some corollaries of the second law	164
	8.11	Thermodynamic temperature scales	167
	8.12	The Carnot function and Kelvin's temperature scales	169
	8.13	The Clausius inequality	170
	8.14	Definition of the entropy of a closed system	172
	8.15	Relation of entropy changes to heat effects attending processes	174
	8.16	Principle of the increase in entropy	174
	8.17	Increase in entropy in an irreversible process	175
	8.18	Review of the properties of entropy	175
	8.19	First- and second-law equation for a closed system	176
References			176

Chapter 9

APPLICATIONS OF THE SECOND LAW ... 178

	9.1	Perfect gas temperature scale	178
	9.2	Variations of entropy with temperature	179
	9.3	Variations of entropy with volume and pressure	180
	9.4	Relations for reversible, adiabatic processes	180
	9.5	Relations for the calorimetric coefficients	181
	9.6	Changes in entropy attending finite changes in state	181
	9.7	Isothermal variations of C_V and C_p	182
	9.8	Relations among heat capacities	182
	9.9	Measurement of heat capacity	183
	9.10	Variation of heat capacity with temperature	184
	9.11	Entropy of phase transformation	184
	9.12	Joule and Joule—Thomson coefficients	184
	9.13	Correction of the indications of gas thermometers to the Kelvin scale	185
	9.14	Perfect gases	186
	9.15	Reversible, adiabatic processes in perfect gases	188
	9.16	Reversible, polytropic processes in perfect gases	189
	9.17	Mixtures of perfect gases	189
	9.18	Entropy of mixing perfect gases	190
	9.19	The Gibbs paradox	192
	9.20	Standard molar entropy of a pure substance	193
	9.21	Entropy changes attending chemical changes in state	194
References			196

Chapter 10

WORK CONTENT AND FREE ENERGY ... 197

10.1	Definition of work content and free energy	197
10.2	Relation of ΔA to the work of certain processes	198
10.3	Relation of ΔG to the work of certain processes	201
10.4	Available energy	203
10.5	Steady-flow processes	203
10.6	Variations of work content and free energy with volume or pressure and temperature.	204
10.7	The chemical potential	205
10.8	Changes in work content and free energy attending finite changes in state	207
10.9	Alternative equations for ΔA and ΔG attending finite changes in state	208
10.10	Integrated equations for the variation of free energy with temperature at constant pressure	210
10.11	Phase diagrams	210
10.12	The Clapeyron equation	212
10.13	Effect of temperature on enthalpy of phase transformation in a univariant system.	214
10.14	Sublimation and vaporization curves	214
10.15	Fusion and transition curves	215
10.16	Effect of pressure on vapor pressure	216
10.17	Effect of temperature on the vapor pressure of a phase at constant pressure.	217
10.18	Perfect gases	218
10.19	Standard molar enthalpy and free energy of a pure substance	219
10.20	Free energy changes attending chemical changes in state	222
10.21	Relations of ΔG and ΔA to chemical changes in state	223
10.22	Equations for h^0, s^0, and g^0 referred to their values at 1 atmosphere and 0 K.	225
10.23	Thermodynamic relations of systems producing non-expansion work	228
References		231

Chapter 11

THE THIRD LAW OF THERMODYNAMICS 232

11.1	Historical introduction	233
11.2	Principle of the unattainability of the absolute zero	239
11.3	Entropy of mixing	239
11.4	The third law of thermodynamics	240
11.5	Some consequences of the third law	241
11.6	Phase diagrams at low temperatures	243
11.7	Heat capacities at low temperatures	246
11.8	Einstein's heat-capacity equation	246
11.9	Debye's heat-capacity equation	247
11.10	Electronic heat capacity of a metal	249
11.11	Difference between c_p and c_V of a solid.	249
11.12	Heat capacity c_p of a solid at low temperatures.	249
11.13	Calorimetric entropies	250
11.14	Spectroscopic entropies	252
11.15	Entropies of polymorphic forms at 0 K	253
11.16	Practical difficulties in determining spectroscopic entropies: internal rotation.	254
11.17	Practical difficulties in determining calorimetric entropies	255
	11.17A Frozen-in disorder at 0 K	256
	11.17B Incorrect extrapolation to 0 K	257
11.18	Entropies of solid solutions at 0 K	257

11.19 Entropies of liquid solutions of ^{4}He and ^{3}He below 1 K 258
11.20 Entropies of glasses at 0 K. 258
11.21 Entropies of solids with configurational disorder at 0 K 261
11.22 Rotational heat capacity of hydrogen 264
11.23 Paramagnetic substances. 268
11.24 Contributions to entropy not included in tabulated values 269
      11.24A Entropy of isotope mixing. 269
      11.24B Nuclear spin entropy. 270
References 270

Chapter 12

CRITERIA OF EQUILIBRIUM AND STABILITY 275

12.1 Entropy and energy criteria of equilibrium and stability 275
12.2 Possible variations 276
12.3 Equivalence of the entropy and energy criteria of equilibrium 276
12.4 Kinds of equilibrium states 277
      12.4A Stable equilibrium 278
      12.4B Neutral equilibrium 279
      12.4C Unstable equilibrium 279
      12.4D Metastable equilibrium 279
12.5 Sufficiency of the entropy and energy criteria of equilibrium 280
      12.5A Stable equilibrium 281
      12.5B Neutral equilibrium 281
      12.5C Unstable equilibrium 282
      12.5D Metastable equilibrium 282
12.6 Necessity of the entropy and energy criteria of equilibrium. 283
12.7 Uniformity of temperature in an isolated system in equilibrium 284
      12.7A Thermal equilibrium in a heterogeneous system 284
      12.7B Effect of non-uniformity of temperature in the varied state 284
12.8 Uniformity of pressure in a simple, isolated system in equilibrium 285
12.9 Three modifications of the energy criterion of equilibrium 286
      12.9A A system of uniform temperature 286
      12.9B A simple system of uniform pressure 287
      12.9C A simple system of uniform temperature and pressure. 287
12.10 Two work-content criteria of equilibrium 288
      12.10A A work-content criterion of equilibrium. 288
      12.10B A modification of the work-content criterion of equilibrium . . . 289
12.11 Two enthalpy criteria of equilibrium. 290
      12.11A An enthalpy criterion of equilibrium 291
      12.11B A modification of the enthalpy criterion of equilibrium 291
12.12 Two free-energy criteria of equilibrium 292
      12.12A A free-energy criterion of equilibrium 292
      12.12B A modification of the free-energy criterion of equilibrium 293
12.13 Résumé of criteria of equilibrium. 294
References 294

Chapter 13

OPEN SYSTEMS 296

13.1 The first- and second-law equation for open, simple phases 296
13.2 Components and species of a phase. 297

13.3	Chemical potentials of components and species.		298
	13.3A	The chemical potential of a component of a phase is independent of the choice of components	298
	13.3B	The chemical potential of a constituent of a phase when considered to be a species is equal to its chemical potential when considered to be a component	299
13.4	Actual and possible components of a phase		299
13.5	Components of a closed, heterogeneous system.		300
13.6	Examples of open systems.		300
	13.6A	Physical processes	300
	13.6B	A heterogeneous chemical reaction	301
	13.6C	A homogeneous chemical reaction.	301
13.7	Chemical potentials in a phase at equilibrium		303
13.8	The summation $\Sigma_i \mu_i dn_i$		303
13.9	The first- and second-law equation for open phases producing non-expansion work		304
13.10	The first-law equation for open systems.		305
13.11	Applications of the first-law equation for open systems		307
	13.11A	Relation between ΔE and h_B	308
	13.11B	Flow of a gas into a system	308
	13.11C	Flow of a gas out of a system	309
13.12	The first-law steady-flow equation		310
13.13	Applications of the first-law steady-flow equation to physical processes.		311
	13.13A	Adiabatic steady-flow processes	311
	13.13B	Steady-flow heating and cooling processes	312
	13.13C	The isothermal porous-plug experiment	312
13.14	Applications of the first-law steady-flow equation to chemical processes		312
13.15	The second-law equation for open systems.		313
13.16	An application of the second-law equation for open systems		314
13.17	The second-law steady-flow equation		316
13.18	The combined first- and second law equation for open, simple phases		316
13.19	Fundamental equations		317
References			320
INDEX OF NAMES			321
SUBJECT INDEX			325

Chapter 1

DEFINITIONS

Thermodynamics is a phenomenological discipline dealing with the interactions of systems with their surroundings, and with the description of and the relations among the macroscopic properties of systems. The values of these interactions and gross properties are determined by macroscopic measuring devices, all of which have the following properties: (1) their spacial resolution is such that they make measurements over regions of space which are very large compared to molecular dimensions (10^{-10} m) and therefore contain many molecules; (2) their time constants are such that they make measurements over time intervals that are very long compared to molecular times (10^{-12} sec); and (3) their inherent errors are very large compared to the contribution of individual molecules to the property under consideration.

Thermodynamics is a logical deductive science based on a small number of postulates called laws. These laws summarize accurately our knowledge of the macroscopic behavior of systems in certain respects, but make no explicit reference to the structure of matter. The final results of thermodynamics are equations of great generality. The actual magnitudes of the quantities appearing in these equations are not predicted; they must be derived from direct measurement, or from the results of statistical thermodynamics which involves molecular theory explicitly.

1.1. System, boundary, surroundings, medium. A thermodynamic system is the part of the physical universe under investigation. It usually contains one or more chemical substances, but it may consist only of one or more of the following: radiation, an electric field, a magnetic field.

The system is confined by a boundary that separates it from the rest of the physical universe, called the surroundings. The boundary usually consists of the inner walls of the vessel containing the system. However, it may be composed entirely or in part of mathematical surfaces. For example, the boundary of a fixed mass of fluid flowing in a horizontal pipe is, in part, the inner surface of the pipe and, in part, two imaginary vertical planes that separate the mass under consideration from the fluid ahead of and that behind it. The description of a thermodynamic system necessarily includes certain information concerning its boundary. The precise location of the boundary and the constraints imposed by the boundary on the interactions of the system with its surroundings are matters of great importance and must be given. Unless otherwise stated we shall always suppose that the boundary does not react chemically with any of the substances composing the system.

The part of the surroundings in the immediate neighborhood of a system is sometimes called the medium. It is large when compared with the system but small compared with the surroundings. The system and medium constitute a composite system which is usually considered to be enclosed in an exterior boundary.

1.2. Closed and open systems. A system is called closed when no mass crosses its boundary, and open when mass flows across its boundary.

1.3. Isolated system. An isolated system has no observable interaction with its surroundings. The boundary of an isolated system shields it from all external influences so that events occurring in the system are not affected by changes taking place in the surroundings. Such a boundary is an abstraction. However, real boundaries approaching this ideal behavior more or less closely are known for all types of interaction except gravitation.

1.4. Thermodynamic properties of a system. The properties of a system that are of interest in thermodynamic discussions are those physical attributes that are perceived directly by the senses, or are made perceptible by certain macroscopic experimental methods of investigation. Thus, thermodynamics deals with macroscopic rather than microscopic properties. We recognize two classes of properties: (1) numerical properties, such as pressure and volume, to which a numerical value can be assigned by direct or indirect comparison with a standard, or by combining certain measurements in accordance with definite rules; and (2) non-numerical properties, such as the kinds of substances composing a system and the states of aggregation of its parts.

1.5. State of thermodynamic equilibrium. A system is in a state of thermodynamic equilibrium, or more simply in an equilibrium state, when each of its thermodynamic properties is time independent and when there are no fluxes within the system or across its boundary. For example, there must not be any transfer of matter or electricity from one place to another, within or across the boundary.

In an equilibrium state "caused by the balance of the active tendencies" [1] of a system, all agents tending to produce change in the properties of the system are so balanced that an infinitesimal variation in the value of each agent, either in the positive or negative direction, is sufficient to produce a corresponding change in the state of the system; and restoration of this agent to its former value causes the system to return to its initial state. Hence, such a state of thermodynamic equilibrium can be reached from the two opposite sides with respect to each agent tending to produce change in the system. Of course, it is possible to pass from one equilibrium state to another by changing the external constraints on the system. We shall call this condition a state of dynamic equilibrium.

In a system as initially constituted several different processes may be taking place simultaneously at quite different rates. After a while the system may apparently have reached a state of equilibrium, at least with respect to variations in the values of the properties being studied. However, processes leading to entirely different equilibrium states may be occurring so slowly that they do not produce a detectable displacement in the supposed equilibrium state because (1) the investigator cannot observe the system for an indefinitely long time, and (2) all of his measuring instruments are subject to inherent errors that may mask the effect of a small change in the properties of the system. Examples of processes which may take place quite slowly are: (1) the sliding of one solid over another when they are in contact; (2) the diffusion of a constituent of a solid or of a viscous liquid solution; (3) a chemical or nuclear reaction; (4) the transformation of a solid from one crystalline form to another; (5) the relief of strains in a solid; and (6) the growth of crystal size in a solid.

Insistence on a strict interpretation of the definition of equilibrium would rule out the application of thermodynamics to practically all states of real systems. Fortunately thermodynamic results can be applied to measurements on systems that are not in equilibrium with respect to all processes occurring therein. Of course, we must know from an experimental study of the given system, or from experience with similar systems, which processes occur so slowly that their effects on the given state may safely be ignored. In the study of the properties of a system in a supposed equilibrium state, the experimental investigator considers the effects of all processes which, he believes, have a perceptible effect on the properties under investigation; and he proves experimentally that this state can be approached from the two opposite sides by each such process. He neglects the effects of very slow processes on the state of the system. The success, or failure, of this procedure when applied to a particular equilibrium state is determined by the agreement, or lack of agreement, between thermodynamic deduction and experiment.

In his discussion of equilibrium states of a system, Gibbs invoked passive forces that prevent certain changes in state no matter how the initial state or external conditions are varied, within limits which, however, allow finite variations in the values of these forces [1]. Hence we can apply the definition of an equilibrium state to a system without regard to very slow processes by associating a passive force with each such agent tending to produce change.

In the study of the properties of systems in equilibrium states the careful experimental investigator specifies: (1) the system and the values of a sufficient number of its properties to fix its state with the requisite accuracy for the proposed study; (2) the location of the boundary and its effect on the interactions between the system and its surroundings; and (3) the variations in state with respect to which the equilibrium is being discussed. Furthermore, he proves that the given state can be approached from the two opposite sides with respect to each agent tending to produce the variations in state mentioned in (3) above; and he checks his measurements by means of equations derived from thermodynamics.

The measured pressure of a closed system composed of liquid nitric oxide (NO) and its vapor, enclosed in a rigid, thermally conducting boundary, attains a constant value shortly after it has been immersed in a thermostat maintained at a fixed temperature. The same value of pressure is observed whether we approach the given temperature from a higher or from a lower value. Yet we have reason to believe from the extrapolation of other data that nitric oxide is decomposing into nitrogen and oxygen, but at an exceedingly small rate. We can either consider that the decomposition of NO is prevented by a passive force, or that the properties of NO are idealized to the extent that no decomposition takes place in the temperature range under consideration. Under either assumption the system attains an equilibrium state with respect to the transfer of NO from one phase to the other and to the establishment of uniform pressure and temperature throughout the system shortly after it has been placed in the thermostat. Measurements of vapor pressures over a range of temperatures can be correlated with other equilibrium measurements on this system by a thermodynamic equation not involving the dissociation of NO.

In contrast to the example just cited, the equilibrium between nitrogen tetroxide (N_2O_4) and nitrogen dioxide (NO_2) is established rapidly from each side. We could not correlate measurements of vapor pressures in this system by means of thermodynamic equations if we supposed it to be composed of an idealized N_2O_4 which did not dissociate under conditions where appreciable quantities of each species are actually present.

Usually we can easily distinguish between a process whose effect on the state of a system must be taken into account and one whose effect can be ignored. Occasionally the rate of a process lies so near the limit at which its effect can be neglected that a more detailed study of the system is required. Such is the case for the two systems now to be considered. Each is used to define a fixed temperature on the International Practical Temperature Scale [2] of 1948, which will be discussed in Chapter 2. Originally it was believed that the temperatures of equilibrium of ice, liquid water, and water vapor, 0.01°C (Int.), and that of liquid sulfur and its vapor under a pressure of one standard atmosphere, 444.6°C (Int.), were established rather quickly, in a matter of several hours or, at most, in a day. And so they were, within the accuracy of measurement of temperature at that time. Later investigations made under more carefully controlled conditions with more sensitive thermometers showed [3] that: (1) the triple point of water increased on the average by 1.7×10^{-4} °C the first day, 0.3×10^{-4} °C the second day, and 0.1×10^{-4} °C per day for about seven more days, thereafter remaining constant to within 0.1×10^{-4} °C; and (2) the sulfur point usually dropped by a total of 10×10^{-3} °C over a period of one to ten days (depending on the amount of trace impurities in the sample), after which it remained constant to 1×10^{-3} °C. The change of the triple point of water has been attributed to the growth in crystal size of ice (thereby reducing the effect of capillarity) and to the relief of strains in the crystal; and the change in the boiling point of sulfur has been attributed to the slowness of the attainment of equilibrium among the polymorphs of sulfur in the liquid and perhaps in the vapor. Further study of these systems would be necessary if we should desire to define the temperature scale more accurately than 0.1×10^{-4} °C at the triple point of water, and 1×10^{-3} °C at the sulfur point.

1.6. State variables, extensive and intensive properties, independent and dependent variables. Each numerical property of a system is called a state variable.

Consider a group of λ identical systems B. The numerical value of an extensive property F of the composite system λB is λ times the corresponding value of F for each system B

$$F(\lambda B) = \lambda F(B) \tag{1.1}$$

The numerical value of an intensive property f of the system (λB) is equal to the value of f for each system B

$$f(\lambda B) = f(B) \tag{1.2}$$

Volume and mass are extensive properties of a system; pressure and temperature are intensive properties.

Any set of state variables, all of whose numerical values must be specified to determine the state of a system, constitutes a set of independent variables. The remaining state variables are called dependent variables.

1.7. Specification of the state of a simple system. The state of a system is determined with sufficient accuracy for most thermodynamic studies by the specification of the values of a relatively small number of properties. A simple system is defined as one whose state is uninfluenced by the following effects: (1) external gravitational, electric, and magnetic field intensities; (2) shape (capillary tensions); (3) distortion of solid crystals [4]; and (4) variations of the isotopic compositions of the constituent substances from their normal values. Fortunately, most of the systems of thermodynamic interest may be regarded as simple systems for all ordinary laboratory conditions and procedures.

In general, we fix the state of a homogeneous simple system composed of one substance by specifying the substance, mass, aggregation state, pressure, and temperature, for example, $2H_2$(g, 1 atm, 25°C). Here the symbol H_2 designates not only the substance hydrogen, but also one formula weight expressed in some unit of mass, such as the gram or the pound.

If the system is a homogeneous mixture of two or more substances (a solution) we must give the composition in addition to the quantities specified above. For example, $HCl \cdot 5.551H_2O$ (l, 1 atm, 25°C) designates a solution of one formula weight of HCl in 5.551 formula weights of H_2O at a given pressure and temperature. In many thermodynamic studies we are interested in only one of the constituents of a solution, for example the HCl in the above system. In this case we would write HCl (0.1 wf, 1 atm, 25°C). This designates the substance HCl in the dissolved state, in a sufficient amount of solution of HCl in water of 0.1 weight formal concentration (0.1 formula weights of HCl in 1 kg of water) to contain one formula weight of HCl. The solvent is always understood to be liquid water unless otherwise stated.

We cannot apply thermodynamics to a system in an equilibrium state unless we have specified the state with sufficient exactness to fix the properties of interest with the required accuracy.

The following proposition has been amply confirmed by experiment. We shall call it a thermodynamic postulate. A simple homogeneous system can exist in only a very small number of equilibrium states when its pressure, density, and composition (expressed in units of mass) are fixed. In general, this number of states is unity. There are a few exceptions where two states can exist over limited ranges of pressure and density. For example, liquid water under a pressure of one atmosphere has a maximum density at about 4°C, so that in the neighborhood of this temperature there are two states having the same pressure, density, and mass: one at a temperature above and one at a temperature below 4°C.

1.8. Change in state, chemical reaction. Let the external constraints acting on a system in a fixed initial state be varied so that the system undergoes a change in its state to some final fixed value. The change in state of the system is completely defined when its initial and its final states are specified. For example, we write

$$H_2O \text{ (l, 1 atm, 100°C)} = H_2O \text{ (g, 2 atm, 150°C)} \tag{1.3}$$

to indicate that the initial state of the system is one formula weight of liquid H_2O at 1 atm and 100°C, and its final state is one formula weight of gaseous H_2O at 2 atm and 150°C. The method of accomplishing this result does not appear in the change in state. If we desire to impart this additional information, we must describe the process, which is defined in Section 1.9.

The importance of the change in state arises from the fact that all thermodynamic quantities concerned with any change in a system depend in part on the change in state; and the increments of many of these quantities depend only on the change in state, not on the process. We have called the latter quantities state variables. The first step in the solution of any problem in thermodynamics concerned with change is to write the change in state. Unless this can be done, the results of thermodynamics cannot be applied to the change under investigation.

We frequently associate with the notation H_2O (l, 1 atm, 100°C) of eqn. (1.3) any extensive property of this system, for example, its volume. We can then specify, next to the right-hand side of the equation, the increase in this property accompanying the change in state; for example $\Delta V = a$ m^3, where a is the volume of the system in its final state minus its volume in its initial state.

Changes in state may be added and subtracted in the same manner as the accompanying changes in volume. For example, the change in state in eqn. (1.3) may be regarded as the sum of the following three changes:

$$H_2O \text{ (l)} = H_2O \text{ (g)} \quad (1 \text{ atm}, 100°C) \tag{1.4}$$

$$H_2O \text{ (g, } 100°C) = H_2O \text{ (g, } 150°C) \quad (1 \text{ atm}) \tag{1.5}$$

$$H_2O \text{ (g, 1 atm)} = H_2O \text{ (g, 2 atm)} \quad (150°C) \tag{1.6}$$

Here a property written to the right of an equation applies to each term, but does not imply that this property is constant during the change.

When writing a change in state involving a chemical reaction, we must have clearly in mind the distinction between a chemical equation and a change in state. A chemical equation expresses only the relative amounts of the substances taking part in the reaction. For example, the equation

$$3H_2 + N_2 = 2NH_3 \tag{1.7}$$

says that whenever NH_3 is produced from H_2 and N_2, the relative numbers of formula weights of these substances involved are in the proportion

$$H_2 : N_2 : NH_3 = 3 : 1 : 2 \tag{1.8}$$

Equation (1.7) does not imply a particular experiment in which two formula weights of NH_3 are formed. In theoretical thermodynamic discussions, we frequently write eqn. (1.7) in the form

$$-3H_2 - 1N_2 + 2NH_3 = 0 \tag{1.9}$$

and represent it by the notation

$$\sum_{1\ i}^{3} \nu_i B_i = 0 \tag{1.10}$$

Here ν_i is the stoichiometric coefficient of B_i in the chemical equation; it has a negative value for a reactant and a positive value for a product. Thus, if we identify B_1, B_2, and B_3 with H_2, N_2, and NH_3, respectively, we find

$$\nu_1 = -3, \nu_2 = -1, \nu_3 = +2 \tag{1.11}$$

In writing chemical equations, we enter the correct molecular formula (see Section 1.17) of the substances involved, and we frequently indicate the aggregation state (see Section 1.14) of each substance, but we do not give all of the information necessary to fix the state of each.

For comparison purposes we shall now write a chemical change in state involving the chemical equation (1.7):

$$0.15\ H_2\ (g) + 0.05\ N_2\ (g) = 0.1\ NH_3\ (g)\ (1\ atm, 400°C) \tag{1.12}$$

or

$$-0.15\ H_2\ (g) - 0.05\ N_2\ (g) + 0.1\ NH_3\ (g) = 0\ (1\ atm, 400°C) \tag{1.13}$$

The initial state of this system is 0.15 formula weights of pure H_2 gas and 0.05 formula weights of pure N_2 gas, each at 1 atm and 400°C, and the final state is 0.1 formula weights of pure NH_3 gas at 1 atm and 400°C. Thus the exact amount of NH_3 formed is designated, as well as the state of each substance entering into the reaction. Equation (1.13) can be written in the form

$$\sum_{i}^{3} \nu_i B_i(g) = 0\ (1\ atm, 400°C) \tag{1.14}$$

1.9. Path, process, cycle. The path of a change in state is defined by giving the initial state, the sequence of intermediate states in the order traversed by the system, and the final state. For example, the change in state

$$N_2\ (g, 10\ atm) = N_2\ (g, 1\ atm)\ (25°C) \tag{1.15}$$

may be brought about by a slow isothermal expansion of the gas along a path given as $p = p(V)$ where $p(V)$ is a known function of volume, or given in the form of a plot of p against V. Interpreted strictly, the definition of path requires the change to take place at an infinitesimal rate; otherwise a path cannot be identified, since gradients and fluxes are produced in the system, and the state of the system is not determined except at the beginning and the end of the change. Evidently there are an infinite number of paths connecting a given initial and a given final state of a system, corresponding to the various ways in which the change can be accomplished through a succession of equilibrium states. Furthermore, there are many ways a particular change can be brought about by processes which proceed so rapidly that no path exists. Frequently we speak, with some loss of accuracy, of a path of a change in state that proceeds at a finite rate.

Any series of events involving a change in the properties of a system is called a process. The description of a process consists in giving the change in state of the system, the path of this change if it exists, and an account of the interactions of the system with its surroundings.

A cycle is a process which, after having produced an initial change in the properties of a system, returns the system to its initial state.

1.10. Quasistatic process. A process which proceeds at an infinitely slow rate is called quasistatic. Hence during such a process the system is infinitesimally near to a state of thermodynamic equilibrium at all times. We can always associate a path with a quasistatic process, which can be given in the form of a functional relation among state variables or exhibited in a graph. As so often happens in thermodynamics, we do not interpret the terms "infinitesimally slow rate" and "infinitesimally near to a state of thermodynamic equilibrium" strictly. Thus we may, and do, speak of the quasistatic expansion of nitric oxide gas at a constant temperature, without taking into account its dissociation into its elements.

1.11. Reversible and irreversible processes, cycles. A process is called reversible if, after its conclusion, the system may be restored to its initial state along a path differing infinitesimally from that of the direct process, and the surroundings restored to their initial condition except for alterations whose ratios to the corresponding alterations produced in the direct path are infinitesimals. A reversible process is always quasistatic. During a reversible process the active tendencies of the system must be so balanced internally and externally that an infinitesimal variation in an external agency can start the reversal of the process at any stage. For example, among the conditions which must be fulfilled in the isothermal reversible expansion of a fixed mass of a gas, contained in a cylinder fitted with a moveable piston, is the following: the difference between the force exerted by the gas on the piston and the externally applied force must at all times be infinitesimal, whatever the direction of motion of the piston.

A process which does not fulfill all of the conditions stated in the last paragraph is irreversible. Such a process may be quasistatic, or it may proceed at a finite rate. During an irreversible process there is a finite imbalance in the agencies tending to produce the change. This lack of balance can occur in the system, or in its surroundings, or in both. We employ such terms as friction, resistance, turbulence, and hysteresis in connection with irreversible processes.

A reversible cycle is composed entirely of reversible processes.

1.12. Phase. The term phase is used in thermodynamics in the discussion of systems in equilibrium. A phase is a part of a system whose intensive properties are independent of position, in the absence of external long-range fields of force. These properties may vary continuously with position under the action of such fields. A phase may consist of one continuous mass; or it may be composed of a group of discrete parts, provided only that they are in physical contact or are connected by other phases containing the same constituents. Evidently, when a phase consists of discrete particles, they must not be so small and variable in size that properties, for example pressure, vary appreciably from particle to particle; and, if solid crystals, they must not be in different states of strain.

It is clear from the foregoing paragraph that the concept of phase depends on the detail with which we wish to describe the system. Thus, for example, if we focus our attention on the optical properties of a system, two discrete parts of a system would be considered different phases if each of their intensive thermodynamic properties has the same value, except for those symmetry properties that transform one part into the other by inversion or mirroring, but would be considered parts of the same phase if one transforms into the other under a pure rotation [5]. On the other hand, if we restrict our attention to ordinary thermodynamic properties, discrete parts of a system which differ only in their symmetry properties can be considered to be parts of the same phase.

In the absence of the effects mentioned above, a phase is a part of a system which is uniform throughout its entire mass, both in chemical composition and in physical state [6].

A system composed of crystals of sodium chloride, a saturated solution of this substance in water, and water vapor in equilibrium, has three phases. There are two phases in a system at equilibrium composed of two liquid mixtures of water and ethyl alcohol of the same composition, contained in separate open vessels but connected by a vapor phase containing both substances. A mixture of right- and left-handed α-quartz crystals is composed of two phases if we are considering optical properties, but one phase for ordinary thermodynamic discussions.

The boundary of a phase or each part thereof is not a mathematical surface, but rather a thin film in which there is a large gradient in intensive properties in the direction normal to the surface. We call this the surface of discontinuity "without implying that the discontinuity is absolute, or that the term distinguishes any surface with mathematical precision" [7]. Until we introduce the effect of surfaces on equilibrium, we shall suppose that the properties of a phase are not affected by the presence of adjacent phases and that their densities are constant up to a mathematical surface which forms the boundary of the phase.

In his discussion of *simple systems*, Gibbs [6] uses the term coexistent for "phases which can exist together, the dividing surface being plane, in an equilibrium which does not depend upon passive resistances to change".

1.13. Homogeneous and heterogeneous systems. A homogeneous system is composed of a single phase; a heterogeneous system contains more than one phase.

Each phase of a closed heterogeneous system is itself an open system if one or more of its constituents can pass from one phase to another. Under these conditions, we may choose to apply the equations of thermodynamics derived for closed systems to the system as a whole, or we may apply the equations derived for open systems to each of the phases. The latter procedure gives the most information about the equilibrium properties of the individual phases.

1.14. States of aggregation. The term aggregation state is applied to a phase, and also to a substance in a mixture, to denote physical condition as opposed to chemical composition. With one exception (crystalline solids) it is not possible to present definitions that distinguish one such state from another sharply for all situations that can occur. However, the meanings of certain terms are generally accepted.

Aggregation states of phases can be divided into two classes: fluids and solids. A fluid cannot sustain even a small shearing stress when in a state of equilibrium: no shear can persist without flow occurring. A solid, unless in the form of a fine powder, can sustain a shearing stress under equilibrium conditions. To be sure, both fluids of high viscosity and solids commonly described as soft do yield more or less slowly under shear, but usually at quite different rates.

Aggregation states of substances in mixtures are also divided into two classes: the dissolved state and the disperse state. A substance in the dissolved state is a constituent of a single phase containing aggregates no larger than molecules of relatively small molecular weights. In the disperse state a substance, called a colloid, is disseminated throughout a gaseous, liquid, or solid medium in the form of small particles, roughly 5×10^{-7}—5×10^{-9} m in diameter, composed of aggregates of molecules or of single molecules of very large molecular weight.

Liquids, gases, and vapors constitute the class of physical states known collectively as fluids. In general we can distinguish a liquid from a gas (or vapor) in a heterogeneous system containing both phases by using the following criteria: under the action of the earth's gravitational field a liquid tends to collect for the most part in one portion at the bottom of the available space, thus giving rise to a surface which changes its form when the container is tilted; a gas (or vapor) fills the remaining available space completely*. On the other hand there is no sharp distinction in the meanings of the terms liquid and gas (or vapor) when applied to a homogeneous fluid system in the region of its liquid—vapor critical point: the term fluid is still appropriate.

A gas in equilibrium with a liquid phase is frequently called a vapor, and this term is sometimes applied to a homogeneous gas phase when below its liquid—vapor critical temperature. We shall use the terms gas and vapor interchangeably when both are applicable.

The term solid includes crystalline solids and amorphous solids. A crystalline solid assumes, on formation, a symmetrical geometrical form bounded by plane surfaces with precisely the same angles between similar faces. It has an internal structure characterized by long-range order, as revealed by its optical properties. When the temperature of a crystalline phase is increased slowly at a constant pressure, melting begins at a definite and reproducible temperature. No critical point involving one crystalline solid and another such phase or a fluid phase has been found experimentally. This means that the distinction between any one crystalline phase and all other possible aggregation states can be maintained unambiguously.

---

*In a mixture of H_2 and He at $-253°C$, the density of the vapor phase is greater than that of the liquid phase at pressures above 49 atm, so that the gas phase is observed to sink in the liquid as the pressure is increased through 49 atm at this temperature. Such exceptional behavior in the liquid—vapor critical region of binary mixtures may be explained in terms of van der Waals' equation of state. See ref. 8.

Amorphous solids do not assume, on formation, the geometrical forms which are characteristic of crystals. Their internal structures do not have the long-range order of crystals but may have the short-range order of liquids. Although classed as solids, they suffer a permanent change in shape under a relatively light load, if it is applied for a sufficiently long time. An amorphous solid does not have a definite melting point; it progressively and continuously softens with increase in temperature until its viscosity is sufficiently low for it to be classed as a liquid. Thus, amorphous solids have in some respects the properties characteristic of a liquid, and in some instances they are called undercooled liquids, or glasses.

Certain crystalline solids melt to a turbid phase that has the low viscosity of a liquid, but also tends to form the definite geometrical shape (although with somewhat rounded contours) and exhibits the optical properties (for example, birefringence) characteristic of a crystalline solid. The terms liquid crystals, anisotropic liquid, and mesomorphic phase are applied to this aggregation state. A liquid crystalline phase changes to an isotropic liquid at a definite temperature, as does a crystalline solid.

Liquids and solids are sometimes collectively called condensed phases to distinguish them from a gas (or vapor) phase.

A system may contain any number of coexistent liquid phases and solid phases. In the absence of internal nonpermeable or semipermeable partitions a system has, in general, only one gas (or vapor) phase*.

We shall use the letter (g) to denote a gaseous phase, (l) a liquid, and (s) a crystalline solid. A crystalline solid may exist in different crystalline modifications, and this can be indicated by appending an appropriate abbreviation. For example, S(r) and S(m) designate rhombic and monoclinic sulfur, respectively. The dissolved state is indicated by giving the concentration of the substance, for example HCl (0.1 wf).

1.15. Pure substances and mixtures. Products obtained from natural or synthetic sources, when subjected to one or more physical methods of fractionation (such as mechanical separation, solvent extraction, selective adsorption, distillation, crystallization, or diffusion), yield substances whose properties are not appreciably changed by further application of any of these resolving methods. Such substances are termed pure. Each is characterized by a set of properties which differ by finite amounts from the corresponding properties of all other pure substances. Delicate tests of the purity of a liquid are its freezing point at a fixed pressure, its boiling point at a fixed pressure, and its vapor pressure at a fixed temperature, and also the constancy of these

---

*Two coexistent gas phases in the critical region of certain binary systems were predicted on the basis of van der Waals' equation of state [9a], and have been found experimentally for the gas mixtures NH_3-N_2 and NH_3-CH_4 [9b]. The curious case of two coexistent fluid phases (both of which may be classed as gaseous), in a system containing an internal rigid partition which is permeable to all of the components of the system, is mentioned by J. W. Gibbs (ref. 1, p. 84).

values as the ratio of the mass of the liquid to that of the other phase involved is varied over wide limits.

Mixtures are composed of two or more pure substances and may be classed as physical mixtures and solutions. The former are heterogeneous, the latter are homogeneous. The properties of solutions can be varied practically continuously, within limits, by changing the relative proportions of their constituents.

Solutions exist in the gaseous, liquid, and solid aggregation states. Gaseous solutions are usually referred to as gas mixtures, so that the term solution usually designates a condensed phase.

1.16. Elements and compounds. The action of sufficiently powerful (but not too violent) physical or chemical resolving influences on most pure substances converts them into a relatively small number of other pure substances which are not further decomposed by similar processes. The latter are called elementary substances, while the former are called compound substances. Each is considered to be composed of a small number of kinds of matter called elements, each elementary substance being composed of one such element and each compound substance of two or more such elements.

Most elements consist of mixtures of two or more isotopes, which differ in atomic weight but not in atomic number. The various isotopes of an element differ very little in their chemical properties, and not greatly in their physical properties, except for those properties which are strong functions of atomic mass. In general, an element derived from natural sources has a quite constant isotopic composition. These facts account for the observed constancy of the properties of so-called pure substances, which actually consist of mixtures of molecular entities differing in isotopic composition but not in chemical structure. Such a mixture is regarded as a single chemical substance, in the absence of nuclear reactions, or of processes especially designed to separate isotopes.

1.17. Extra-thermodynamic laws. In thermodynamics, changes in state are written to conform with the following laws of matter.

1. Mass is conserved in all changes in state for which the effects of the special theory of relativity may be neglected. In all such changes in state, each compound is conserved when no chemical reaction occurs, and each element is conserved in the absence of nuclear reaction.

2. Net electric charge is conserved in all changes in state.

3. In a change in state, a component of one phase of a system cannot be transported to another phase containing this substance, unless the two phases are in contact or are connected by other phases each containing the substance or its constituents [1].

The application of thermodynamics to systems in which chemical reactions occur is simplified by the use of chemical units of mass. Originally, the atomic weights of elements and the molecular weights of elementary and compound substances were referred to the basis O = 16, where O is the naturally-occurring isotopic mixture of oxygen; and they were arrived at by application of the following laws: (1) the law of definite proportions; (2) the law of combining weights; (3) Gay-Lussac's law of combining volumes; (4) Avogadro's law; (5) the dilute gas law; and (6) the dilute solution law. Today atomic weights of elements are expressed on the basis of the assignment $^{12}C = 12$, where ^{12}C is one of the isotopes of carbon; and they are usually computed from isotopic masses and relative natural abundances as revealed by mass-spectrographic measurements.

From an analysis of a substance and a table of atomic weights we can determine the substance's empirical formula, which expresses its composition in the terms involving the smallest possible integers. In general, the molecular formula of an elementary or compound substance in a given state is written to agree with its molecular weight in that state if known; otherwise the empirical formula is used.

1.18. Chemical units of mass. The amount of an element whose mass in grams is numerically equal to its atomic weight is called one gram-atom.

Three chemical units of mass are used for elementary and compound substances: formula weight, mole, and equivalent weight. Each is the number of grams (or other mass unit) of the substance equal to the weight corresponding to a certain molecular formula. For the formula weight a definite formula is specified, for example, $Ba(OH)_2$, or $Ba(OH)_2 \cdot 8H_2O$. The term mole is appropriate when there is general agreement on the molecular formula to be used, so that a definite formula need not be specified, for example, O_2, H_2O. For the equivalent weight the formula corresponds to the weight of a substance which reacts with one gram-atom of hydrogen, or with that weight of any other substance which itself reacts with one gram-atom of hydrogen; and, in general, it refers to a designated reaction. For example, one equivalent weight of sulfuric acid when used as an acid is $H_2SO_4/2$, while its equivalent weight when used as an oxidizing agent may be $H_2SO_4/2$, $H_2SO_4/6$, or $H_2SO_4/8$, depending on whether the sulfur is reduced from H_2SO_4 to H_2SO_3, S, or H_2S, respectively.

1.19. Species and components. A species of a phase is the aggregate of all of the constituent molecular or atomic entities which have the same chemical structure. Entities differing only in isotopic composition belong to the same species, unless processes such as nuclear reactions or isotopic fractionation are considered to occur. For example, the group of all molecules in liquid water identified by the formula H_2O is normally considered to constitute a single species, although the group is actually composed of a collection of many kinds of particles differing among themselves in isotopic composition with respect to both hydrogen and to oxygen: $^1H_2^{16}O$, $^1H_2^{17}O$, $^1H_2^{18}O$, $^2H_2^{16}O$, $^1H^2H^{16}O$, etc. Similarly, all of the molecules having the formula H_4O_2 constitute a single species irrespective of their isotopic compositions;

and this is true for the molecules H_6O_3, etc., and also for the ions OH^-, and H_3O^+, etc.

The number of components of a phase is the minimum number of pure chemical substances whose relative quantities suffice to express the composition of the phase in the given state, and in every state which we regard as possible for the phase to reach. Hence the values of the differentials of the masses of the components of a phase must be independent, and capable of expressing every variation of composition which we regard as possible [10]. When these conditions are satisfied, it is immaterial which of the constituent substances are chosen as the components. For example, when water is the constituent of a phase, all of the species or group of species in equilibrium which have the overall composition (not simply the molecular formula) represented by the formula H_2O, can be considered to form a single component, namely the component substance water. Thus, the single component substance water is constituted of the species H_2O, H_4O_2, H_6O_3, etc., and of the groups of ions $H_3O^+ + OH^-$, $2H_3O^+ + O^{2-}$, etc., the members of each group being taken collectively.

The number of components of a phase is the number of species diminished by: (1) the number of independent chemical reactions at equilibrium among these species, (2) the condition of electrical neutrality if ions are present (one equation), and (3) the number of independent arbitrary conditions of restraint imposed on the composition of the phase by the investigator. Thus

$$c = m - r - a - b \tag{1.16}$$

where c = number of components of the phase, m = number of species identified in the phase, r = number of independent chemical reactions at equilibrium among these species, a = zero for a non-electrolyte phase and unity for an electrolyte phase, and b = number of independent arbitrary conditions of restraint imposed on the composition of the phase. Fortunately, we need not identify all conceivable species in order to apply eqn. (1.16). We can, if we choose, omit any species which is formed by means of an equilibrium chemical reaction from other species already listed, since its inclusion would increase each of the terms m and r of eqn. (1.16) by unity, and hence would not change the value of c.

Consider a gas phase containing the species H_2, N_2, and NH_3. Among these species there is a possible chemical reaction

$$3H_2 + N_2 = 2NH_3 \tag{1.17}$$

Under conditions where this reaction proceeds at such a small rate that it does not have an appreciable effect on the state of the phase during the time it is under observation (see Section 1.5), we must evidently conclude that the phase has three components: H_2, N_2, and NH_3. Under conditions where the rate of the reaction is appreciable, so that equilibrium with respect to the reaction prevails, there are two components in the system, in the absence of any arbitrary conditions of restraint on its composition. We may select

H_2 and N_2, or H_2 and NH_3, or N_2 and NH_3 as the two components. At an intermediate temperature, where the rate of the reaction (1.17) lies near the limit of perceptibility, the decision as to whether the phase is composed of two or three components would be based on the relation of the relaxation time of the attainment of equilibrium to that of the measuring instrument employed to fix the state of the phase. If, in the case where the reaction proceeds at an appreciable rate, we should impose the arbitrary constraint that the ratio of the number of gram atoms of H to the number of gram atoms of N is always 3 to 1, then the phase can be considered to be composed of a single component, namely NH_3.

Suppose that in a phase initially consisting of an aqueous solution of sodium chloride, we identify the following species

$$H_2O, H_4O_2, NaCl, NaOH, HCl, Na^+, Cl^-, H^+, OH^- \qquad (1.18)$$

There are five chemical reactions at equilibrium:

$2H_2O = H_4O_2$, $NaCl = Na^+ + Cl^-$, $NaOH = Na^+ + OH^-$, $HCl = H^+ + Cl^-$,

$$H_2O = H^+ + OH^- \qquad (1.19)$$

The condition of electrical neutrality is

$$(Na^+) + (H^+) = (Cl^-) + (OH^-) \qquad (1.20)$$

where the symbol (Na^+) represents the number of formula weights of (Na^+) associated with 1 kg of water. If we add the further arbitrary condition of restraint that the phase never loses or gains NaOH or HCl, except in the ratio 1 NaOH to 1 HCl, we have the additional relation

$$\Sigma(Na) = \Sigma(Cl) \qquad (1.21)$$

where the symbol $\Sigma(Na)$ denotes the total number of formula weights of Na from all sources $(NaCl + NaOH + Na^+)$ associated with 1 kg of water, and $\Sigma(Cl)$ denotes the corresponding quantity $(NaCl + HCl + Cl^-)$ for Cl. Then, from eqn. (1.16)

$$c = 9 - 5 - 1 - 1 = 2 \qquad (1.22)$$

and the phase has two components, say H_2O and NaCl. On the other hand, if we do not impose the arbitrary condition of constraint (1.21), the phase has three components, say H_2O, NaOH, and HCl.

Let us now choose to add the two additional species H_6O_3 and H_3O^+ to the list (1.18). Then two more equilibrium equations

$$3H_2O = H_6O_3, \quad H_2O + H^+ = H_3O^+ \qquad (1.23)$$

must be added to the list (1.19). Thus the value of c remains unchanged.

Let us now turn to the consideration of a closed heterogeneous system. The number of components of such a system is the minimum number of pure chemical substances whose relative quantities suffice to express the compositions of every phase of the system in the given state, and in every state which we regard as possible for the system to reach. In applying this definition we must remember that each phase of a closed heterogeneous system is itself an open homogeneous system. The conditions of restraint imposed on the system will determine whether or not we regard it to be possible for any of the phases initially present to disappear, and for any phases not initially present to appear. Once we have found the number of components of the system, we can select the actual components without reference to the condition that the variations of the differentials of their masses shall be independent in each phase.

The number of components of a heterogeneous system is the sum of the components chosen for the constituent phases (no component being counted more than once) diminished by all independent chemical reactions at equilibrium among these components [11]. Consider the following system at equilibrium

$$CuSO_4 \cdot 5H_2O(s), CuSO_4 \text{ (saturated aqueous solution)}, H_2O(g) \tag{1.24}$$

On applying the definition of the number of components of a phase, we find that the solid phase has one component, $CuSO_4 \cdot 5H_2O$, the liquid phase has two, say $CuSO_4$ and H_2O, while the gaseous phase has one, H_2O. The sum of these components is three. But there is one chemical equilibrium among these substances, namely

$$CuSO_4 + 5H_2O = CuSO_4 \cdot 5H_2O \tag{1.25}$$

Hence the system has two components. This is true whether or not we consider it possible for variations in the state of the system to cause one of the phases to disappear, or new solid phases to form. Usually we would select $CuSO_4$ and H_2O as the two components, since with this choice we would never have to assign a negative mole number to a component in any phase (see Section 1.20). However, we note that, under these conditions, the variations of the masses of $CuSO_4$ and H_2O are not independent in the crystalline phase but must always be in the ratio of 1 to 5.

1.20. Composition of a system expressed in terms of its species and its components. Consider first a closed homogeneous system in which there is one chemical reaction at equilibrium. The system has m species and $c = m - 1$ components. Let n_{i_s} ($i = 1, 2, ..., m$) be the mole number (number of formula weights) of the constituent B_i when it is considered to be a species of the system in any state; and let n_i ($i = 1, 2, ..., c = m - 1$) be the corresponding mole number of the same constituent when it is considered to be a component of the system in this state. Thus the species B_m is not a component and the value of n_m is identically zero. The equation for the chemical reaction is

$$\sum_{1}^{m} \nu_i B_i = 0 \tag{1.26}$$

The stoichiometric coefficient ν_i is zero for any species which does not take part in the reaction. Let the phase, while remaining closed, undergo a real or virtual variation in state whereby n_{is} becomes $n_{is} + \delta n_{is}$. Then, from eqn. (1.26) we find

$$\delta n_{1s}/\nu_1 = \delta n_{2s}/\nu_2 = \ldots = \delta n_{ms}/\nu_m = \delta \alpha \tag{1.27}$$

where $\delta\alpha$ is a measure of the extent of the reaction [12]. Integration of these relations from a fiducial composition n_{is}^0 ($i = 1, 2, \ldots, m$), that is, some fixed composition where the extent of the reaction is arbitrarily placed equal to zero, to a variable composition n_{is} ($i = 1, 2, \ldots, m$) where it has the value α gives the relations

$$n_{is} = n_{is}^0 + \nu_i \alpha \qquad (i = 1, 2, \ldots, m) \tag{1.28}$$

Now, as we have said, the number of components of the system under consideration is one less than the number of constituents. The composition of the system in a given state, when expressed in terms of components, depends on the selection of the c components from among the m species. Let the species not appearing in the list of components be B_m, the only limit on its selection being that ν_m in the chemical reaction (1.26) must not be zero. Suppose we know the values of the mole numbers n'_{is} ($i = 1, 2, \ldots, m$) of the species of the system in a given state, the single prime sign designating this particular composition. The corresponding extent of reaction α' for this state depends on the fiducial composition n_{is}^0 ($i = 1, 2, \ldots, m$) from which we choose to measure α. Let us choose this composition by supposing that the reaction (1.26) proceeds to such an extent in one direction or the other that the mole number of the species B_m becomes zero. Hence, in eqn. (1.28) $n_{ms}^0 = 0$, and we can compute the values of α' and of the $m-1$ fiducial mole numbers n_{is}^0 $(1, 2, \ldots, m-1)$ from the following relations

$$n_{ms}^0 = 0 \tag{1.29a}$$

$$\therefore n'_{ms} = \nu_m \alpha', \quad \alpha' = n'_{ms}/\nu_m \tag{1.29b}$$

$$n_{is}^0 = n'_{is} - \nu_i \alpha' \qquad (i = 1, 2, \ldots, m-1) \tag{1.29c}$$

The state determined by placing n_{ms}^0 equal to zero need not be an equilibrium state of the system*.

We now identify the fiducial values n_{is}^0 ($i = 1, 2, \ldots, m-1$) computed from eqns. (1.29) with the mole numbers n_i ($i = 1, 2, \ldots, c$) of the c components of the system, n_m being identically zero since B_m is not a component. Thus, eqns. (1.29) may be written in the form

---

*In eqns. (1.27) and throughout this book, the symbol δ is used to denote a possible variation in the state of a system as this term was used by J. W. Gibbs (ref. 1, p. 57). The meaning of the term is discussed in detail later. It will be sufficient to say here that a possible variation is one that: (1) starts from a state of equilibrium, (2) does not violate the laws of thermodynamics and of matter, and (3) is consistent with the stated conditions of restraint. A possible variation need not proceed to an equilibrium state.

$$n'_{ms} = \nu_m \alpha', \; \alpha' = n'_{ms}/\nu_m \tag{1.30a}$$

$$n_i = n'_{is} - \nu_i \alpha' \qquad (i = 1,2,...,c = m-1) \tag{1.30b}$$

Evidently the quantities n_i are the numbers of formula weights of the c components introduced into the system when it was formed. We shall measure the extent α of the reaction from a state, whether real or hypothetical, having this composition. The values of the mole numbers n_{is} of the m species for any accessible state of the closed system are then given by the relations

$$n_{is} = n_i + \nu_i \alpha \qquad (i = 1,2,...,m) \tag{1.31a}$$

$$n_m \equiv 0 \tag{1.31b}$$

We shall now apply these results to the ammonia synthesis reaction. Consider a closed gaseous phase initially composed of the species

$$n'_{1s} = 12 \text{ moles } H_2, \; n'_{2s} = 5 \text{ moles } N_2, \; n'_{3s} = 6 \text{ moles } NH_3 \tag{1.32}$$

Among these there is one chemical reaction at equilibrium

$$3H_2 + N_2 = 2NH_3 \tag{1.33}$$

In Table 1.1 the values of α' and of the compositions of the phase in terms of its components, as computed from eqns. (1.30), are listed for each of three sets of components: H_2 and N_2, H_2 and NH_3, and N_2 and NH_3. It will be noted that the designation of H_2 and NH_3 as the components of the phase leads to a negative value for the mole number of the component H_2. Although this may be an inexpedient choice of components for some computational purposes, it is not thermodynamically incorrect.

TABLE 1.1

COMPOSITION OF A CLOSED SYSTEM IN TERMS OF ITS SPECIES AND ITS COMPONENTS

Components	α'	Composition expressed in terms of components		
		Moles H_2 ($\nu_{H_2} = -3$)	Moles N_2 ($\nu_{N_2} = -1$)	Moles NH_3 ($\nu_{NH_3} = 2$)
Phase (1.32): 12 moles H_2, 5 moles N_2, 6 moles NH_3 as species				
H_2 and N_2	3	12—(—3)(3) = 21	5—(—1)(3) = 8	6—(2)(3) = 0
H_2 and NH_3	—5	12—(—3)(—5) = —3	5—(—1)(—5) = 0	6—(2)(—5) = 16
N_2 and NH_3	—4	12—(—3)(—4) = 0	5—(—1)(—4) = 1	6—(2)(—4) = 14
Phase (1.34): 6 moles H_2, 3 moles N_2, 10 moles NH_3 as species				
H_2 and N_2	5	6—(—3)(5) = 21	3—(—1)(5) = 8	10—(2)(5) = 0
H_2 and NH_3	—3	6—(—3)(—3) = —3	3—(—1)(—3) = 0	10—(2)(—3) = 16
N_2 and NH_3	—2	6—(—3)(—2) = 0	3—(—1)(—2) = 1	10—(2)(—2) = 14

With the phase remaining closed, let its state be varied so that four additional moles of NH_3 are formed. The new composition of the phase in terms of its species becomes

$n'_{1s} = 6$ moles H_2, $n'_{2s} = 3$ moles N_2, $n'_{3s} = 10$ moles NH_3 (1.34)

We see from Table 1.1 that the states (1.32) and (1.34) have the same compositions when each is expressed in terms of a particular set of components, although the extents α' of the reaction are different in the two cases. We can employ eqns. (1.31) to compute all physically possible compositions of the system, including the two examples cited.

Let us now consider a closed homogeneous system containing the species B_i ($i = 1,2,...,m$) among which there are r independent chemical reactions at equilibrium

$$\sum_i^m \nu_i^k B_i = 0 \qquad (k = \text{I, II},...,r) \qquad (1.35)$$

The condition [14] that these reactions be independent is that the following matrix shall have the rank r

$$\begin{Vmatrix} \nu_1^{\text{I}} & \nu_2^{\text{I}} & ... \nu_m^{\text{I}} \\ \nu_1^{\text{II}} & \nu_2^{\text{II}} & ... \nu_m^{\text{II}} \\ ... & ... & ... \\ \nu_1^r & \nu_2^r & ... \nu_m^r \end{Vmatrix} \quad (r < m) \qquad (1.36)$$

We follow the same general procedure as that used above for the case of a single chemical reaction. There are m species and $c = m - r > 0$ components of the system. Let n_{is} ($i = 1,2,...,m$) be the mole number of the constituent B_i when it is considered to be a species of the system in any state; and let n_i ($i = 1,2,...,c = m-r$) be the corresponding mole number of the same constituent when considered to be a component of the system in this state. Thus the r species B_i ($i = c+1, c+2,...,m$) are not components, and the corresponding values of n_i are identically zero. Instead of the single set of equations (1.27) we now have the r sets

$$\delta n_{1s}^k / \nu_1^k = \delta n_{2s}^k / \nu_2^k = ... = \delta n_{ms}^k / \nu_m^k = \delta \alpha^k \qquad (k = \text{I, II},...,r) \qquad (1.37)$$

Here δn_{1s}^k is the number of moles of the species B_1 produced when the k-th chemical reaction proceeds to the extent $\delta \alpha^k$. The relations corresponding to eqns. (1.28) are

$$n_{is} = n_{is}^0 + \nu_i^{\text{I}} \alpha^{\text{I}} + \nu_i^{\text{II}} \alpha^{\text{II}} + ... + \nu_i^r \alpha^r \qquad (i = 1,2,...,m) \qquad (1.38)$$

Here n_{is}^0 is the mole number of the species B_i in a fiducial composition where each extent of reaction α^k ($k = \text{I, II},...,r$) is placed equal to zero, and n_{is} is the corresponding quantity when each extent of reaction has the value α^k.

Let us assume that the mole numbers n'_{is} ($i = 1,2,...,m$) of the m species of the system in a given state are known. Proceeding as before, we identify the mole numbers of the constituents in a fiducial state obtained by placing the r values n_{is}^0 equal to zero for $i = c+1, c+2, ..., m$ in eqns. (1.38) as the mole numbers of the c components of the system. The relations corresponding to eqns. (1.29) and (1.30) are

$$n_{is}^0 = 0 \; (i = c + 1, c + 2, \ldots , m) \tag{1.39a}$$

$$n'_{is} = v_i^I \alpha'^I + v_i^{II} \alpha'^{II} + \ldots + v_i^r \alpha'^r \qquad (i = c + 1, c + 2, \ldots , m) \tag{1.39b}$$

$$D\alpha'^k = n'_{c+1, s} A_{c+1}^k + n'_{c+2, s} A_{c+2}^k + \ldots + n'_{ms} A_m^k \qquad (k = I, II, \ldots , r) \tag{1.39c}$$

$$n_i = n'_{is} - v_i^I \alpha'^I - v_i^{II} \alpha'^{II} - \ldots - v_i^r \alpha'^r \qquad (i = 1, 2, \ldots , c) \tag{1.39d}$$

In these relations the known mole numbers of the m species are $n'_{is} \; (i = 1,2,\ldots,m)$, those of the c components are $n_i \, (i = 1,2,\ldots, c = m-r)$, the extents of the r chemical reactions for the given state are α'^k ($k = I, II, \ldots , r$), D is the determinant associated with the coefficient matrix of the second equation of the set (1.39), and A_i^k is the cofactor of the element in the i-th row and k-th column of this matrix, which is

$$D = \begin{Vmatrix} v_{c+1}^I & v_{c+1}^{II} & \ldots & v_{c+1}^r \\ v_{c+2}^I & v_{c+2}^{II} & \ldots & v_{c+2}^r \\ \ldots\ldots\ldots\ldots\ldots\ldots\ldots \\ v_m^I & v_m^{II} & \ldots & v_m^r \end{Vmatrix} \quad (r = m - c) \tag{1.40}$$

It will be noted that this square matrix is formed from the matrix (1.36) by deleting the first $m - r$ columns and changing the rows into columns without change of order. Hence it will always be possible to choose a set of components such that the determinant D is not zero*.

As in the case of the system having only one chemical reaction, we shall measure the extents of the reaction α^k ($k = I, II, \ldots , r$) from the composition $n_i \, (i = 1, 2, \ldots , c)$ computed from eqns. (1.39). Instead of eqns. (1.31) we find the following relations for the mole numbers of the m species B_i for any arbitrary values of the extents α^k of the r chemical reactions:

$$n_{is} = n_i + v_i^I \alpha^I + v_i^{II} \alpha^{II} + \ldots + v_i^r \alpha^r \qquad (i = 1, 2, \ldots , m) \tag{1.41a}$$

$$n_i \equiv 0 \qquad (i = c + 1, c + 2, \ldots , m) \tag{1.41b}$$

So far we have discussed only closed homogeneous systems (closed phases). Equations (1.26)–(1.41) evidently hold for a closed heterogeneous system when we interpret n_{is} and n_i to be, respectively, the numbers of formula weights of B_i as species and as components in the system. Consider a system of seven species in three phases, having the following composition (in round numbers)

---

*This condition is known as Jouguet's criterion (see I. Prigogine and R. Defay, ref. 12, p. 468). When it holds, we can form, from (1.36) at least one $r \times r$ square matrix whose determinant is not zero.

$n_{1s} = 4H_2O(g)$, $n_{2s} = 8H_2(g)$, $n_{3s} = 3CO_2(g)$, $n_{4s} = 5Ar(g)$, (1.42a)

$n_{5s} = 8Fe(s)$, $n_{6s} = 6CO(g)$, $n_{7s} = 10FeO(s)$ (1.42b)

There are two reactions at equilibrium, which may be written

I $FeO(s) + H_2(g) = Fe(s) + H_2O(g)$ (1.43a)

II $CO_2(g) + H_2(g) = CO(g) + H_2O(g)$ (1.43b)

There are $c = 7 - 2 = 5$ components. Now these can and must be so selected that the determinant D, numbered (1.40), is not zero. This requires the list of components to contain Ar, and at least one species selected from each of the following three pairs (two from any one pair): H_2O and H_2, CO_2 and CO, and Fe and FeO. If we choose to regard H_2O, H_2, CO_2, Ar, and Fe as components, the results of applying eqns. (1.39) to the system (1.42) are listed in Table 1.2, where the last column is the number of formula weights of the selected components in the system.

If we select a different set of components, the values entered in the last column of Table 1.2 would be different, but those in the other columns would remain unchanged. If we use a different set of independent chemical reactions, for example

I $FeO(s) + H_2(g) = Fe(s) + H_2O(g)$ (1.44a)

II $FeO(s) + CO(g) = Fe(s) + CO_2(g)$ (1.44b)

TABLE 1.2

COMPOSITION OF A CLOSED HETEROGENEOUS SYSTEM IN TERMS OF SPECIES AND COMPONENTS

System (1.42), equilibrium reactions (1.43), eqns. (1.39) and (1.40)

Substance	i	ν_i^I	ν_i^{II}	n'_{is} (species)	n_i (component)
H_2O	1	+1	+1	4	8
H_2	2	−1	−1	8	4
CO_2	3	0	−1	3	9
Ar	4	0	0	5	5
Fe	5	+1	0	8	18
CO	6	0	+1	6	0
FeO	7	−1	0	10	0

$D = \begin{vmatrix} \nu_6^I & \nu_6^{II} \\ \nu_7^I & \nu_7^{II} \end{vmatrix} = \begin{vmatrix} 0 & +1 \\ -1 & 0 \end{vmatrix} = 1$ from eqn. (1.40)

$A_6^I = 0$, $A_7^I = -1$, $A_6^{II} = +1$, $A_7^{II} = 0$

$\left. \begin{aligned} D \times \alpha'^I &= 6 \times (0) + 10 \times (-1) = -10 \\ D \times \alpha'^{II} &= 6 \times (+1) + 10 \times (0) = +6 \end{aligned} \right\}$ from eqn. (1.39)

and the same choice of components, the composition of the system in terms of components would be unchanged.

REFERENCES

1 J. W. Gibbs, Scientific Papers, Vol. I, Thermodynamics, Longmans, Green and Co., New York, 1906, pp. 57, 58.
2 H. F. Stimson, J. Res. Natl. Bur. Stand., Sect. A, 65 (1961) 139—145.
3 R. J. Berry, Can. J. Phys., 37 (1959) 1230—1248, 38 (1960) 1027—1047; H. F. Stimson, in H. C. Wolfe (Ed.), Temperature, Vol. II, Reinhold, New York, 1955, pp. 141—168.
4 J. W. Gibbs, ref. 1, p. 62.
5 L. Tisza, Ann. Phys., 13 (1961) 1—92, especially pp. 5 and 6.
6 J. W. Gibbs, ref. 1, pp. 57, 63, 96, 358.
7 J. W. Gibbs, ref. 1, p. 219.
8 H. K. Onnes and W. H. Keesom, Commun. Phys. Lab., Univ. Leiden, No. 96 and Suppl. No. 16 (1906—1907); J. D. van der Waals and Ph. Kohnstamm, Lehrbuch der Thermostatik, Vol. II, 2nd edn., J. A. Barth, Leipzig, 1927, pp. 109—112.
9 (a) H. K. Onnes and W. H. Keesom, Commun. Phys. Lab., Univ. Leiden, Suppl. No. 15 (1907); (b) see J. S. Rowlinson, in S. Flügge (Ed.), Encyclopedia of Physics, Vol. XII, Thermodynamics of Gases, Springer, Berlin, 1958, p. 69.
10 J. W. Gibbs, ref. 1, p. 63.
11 J. W. Gibbs, ref. 1, pp. 96—98.
12 Th. DeDonder, L'Affinité, Gauthier-Villars et Cie, Paris, 1927; I. Prigogine and R. Defay, Chemical Thermodynamics, (translated by D. H. Everett), Longmans, Green and Co., London, 1954, p. 10.

Chapter 2

PRESSURE AND TEMPERATURE

We now select for special attention the pressure and temperature of a system, because the values of these two quantities are normally given in the specification of the state of a system. Both are relatively easy to control and measure precisely. Pressure is related to work of expansion and temperature is related to heat, two quantities of basic importance in thermodynamics, which will be treated in Chapter 3.

2.1. Pressure. The pressure p at a point P in the boundary surface of a system in equilibrium is defined as the normal component of mechanical force exerted over unit area of the surface at P. If we define ΔF_n as the normal component of mechanical force exerted over an element of surface ΔA including P, then p is defined as the limit of the ratio $\Delta F_n/\Delta A$ as ΔA becomes sufficiently small. Thus,

$$p = \lim_{\Delta A \to 0} \Delta F_n/\Delta A = dF_n/dA \tag{2.1}$$

The above definition of pressure and eqn. (2.1) also hold for an interior point of a phase.

Since a fluid in equilibrium cannot sustain a shearing stress (see Section 1.14), the only force acting over an element of area dA at a point in the bounding surface of the fluid is normal to the surface. Moreover, the force acting over an area dA at an interior point of a fluid phase in equilibrium is normal to the area and independent of its orientation. Thus, the pressure at every point in the surface and in the interior of a fluid in equilibrium depends only on the position of dA and not on its orientation, and the force acting over dA is normal to it. This type of pressure is sometimes called hydrostatic, and it is what we shall mean by the term pressure.

In the absence of fields of force (such as a gravitational field) and internal rigid walls, the pressure of a fluid at equilibrium is uniform over its entire boundary and throughout its interior.

The pressure p at a point P in a fluid at equilibrium situated in the Earth's gravitational field varies with the vertical height z of P above a fixed horizontal plane, according to the relation

$$dp/dz = -\rho g \tag{2.2}$$

where ρ is the density of the fluid and g the acceleration due to gravity, both measured at z. [If z is taken to be the vertical distance of the point

P *below* a fixed horizontal plane, the negative sign in eqn. (2.2) is replaced by a positive sign.] The above expression holds for the effect of any gravitational field on p, if we interpret g as the gravitational field intensity G at point P. In this case, we ordinarily take z to increase in the direction of the field intensity and use the positive sign on the right-hand side of the equation.

The pressure p at a point in a fluid phase situated in an external centrifugal field of force produced by rotation of the phase around a vertical axis with an angular velocity ω varies with the radial distance r of P from the axis of rotation, according to the equation

$$dp/dr = \rho\omega^2 r \tag{2.3}$$

where ρ is the density of the fluid at r.

The density of a fluid is a function of its pressure, temperature, and composition. In the usual applications of eqns. (2.2) and (2.3) we compute the density of the fluid from its equation of state, on the assumption that its temperature is uniform. Frequently we assume that a liquid is incompressible and hence has a constant density, and that a gas (or vapor) obeys the perfect gas equation of state [see eqn. (2.41).]

The pressure of a system is considered to be positive when it is directed outward, and negative when directed inward. From the definition of pressure, eqn. (2.1), it is evident that pressure is an intensive state property of a system.

In the CGS system, the unit of pressure is the dyne cm^{-2}; in the MKS system it is the newton m^{-2}. Thermodynamic measurements are also reported in a number of other units. Those in common use are listed in Table 2.1, together with their conversion factors to newton m^{-2}, and, in some cases, to other units.

TABLE 2.1

CONVERSION FACTORS FOR PRESSURE UNITS

[a]1 dyne cm^{-2}	= 1 × 10^{-1} newton m^{-2}
1 kgf cm^{-2}	= 9.80 665 × 10^4 newton m^{-2}
1 lbf in^{-2}	= 6.894 7573 × 10^3 newton m^{-2}
[a]1 bar	= 1 × 10^5 newton m^{-2}
[a]1 standard atm	= 1.01 325 × 10^5 newton m^{-2}
1 standard atm	= 1.469 5949 × 10^1 lbf in^{-2}
[a]1 torr	= (760)$^{-1}$ standard atm
1 torr	= 1.333 2237 × 10^2 newton m^{-2}
[a]1 standard m of Hg	= 1 × 10^3 torr
1 standard m of Hg	= 1.333 2237 × 10^5 newton m^{-2}

[a]Exact by definition.
Note: 1 kgf and 1 lbf are, respectively, the weights of 1 kg mass and 1 lb mass (0.4535 9237 kg mass) at standard gravity (9.80 665 m sec^{-2}).

2.2. Temperature. Temperature is the degree of hotness or coldness of a system measured on some scale. Our naive idea of this concept is derived from the sensation of hotness or coldness we experience when we examine a system by our senses, particularly the senses of touch and sight. We associate the term hot with high temperature and the term cold with low temperature. Our naive idea of temperature is subjective, and it is too qualitative and unreliable for our present purpose. Before developing an exact concept of temperature as this term is used in thermodynamics and describing an objective method of measuring it quantitatively, we must define two terms with great care: thermal contact and thermal equilibrium.

2.3. Thermal contact and thermal equilibrium. Consider a composite system consisting of two simple systems, A and B, not initially in contact. The systems have approximately the same size. Each is enclosed in a rigid, non-permeable boundary* composed of the same material and has no interior partitions. The composite system is enclosed in a boundary that isolates it from its surroundings. Inside this latter boundary there is a removable partition that isolates one system from the other. In each compartment so formed, the space exterior to the boundary of the system of interest is evacuated. System A is homogeneous and is composed of a pure gaseous substance (such as helium) at a density ρ and an initial pressure p'_A; B is also homogeneous, and consists of the same gas at the same density ρ but at an initial pressure p'_B different from p'_A. Now let the isolating partition between A and B be removed and the systems so displaced that finite areas of their individual boundaries are in physical contact, while both systems remain inside the isolating boundary. Experimentally, we always observe one or other of the following results: either the pressures of both systems change with time and finally become equal, or the pressure of each system remains essentially constant.

If the pressures of A and B each change until each system reaches a time-independent state in which the final pressures p''_A and p''_B are essentially equal, then: (1) the two systems are by definition in thermal contact; (2) the material composing their boundaries is termed a thermal conductor; and (3) the boundaries themselves are called diathermal. The two systems A and B are by definition in thermal equilibrium after the final time-independent states have been reached. Many quantitative experiments have shown that the final equilibrium states of A and B are uniquely determined by their initial states.

On the other hand, if the initial pressures p'_A and p'_B remain essentially unchanged with time during the experiment, then: (1) the two systems are not in thermal contact; (2) the material composing their boundaries is termed a thermal insulator; and (3) the boundaries themselves are called

---

*If the material composing the boundaries of A and B is not itself rigid, means must be provided to maintain the volumes of each system essentially constant during the experiment.

adiabatic. Under these conditions, we say that the systems A and B did not come into thermal equilibrium.

Systems which come into thermal equilibrium when placed in thermal contact are said to be thermal conductors. Helium and all gaseous systems are thermal conductors, and hence we may use such systems to distinguish between diathermal and adiabatic boundaries. Many careful investigations involving different types of systems and different kinds of properties have led us to the conclusion that boundaries which permit thermal contact between two specified systems also permit thermal contact between any two systems whatever.

It may happen that a condensed system (such as a piece of asbestos) does not come into thermal equilibrium with its surroundings, or it approaches this condition at a slow rate, when enclosed in a diathermal boundary. Such a system is called a thermal non-conductor, or a poor thermal conductor.

Thermal contact between a system and its surroundings is determined solely by properties of the material composing the boundary. The attainment of thermal equilibrium between a system and its surroundings involves the additional requirement that the system be a thermal conductor.

No real boundary is perfectly adiabatic, and no real system is completely thermally non-conducting. The classification of a boundary as diathermal or adiabatic and of a system as a thermal conductor or a non-conductor with respect to a particular experiment depends on whether the time required to reach thermal equilibrium between the system and its surroundings is short or long in comparison with the time required to complete the measurements of interest.

We have treated the boundary of a system as though it were a material substance capable of a separate existence. It may be simply a mathematical surface, such as the surface of a block of copper, or the interface between a liquid and a vapor phase in a heterogeneous system.

2.4. Zeroth law of thermodynamics. Consider three simple, thermally conducting systems A, B, and C. Each is enclosed in a rigid, diathermal, non-permeable boundary, and each is in a definite initial state. The composite system consisting of A, B, and C is enclosed in a boundary that isolates it from its surroundings. This latter boundary is provided with removable partitions which isolate each system from the other. The space exterior to the boundary of each system of interest is evacuated. We proceed in the same manner as that described in the last section, and place system C in thermal contact first with A and then with B. We always find experimentally that, if system C has the same time-independent state when it is in thermal equilibrium successively with A and with B, then A and B are initially in thermal equilibrium when placed in thermal contact with each other.

Thus, two systems, each in thermal equilibrium with a third system, are in thermal equilibrium with each other. This has been called the zeroth law of thermodynamics.

2.5. Definition of temperature. The experiments described in the last two sections bring to our attention the property of a system called temperature. Temperature is a state property of a thermally-conducting system which determines whether or not it would be in thermal equilibrium with another system if the two were placed in thermal contact. By definition, two systems in thermal equilibrium have the same temperature. Under these conditions we say that the temperature of one system is equal to the temperature of the other.

We would have found the same result in the experiment described in Section 2.4 if system B were an exact replica of A. Thus we may say that the temperature of A is equal to the temperature of the composite system A + B, that is, of the system 2A. Hence, from the zeroth law of thermodynamics, we find that temperature is an intensive state property of a system.

When two thermally-conducting systems having unequal temperatures are placed in thermal contact, we always find that each undergoes a change in state, and at least one system experiences a change in temperature. Other effects observed in either system or in both include one or more of the following: (1) an alteration of pressure if the volume of the system is held constant, or of volume if the pressure is maintained constant; (2) the appearance or disappearance of one or more phases; and (3) the redistribution of matter among the phases of a system (if the system is heterogeneous).

2.6. Thermometry. In its simplest terms, a thermometer is a system that registers its own temperature by means of some type of read-out device, and does not appreciably change the temperature or state of the system under investigation when thermometer and system are brought into thermal equilibrium.

A proper thermometer is so designed that its temperature depends on a single independent property P. An empirical scale of temperature is established for the thermometer by selecting a suitable arbitrary function

$$t = t(P) \tag{2.4}$$

connecting its temperature t and the thermometric property P.

The function $t(P)$ must be a continuous, single-valued, and monotonic function of P. These restrictions are necessary to ensure that: (1) there are no gaps in the proposed scale; (2) two systems at temperatures corresponding to the same value of P of the thermometer will be in thermal equilibrium when placed in thermal contact; and (3) two systems at temperatures corresponding to different values of P of the thermometer will not be in thermal equilibrium when placed in thermal contact. These are by no means trivial restrictions. For example, a liquid-in-glass thermometer with water as the thermometric fluid would not be a proper thermometer over the entire liquid range of water because liquid water under a constant pressure has a maximum in its density vs. temperature curve, at a temperature just above its freezing point. Many experimental investigations have shown that the pressure of a simple gas, such as He, Ne, Ar, H_2, or N_2, maintained at a constant density, and also the density of such a gas maintained at a constant pressure, are suitable thermometric properties throughout the entire temperature range over which the system remains homogeneous.

The functions $t(P)$ are so chosen that all proper thermometers agree on the direction of increasing temperature, namely, so that the algebraic value of t increases from cold to hot. Also, all proper thermometers must agree that a fixed temperature t_A is greater than, equal to, or less than a second fixed temperature t_B, as the case may be. However, the thermometers do not necessarily agree on the numerical value of either t_A or t_B.

Interpreted strictly, the method of determining temperature presented above is applicable only to a system in a state of dynamic equilibrium. In practice, we employ it for systems in quasistatic states, and even, with some reservations as to the interpretation of the observations, to systems whose states are changing at finite rates. In the latter case, the interpretation of the indication of the thermometer at each instant of time as a measure of a thermodynamically significant temperature of the system at some particular time is subject to qualifications which involve the relation of the relaxation time of the attainment of thermal equilibrium between the system and its surroundings to the relaxation times of the attainment of uniform temperatures throughout the system and the thermometer. We should also bear in mind that it takes a finite, though short, time for an isolated nonequilibrium system to relax to a state where a temperature can be defined at each point. The results of radiation techniques for determining temperature must be interpreted with this fact in mind.

One of the consequences of the second law of thermodynamics is the establishment of an absolute scale of temperature, called the thermodynamic scale (see Sections 8.11 and 8.12). It is independent of the properties of any system, and has a zero point fixed by the second law. This is the scale we shall ultimately employ. In the present stage of our development of thermodynamics we are limited to empirical temperature scales based on the indications of a particular thermometer, which we traditionally convert to temperatures by use of a linear relation $t(P)$. The choice of a linear relation between t and P is completely arbitrary. It has the great advantage that it results in a simple, straightforward procedure for establishing a temperature scale for the thermometer in question, as shown below. Thus we write

$$t = t(P) = a + bP \tag{2.5}$$

which may also be written as

$$P = P(t) = a' + b't \tag{2.6}$$

where a′ and b′ bear a simple relationship to a and b. Sometimes we place a and a′ equal to zero in the above expressions, and call the scale so defined an absolute temperature scale. The determination of the values of the parameters a, b and a′, b′ of these equations for a given thermometric system is called calibration. One method of determining the temperature of a system on the empirical scale of a given thermometer is outlined below.

Two fixed and reproducible temperatures t_1 and t_2 are selected as defining fixed points of the proposed temperature scale. Each is assigned a numerical value, and each is maintained by a system whose temperature is either completely independent of external conditions or depends only on the applied pressure. Examples are: (1) the temperature of equilibrium of ice, liquid water, and water vapor (the triple point of water); (2) the temperature of equilibrium of ice and liquid water under a pressure of one standard atmosphere (the ice point); and (3) the temperature of equilibrium of liquid water and its vapor under a pressure of one standard atmosphere (the steam point). The values P_1 and P_2 of the thermometric property of the thermometer are determined when it is in thermal equilibrium with the systems at t_1 and at t_2, respectively. The mean coefficient α of thermal increase in the thermometric property P from t_1 to t_2 is defined by the equation

$$\alpha = \frac{1}{t_2 - t_1}\left(\frac{P_2}{P_1} - 1\right) \tag{2.7}$$

The value P_t of the thermometric property is then measured while the thermometer is in thermal equilibrium with the system whose temperature we wish to measure. The temperature t of the system on the scale of the thermometer is given by the expression

$$t = t_1 + \frac{1}{\alpha}\left(\frac{P_t}{P_1} - 1\right) \tag{2.8}$$

2.7. Temperature scales. A number of temperature scales are in current use for reporting scientific work. All have undergone more or less significant revision since 1948 by the international body known as the General Conference of Weights and Measures.

The Kelvin temperature scale (T, K) is an absolute thermodynamic scale based on the second law of thermodynamics. As mentioned earlier, this scale is independent of the properties of any system and has a zero point fixed by the second law. We are, of course, allowed to fix the size of the unit of thermodynamic temperature, the kelvin. At the Ninth General Conference on Weights and Measures in 1948 [1], the size of this unit was defined, as heretofore, by the statement that the interval from the ice point, T_0, to the steam point, T_{100}, is exactly 100 kelvin. The value of T_0 on the scale was subject to experimental determination. In 1954, the Tenth General Conference [1] adopted a resolution in which the size of the kelvin was fixed in a different manner, namely, by assigning the exact value 273.16 K to the triple point of water, the other value being, of course, the natural zero point of the scale. On this basis, the ice and steam points have no special significance, and their values are subject to experimental determination. One group of investigators (mentioned at the end of this section) found the interval from the ice point to the steam point to be 99.994 kelvin on the basis of the 1954 definition of the size of the unit; it is exactly 100 kelvin on the basis of the earlier definition.

The Celsius, formerly known as the Centigrade, scale ($t°C$) differs from the Kelvin scale only in that its zero point is displaced. At the Ninth General Conference in 1948, the ice point was taken to be 0°C, as heretofore. In 1954, the Tenth General Conference adopted a resolution defining 0°C to be exactly 0.01 deg. below the triple point of water. The latter is the only fixed point whose temperature is known exactly on both the Celsius (+ 0.01°C) and on the Kelvin (273.16 K) scales. The new zero point of the Celsius scale is quite close to the ice point, but there is experimental evidence that it may be 1×10^{-4} to 2×10^{-4} deg. below this fixed point. The relations between the Celsius and the Kelvin scales in 1948 and in 1954 are shown by the following equations.

$$t°C\,(1948) = T\,K\,(1948) - T_0\,K\,(1948), \qquad T_{100} - T_0 = 100 \text{ deg.}$$
The value of $T_0\,K\,(1948)$ is to be determined experimentally on this scale.
(2.9)

$$t°C\,(1954) = T\,K\,(1954) - T_0\,K\,(1954), \qquad T_0\,(1954) = 273.15\,K.$$
The ice and steam points do not define fixed points on this scale. (2.10)

Throughout the text, we will use the unmodified terms Kelvin and Celsius to denote thermodynamic temperatures.

Because of the difficulty of measuring temperatures on either of the above thermodynamic scales, and the resulting discordant scales used in different countries, the Seventh General Conference on Weights and Measures [2] adopted an International Temperature Scale in 1927. This scale was revised at the Tenth General Conference in 1948 [3]; and there was a "Text Revision" of the scale at the Eleventh General Conference in 1960 [4]*, at which its name was changed to the International Practical Temperature Scale (IPTS). The IPTS is a Celsius scale ($t°C$, Int.) designed to agree with the thermodynamic Celsius scale as closely as contemporary knowledge permits, but it is not itself a thermodynamic scale (as will be shown below). It is based on six reproducible temperatures, to which numerical values are assigned (see Table 2.2). Interpolation between the defining fixed points is accomplished by the following equations [4], which describe the relation between temperature and the thermometric properties of certain designated thermometers over fixed temperature

---

*The pretense is made that the action of the General Conference in 1960 was not a revision of the International Temperature Scale of 1948 "but merely a revision of its text." Hence the scale was called the IPTS of 1948. Fortunately the IPTS was not changed appreciably, but the change in the size of the Kelvin degree (in 1954) was more significant.

TABLE 2.2

DEFINING FIXED POINTS ON THE INTERNATIONAL PRACTICAL TEMPERATURE SCALE

Exact values assigned. The pressure is 1 standard atmosphere, except for the triple point of water

Fixed point	Temperature °C (Int., 1948)
Temperature of equilibrium between liquid oxygen and its vapor (oxygen point)	−182.97
Temperature of equilibrium between ice, liquid water, and water vapor (triple point of water)	+0.01
Temperature of equilibrium between liquid water and its vapor (steam point)	100
Temperature of equilibrium between liquid sulfur and its vapor (sulfur point)	444.6[a]
Temperature of equilibrium between solid silver and liquid silver (silver point)	960.8
Temperature of equilibrium between solid gold and liquid gold (gold point)	1063

[a]In place of the sulfur point, it is recommended that the temperature of equilibrium between solid zinc and liquid zinc (zinc point) with the value 419.505°C (Int., 1948) should be used.

intervals. These equations are not linear, since we are interested here in reproducing, as closely as possible, a thermodynamic scale, rather than the arbitrary scale of the particular thermometer in question (which we considered in Section 2.6).

From 0 to 630°C (the temperature of equilibrium between solid antimony and liquid antimony) the temperature t is defined by the equation

$$R_t = R_0 (1 + At + Bt^2) \quad (2.11)$$

where R_t is the electrical resistance at temperature t of the platinum wire resistor of a standard platinum resistance thermometer, and R_0 is the resistance at 0°C. The constants R_0, A, and B are to be determined from the value of R_t at the triple point of water, the steam point, and the sulfur point (or the zinc point). This equation can be written in the Callendar form which shows its relation to eqns. (2.7) and (2.8):

$$t_{pt} = \frac{1}{\alpha}\left(\frac{R_t}{R_0} - 1\right), \alpha = \frac{1}{100}\left(\frac{R_{100}}{R_0} - 1\right) \quad (2.12a)$$

$$t = t_{pt} + \delta \frac{t}{100}\left(\frac{t}{100} - 1\right) \quad (2.12b)$$

where

$$\alpha \equiv A + 100B, \delta \equiv -\frac{10^4 B}{A + 100B} \quad (2.13)$$

Here α is the mean coefficient of thermal increase of resistance of the thermometer from 0 to 100°C, and t_{pt} is the "platinum temperature" corresponding to t, computed in accordance with eqns. (2.7) and (2.8). The term containing the constant δ is a measure of the deviation $t - t_{pt}$ of the platinum scale from the Celsius scale of temperature in the range 0—630.5°C.

From the oxygen point to 0°C, the temperature t is defined by the equation

$$R_t = R_0 [1 + At + Bt^2 + Ct^3 (t - 100)] \tag{2.14}$$

where R_0, A, and B are determined in the manner described in the last paragraph above, and the constant C is then determined from the value of R_t at the oxygen point. Equation (2.14) can be written in the Callendar—Van Dusen form:

$$t = t_{pt} + \delta \frac{t}{100} \left(\frac{t}{100} - 1\right) + \beta \left(\frac{t}{100}\right)^3 \left(\frac{t}{100} - 1\right) \tag{2.15}$$

where t_{pt} and δ are given by eqns. (2.12) and (2.13), and

$$\beta = -\frac{10^8 C}{A + 100B} \tag{2.16}$$

From 630.5°C to the gold point, the temperature t is defined by the equation

$$E = a + bt + ct^2 \tag{2.17}$$

where E is the electromotive force of a standard thermocouple of platinum and platinum—rhodium alloy, when one of the junctions is at 0°C and the other at the temperature t. The constants a, b, and c are to be determined from the values of E at 630.5°C (as measured with a standard platinum resistance thermometer), at the silver point, and at the gold point.

Above the gold point, the temperature t is defined by the equation

$$\frac{J_t}{J_{Au}} = \frac{\left\{\exp\left[\frac{C_2}{\lambda(t_{Au} + T_0)}\right]\right\} - 1}{\left\{\exp\left[\frac{C_2}{\lambda(t + T_0)}\right]\right\} - 1} \tag{2.18}$$

where J_t and J_{Au} are the radiant energies per unit wavelength interval at wavelength λ, emitted per unit time per unit solid angle per unit area of a black body at the temperature t and at the gold point, respectively, C_2 is the second radiation constant with the value $C_2 = 0.01438$ meter-degrees, λ is in meters, and $T_0 = 273.15$ degrees.

The IPTS of 1948 as revised in 1960 was not defined for temperatures below the oxygen point.

The complete revised text defining the IPTS of 1948 is given in ref. 4. Besides the definitions given above, this publication includes recommendations for the construction of standard resistance thermometers and standard thermocouples, tests for purity of the materials of construction, recommendations for constructing the defining fixed points, and a list of secondary reference temperatures.

In the range from the ice point to the sulfur point, intercomparison of the IPTS with the Celsius scale, as realized by two constant-volume nitrogen gas thermometers, gave the following relation [4, 5]

$$t(\text{therm.}) - t(\text{Int.}) = [-0.0060 + (0.01t-1)(0.04106 - 7.363 \times 10^{-5}t)] \times (0.01\,t) \qquad (2.19)$$

where the term t in the right-hand side of the equation is on the IPTS. This equation gives 99.9940°C (therm., 1954) for the steam point. As mentioned above, this value is exactly 100°C (therm., 1948).

At the tenth General Conference on Weights and Measures in 1960, an International Practical Kelvin Temperature Scale (°K, Int.) was defined [4] by the following relation

$$T°K\,(\text{Int.}) = t°C\,(\text{Int.}) + 273.15 \qquad (2.20)$$

where t is the temperature on the corresponding International Celsius scale. This is not the Kelvin scale, which is based on thermodynamic theory, but it was designed to agree with the latter as closely as current knowledge permitted.

The most accurate measurements of the difference between the IPTS (1968) and the thermodynamic Kelvin temperature scale over the range 273.16—730 K have been made at the U.S. National Bureau of Standards [6]. The IPTS (1968) was measured using a standard platinum resistance thermometer. The thermodynamic scale consisted of a constant-volume helium gas thermometer in which the bulb was made of a platinum—rhodium alloy. The mercury manometer used "capacitance sensing of meniscus positions". The results are given in Table 2.3.

TABLE 2.3

DIFFERENCE BETWEEN THE IPTS OF 1968 AND THE THERMODYNAMIC SCALE OVER THE RANGE 273.16—730 K

$$T/K - T_{68}/K_{68} = -120{,}887.784/T_{68}^2 + 1213.53295/T_{68} - 4.3159552$$
$$+ 6.44075647 \times 10^{-3}\,T_{68} - 3.56638846 \times 10^{-6}\,T_{68}^2$$

Application of the above equation to the IPTS-68 boiling point of water and freezing points of tin and zinc gives the results shown in the Table

		Uncertainty	
$t(°C)$	$T/K - T_{68}/K_{68}$	Random (99% confidence limits)	Systematic
100	−0.0252	±0.0018	±0.00054
231.9681	−0.0439	±0.0022	±0.0015
419.58	−0.0658	±0.0028	±0.0028

In the above equations T/K is the thermodynamic temperature on the Kelvin scale: T_{68}/K_{68} or simply T_{68} is the IPTS-68 temperature on the Kelvin scale.

TABLE 2.4

SOME THERMOMETRIC SYSTEMS

Thermometer	Thermometric Property
Pure gas at constant volume and mass	Pressure
Pure gas at constant pressure and mass	Volume
Pure gas at constant temperature and mass	Pressure—volume product
Optical pyrometer	Spectral radiance
Paramagnetic crystal at constant pressure	Magnetic susceptibility
Pure liquid and vapor in equilibrium	Vapor pressure
Annealed wire at constant pressure	Electrical resistance
Annealed thermocouple at constant pressure	Thermal electromotive force
Liquid at constant pressure (confined in a glass)	Volume

2.8. Thermometers. Some thermometric systems and their corresponding thermometric properties are listed in Table 2.4. Many other methods for determining or estimating temperatures have been proposed.

The theory for the computation of thermodynamic temperatures from the indications of certain thermometers, called primary thermometers, has a firm basis in thermodynamic theory. Usually, in addition to measurements of the thermometric property of the thermometer, we must determine experimentally other properties of the system.

Of the systems listed in Table 2.4, the three gas thermometers, the optical pyrometer, the magnetic thermometer, and the vapor pressure thermometer have been employed as primary instruments for the realization of one segment or another of a thermodynamic temperature scale. The gas thermometer, in one or another of the three forms listed in the Table, is the instrument actually employed for the determination of thermodynamic temperatures from about 1 K to the gold point (1063 K); the procedures are outlined later in this section. Thermodynamic temperatures above the gold point are determined by means of optical pyrometers [7], see eqn. (2.18); those below ~1 K are measured using magnetic thermometers [8]. In principle, the vapor pressure thermometer can be used for the realization of the Kelvin scale in those regions where the liquid (or solid) and vapor phases of a pure substance coexist. In practice, this thermometer has been employed (as a primary instrument) only for the purposes of checking the accuracy of the realization of the Kelvin scale by other thermometers [9].

The experimental difficulties of determining thermodynamic temperatures with the primary thermometers has led to the introduction of many secondary instruments. These have one or more of the following advantages: (1) small bulk, at least for the part of the instrument inserted into the system under investigation; (2) short relaxation time for the attainment of thermal equilibrium with the system in question; (3) ease of operation; (4) reproducibility; and (5) precision. Of the instruments listed in Table 2.4, the resistance thermometer, the thermocouple, and the liquid-in-glass thermometer have always been used as secondary thermometers, while the optical pyrometer, the magnetic thermometer, and the vapor pressure thermometer have been so employed in certain temperature ranges. A secondary thermometer is standardized against a primary instrument, and its indications converted into thermodynamic temperature by use of a table, or of an empirical equation with or without a supplementary correction table. This is the procedure employed for the definition of the International Practical Temperature Scale from the oxygen point to the gold point, and for the provisional extension of this scale below the oxygen point.

Let us now consider the methods employed to realize a thermodynamic temperature scale with gas thermometers. Their indications, when converted into temperature by linear relations similar to eqns. (2.5) and (2.6), do not yield thermodynamic temperatures directly, but the theory of correcting their indicated temperatures to this scale have a firm basis in experiment and in thermodynamic principles.

In the idealized constant-volume (or constant-density) gas thermometer [10], a fixed mass of gas is confined in a bulb of invariant volume connected by a capillary of negligible volume to the lower arm of a mercury manometer which is also of negligible gas volume. The connecting capillary and the crown of the mercury meniscus in the short arm of the manometer lie in the same horizontal plane as the center of the thermometer bulb, so that the effect of the Earth's gravitational field may be ignored. Two methods of operation for the determination of temperature have been used.

Prior to 1954, the procedure [10] was to measure the pressures p_0, p_{100}, and p_t of the gas while the bulb of the thermometer was maintained at the ice point t_0, the steam point t_{100}, and in thermal equilibrium with a system at an unknown Celsius temperature t, respectively. The mean coefficient α_v of thermal pressure increase at constant volume of the gas from 0 to 100°C, and the Celsius temperature t_v, corresponding to t, on the constant-volume scale of the thermometric gas at the given fixed density were defined by the relations

$$\alpha_v = \frac{1}{100}\left(\frac{p_{100}}{p_0} - 1\right) \tag{2.21a}$$

$$t_v = \frac{1}{\alpha_v}\left(\frac{p_t}{p_0} - 1\right) \tag{2.21b}$$

Both depend on the thermometric gas used and on its density, that is, on the ice point pressure, p_0, of the series of measurements. In order to determine thermodynamic temperatures, several determinations of α_v and t_v were made for each of several values of density, corresponding to several values of p_0, each of which were pressures lower than one meter of mercury. We shall show later that

$$\lim_{p_0 \to 0} \alpha_v = \frac{1}{T_0}, \quad \lim_{p_0 \to 0} t_v = t, \quad T = T_0 + t \tag{2.22}$$

where T_0 is the Kelvin temperature of the ice point, t is the Celsius temperature of the system under investigation, and T is the corresponding Kelvin temperature. The last expression of the eqns. (2.22) was the definition of the relation of T to t prior to 1954 [see eqn. (2.9)].

The procedure for determining thermodynamic temperatures since 1954 has been to measure the pressures p_{tr} and p_t of the gas while the bulb of the thermometer is maintained at the temperature of the triple point T_{tr} of water (273.16 K), and in thermal equilibrium with a system at the unknown Kelvin temperature T, respectively. The absolute temperature T_v, corresponding to T, on the constant-volume scale of the thermometric gas at the given fixed density is defined by the relation

$$T_v = 273.16 \, p_t/p_{tr} \tag{2.23}$$

where T_v depends on the thermometric gas and its density, or pressure p_{tr}, at the triple point of water. As in the method described in the last paragraph, we determine T_v for several different values of p_{tr} and apply the limiting process of eqn. (2.22):

$$T = \lim_{p_{tr} \to 0} T_v, \, t = T - 273.15 \tag{2.24}$$

where T is the Kelvin temperature of the system under investigation, and t is the corresponding thermodynamic Celsius temperature [see eqn. (2.10)].

We note that, in the method used prior to 1954, the values of both t and T_0 were determined experimentally, and the value of T, corresponding to t, was calculated by adding T_0 to t; in the procedure employed since 1954, the value of T is determined experimentally, and the value of t, corresponding to T, is calculated by subtracting from T the defined value (273.15) of the Kelvin temperature of t_0. These methods are used to conform to the procedures adopted by the General Conference on Weights and Measures, as described in Section 2.7.

The idealized constant-pressure gas thermometer [10] differs from the constant-volume instrument in only one respect. A tee is inserted in the capillary connecting the thermometer bulb to the manometer. One arm of the tee is connected to a capillary of negligible volume, which leads to a pipette maintained at a fixed temperature. The gas volume in the pipette can be varied and measured by the introduction or removal of known masses of liquid mercury. In the operation of the thermometer, the pressure p of the gas is brought exactly to the same fixed value at each bulb temperature by adjustment of the gas volume in the pipette. As in the constant-volume instrument, the mass of gas in the thermometric system is constant during any one series of measurements; but, as the temperature of the thermometer bulb is changed, various amounts of gas are transferred from the bulb to the pipette.

Prior to 1954, the procedure [10] was to measure the volumes v_0, v_{100}, and v_t of the gas in the pipette (always maintained at a fixed temperature, for example the ice point) required to adjust the pressure in the thermometer bulb of volume V_b to the fixed value p, while the bulb was maintained at the ice point t_0, the steam point t_{100}, and in thermal equilibrium with a system at an unknown Celsius temperature t, respectively. Let $\bar{V}_0$, $\bar{V}_{100}$, and $\bar{V}_t$ be the reciprocals of the corresponding densities D_0, D_{100}, and D_t (expressed in any convenient set of units) *of the gas in the thermometer bulb* while the bulb was maintained at t_0, t_{100}, and $t°C$, respectively. The mean coefficient α_p of thermal dilation at constant pressure of the gas from 0 to 100°C, and the Celsius temperature t_p, corresponding to t, on the constant pressure scale of the thermometric gas at the fixed pressure p were defined by the relations

$$\alpha_p = \frac{1}{100} \left(\frac{\bar{V}_{100}}{\bar{V}_0} - 1 \right) = \frac{1}{100} \left(\frac{D_0}{D_{100}} - 1 \right) \tag{2.25a}$$

$$t_p = \frac{1}{\alpha_p} \left(\frac{\bar{V}_t}{\bar{V}_0} - 1 \right) = \frac{1}{\alpha_p} \left(\frac{D_0}{D_t} - 1 \right) \tag{2.25b}$$

Both depend on the thermometric gas used and on the fixed pressure p of the series of measurements. We can express these quantities in terms of the measured volumes by noting that the following relations hold, because the temperature of the pipette, the mass of gas in the thermometric system, and the pressure were each maintained constant throughout a series of runs:

$$V_b D_0 + v_0 D_0 = V_b D_{100} + v_{100} D_0 \tag{2.26a}$$

$$V_b D_0 + v_0 D_0 = V_b D_t + v_t D_0 \tag{2.26b}$$

From these relations and eqns. (2.25) we find

$$\alpha_p = \frac{1}{100} \frac{v_{100} - v_0}{V_b - (v_{100} - v_0)} \tag{2.27a}$$

$$t_p = \frac{1}{\alpha_p} \frac{v_t - v_0}{V_b - (v_t - v_0)} \tag{2.27b}$$

After determining the values of α_p and t_p for each of several values of the fixed pressure p, we applied the limiting procedures of eqns. (2.22):

$$\lim_{p \to 0} \alpha_p = \frac{1}{T_0}, \lim_{p \to 0} t_p = t, \ T = T_0 + t \tag{2.28}$$

where T_0 is the Kelvin temperature of the ice point, and t and T are, respectively, the Celsius and the Kelvin temperatures of the system under investigation.

The procedure for determining thermodynamic temperatures since 1954 has been to measure the volumes v_{tr} and v_t of the gas in the pipette (always maintained at a fixed temperature, for example the triple point of water) required to adjust the pressure in the thermometer bulb of volume V_b to the fixed value p while the bulb was maintained at the temperature of the triple point T_{tr} of water (273.16 K), and in thermal equilibrium with a system at an unknown Kelvin temperature T, respectively. Let $\bar{V}_{tr}$ and

$\bar{V}_t$ be the reciprocals of the corresponding densities D_{tr} and D_t *of the gas in the thermometer bulb*, while the bulb is at the temperatures T_{tr} and T, respectively. The absolute temperature T_p, corresponding to T, on the constant pressure scale of the thermometric gas at the fixed pressure p is defined by the relation

$$T_p = 273.16 \, \bar{V}_t/\bar{V}_{tr} = 273.16 \, D_{tr}/D_t \tag{2.29}$$

The relation corresponding to eqn. (2.26) is

$$V_b D_{tr} + v_{tr} D_{tr} = V_b D_t + v_t D_{tr} \tag{2.30}$$

Equations (2.29) and (2.30) give

$$T_p = 273.16 \, V_b/[V_b - (v_t - v_{tr})] \tag{2.31}$$

The relations corresponding to eqns. (2.28) are

$$\lim_{p \to 0} T_p = T, \quad t = T - 273.15 \tag{2.32}$$

where t is the Celsius temperature corresponding to the Kelvin temperature T.

The practical constant-volume and constant-pressure gas thermometers differ from the idealized instruments in many respects. Among these are the variations of the volumes of the bulb and other parts of the system with pressure and with temperature, the variation of the gas volume in the short arm of the manometer with the location and shape of the mercury meniscus, and the quite appreciable amount of gas in the capillary leading from the bulb to the manometer (and to the pipette in the constant-pressure thermometer), part of which is in a region of large thermal gradient. We must also take into account the effect of the local gravitational field. In the choice of the thermometric gas and in the design of the system, we must take into consideration the effects of variations in the amounts of gas adsorbed on the walls of the system with pressure and temperature, and the thermomolecular pressure effect in the part of the capillary situated in the region of large thermal gradient. The values of pressures and volumes appearing in the equations given above are the corresponding observed values corrected for these effects.

A sketch [11] of a double constant-volume gas thermometer is shown in Fig. 2.1. Each thermometer could be converted into a constant-pressure instrument by connecting a pipette to the capillary N_2 leading from the valve F. Each thermometer system consists of a bulb B connected by the capillaries A' and D to the short arm G of the precision mercury manometer GJ, which is enclosed in an air thermostat K. The correction for the gas in the region of large thermal gradient A' is made by means of the indications of an auxiliary gas thermometer A which, in turn, is connected by the capillary C to the somewhat less precise auxiliary manometer PQ, also enclosed in an air thermostat. The bulb B, the capillary A', and the auxiliary thermometer A are enclosed in interconnected steel cases filled

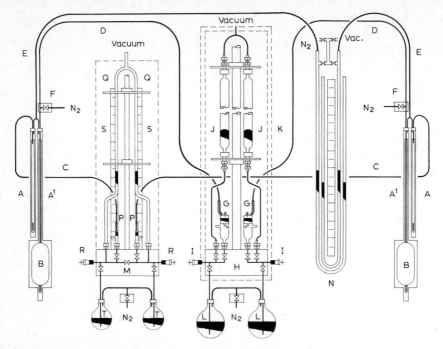

Fig. 2.1. Sketch of a double constant-volume gas thermometer.

during each measurement to the same pressure of gas as that in the thermometer bulb B. The capillary E leads from the steel case to the manometer N.

A third procedure for realizing the thermodynamic temperature scale by means of a gas thermometer has been proposed, and found suitable for the high-temperature and low-temperature portions of the scale. The constant-bulb-temperature gas thermometer is similar in construction to a constant-pressure instrument, but is operated in a different manner. The thermometer bulb is maintained at the temperature whose value is to be determined, while successive portions of gas are withdrawn into the pipette which is held at a fixed reference temperature whose value is known, for example, the triple point of water. The pressure in the system is measured before and after each withdrawal of gas. The low-pressure equation of state of a real gas, see eqn. (2.44), is the basis for computing the Kelvin temperature of the thermometer bulb from the observed pressures and volumes. In the high-temperature version [12] of this instrument, which has been used up to the gold point, a knowledge of the mass of gas in the system is not required, but the values of the second virial coefficients B of the thermometric gas (see Section 2.11) must be determined in a separate experiment at the reference temperature and at each bulb temperature. In the low-temperature version [13], which has been used at liquid-helium temperatures, a knowledge of the second virial coefficient of the gas at the reference temperature and of the universal gas constant R is required, since the number of moles of gas in the thermometer bulb at each pressure is computed. The advantages claimed for this type of thermometer are that the effects of gas adsorption on the walls of the bulb, and of hysteresis in the volume of the bulb, are reduced, since the temperature of the bulb is not changed during a complete determination of temperature.

2.9. Perfect gas temperature scales. As the pressure of a gaseous substance approaches zero, the substance approaches a condition known as the perfect gas state where it conforms to certain simple principles called the laws of perfect gases. We have not, as yet, developed thermodynamic theory sufficiently to define a perfect gas completely from the thermodynamic viewpoint, but we can now discuss two empirical perfect gas temperature scales, which bear the same relation to each other as the Celsius scale bears to the Kelvin scale. The perfect gas scales depend on the laws of Boyle and Gay-Lussac, and they are defined in accordance with eqns. (2.5) and (2.6), with the pressure—volume product of a fixed mass of gas as the thermometric property.

Boyle's law states that the pressure—volume product pV of a fixed mass, m, of a perfect gas is constant at a constant temperature θ, that is,

$$pV = f(\theta) \qquad (m = \text{constant}) \qquad (2.33)$$

Here θ may be measured on any temperature scale. Boyle's law does not specify the function $f(\theta)$.

Gay-Lussac's law (also known as Charles' law) states that all perfect gases undergo the same fractional increase in their pressure—volume product when heated in a closed system from the same initial temperature θ_1 to the same final temperature θ_2, that is,

$$\frac{(pV)_2 - (pV)_1}{(pV)_1} = F(\theta_2, \theta_1) \qquad (m = \text{constant}) \qquad (2.34)$$

or

$$\frac{(pV)_2}{(pV)_1} = 1 + F(\theta_2, \theta_1) = \phi(\theta_2, \theta_1) \qquad (m = \text{constant}) \qquad (2.35)$$

Here the functions F and ϕ depend on the temperature scale on which θ is measured, but are independent of the gas so long as it is in the state of a perfect gas.

By virtue of eqn. (2.35) and in accordance with eqn. (2.5) we define a perfect gas absolute scale $T°A$ (which we shall also call the perfect gas Kelvin scale TK), by means of the relation

$$\frac{T}{T_{tr}} = \frac{T}{273.16} = \frac{(pV)_T}{(pV)_{tr}} \qquad (m = \text{constant}) \qquad (2.36)$$

A corresponding perfect gas Celsius scale is defined by the equation

$$t°C = T°A - 273.15 \qquad (2.37)$$

Whence

$$\frac{t + 273.15}{t_{tr} + 273.15} = \frac{t + 273.15}{273.16} = \frac{(pV)_t}{(pV)_{tr}} \qquad (2.38)$$

Here we have assigned the values $T_{tr} = 273.16$ K and $t_{tr} = 0.01°$C to the temperature of the triple point of water on the Kelvin and the Celsius perfect gas scales, respectively, in agreement with the values assigned on the Kelvin and the Celsius (thermodynamic) scales by the Tenth Conference on Weights and Measures (see Section 2.7). The laws of Boyle and Gay-Lussac assure us that each of the two scales of temperature defined by eqns. (2.36) and (2.38) is independent of the particular gas used, providing it is in the state of a perfect gas. For this reason, the application of the limiting processes of eqns. (2.22), (2.24), (2.28), and (2.32) to the empirical temperatures derived from constant-volume and constant-pressure gas thermometers yield perfect gas Kelvin and perfect gas Celsius temperatures, which are independent of the real gas used as the thermometric fluid.

The reader will not fail to notice that we have used the same symbols, T and t, for the perfect gas scales as for the corresponding thermodynamic scales. This is done intentionally. After the first and second laws of thermodynamics have been presented, we shall define a perfect gas (for thermodynamic purposes) as one that obeys Boyle's and Joule's laws. We shall then show that Gay-Lussac's law is one thermodynamic consequence of these two laws. Another consequence is the identity of perfect gas temperature scales with the corresponding thermodynamic scales. In order to keep the number of symbols used to a minimum, we shall indicate perfect gas Kelvin and Celsius temperatures by writing T K and $t°$C, respectively. Until we have proved the identity of perfect gas scales with the corresponding thermodynamic scales, from Boyle's law and Joule's law and the first and second law equation of thermodynamics, we shall regard the perfect gas scales as simply two empirical scales of temperature.

2.10. Equation of state of a perfect gas. The relation connecting the pressure p, molar volume v, and perfect gas absolute temperature T of a perfect gas rests on two laws and two definitions. These are Boyle's law, Gay-Lussac's law, the definition of the perfect gas absolute temperature scale, and the empirical definition of the mole as a unit of mass for gases. All except the last have been discussed above. The following empirical definition of a mole suffices for expressing the pressure—volume—temperature relations of gases and for the thermodynamic applications of this equation.

One mole, or one molar weight, M, is the number of grams of a gas which has the same value of pv/T as one mole of oxygen when each is in the perfect gas state. Formerly, one mole of oxygen was defined to be exactly 32 grams of the naturally-occurring elementary substance; now it is taken to be 31.9988 grams of this substance, in accordance with the shift of the basis of atomic weights mentioned in Section 1.17. One mole of naturally-occurring oxygen has a possible mass variation of ±0.0002 gram occasioned by variations in its isotopic composition with the source of the sample.

The above definition of a mole states that the value of pv/T for every perfect gas, and hence for every real gas at a sufficiently low pressure, is a universal constant. This quantity is evidently equal to the limiting value of $(pv)_0/273.15$ for one mole of the real gas oxygen as the pressure approaches zero at 0°C (273.15 K). Equation (2.36) can now be written in the following manner

$$\frac{(pv)_T}{T} = \frac{(pv)_{tr}}{273.16} = \lim_{p \to 0}\left[\frac{(pv)_0}{273.15}\right]_{\text{oxygen}} = R \tag{2.39}$$

or

$$pv = RT \tag{2.40}$$

In these relations, R has the same value for all perfect gases, and is called the gas constant.

The molar pressure-volume product $(pv)_0$ of a perfect gas at 0°C is usually tabulated as the standard molar volume v_0 of a perfect gas at 1 atmosphere and 0°C. Values [14] of v_0 and of R are listed in various sets of units in Table 2.5.

Since volume is an extensive property of a system, the volume V of n moles of a gas is nv, where V and v are measured at the same pressure and temperature. Evidently n is the ratio (m/M) of the number of grams m of the gas in the volume V to the mass M of one mole, as defined above. Thus, eqn. (2.40) can be written in the form

$$pV = nRT = \frac{m}{M}RT \tag{2.41}$$

This is the equation of state of a perfect gas. Although we shall call T in the above expression Kelvin temperature, we must bear in mind that it refers only to an empirical scale known as the perfect gas absolute scale, until we show that it has a more fundamental significance. In particular, we must remember that the derivative of perfect gas temperature with respect to the corresponding thermodynamic temperature cannot be assumed to be unity until so proven by thermodynamic arguments.

TABLE 2.5

VALUES OF THE STANDARD MOLAR VOLUME OF A PERFECT GAS, v_0, AND OF THE GAS CONSTANT R IN VARIOUS UNITS

v_0	Units	R	Units[a]
22,413.6	cm^3 $mole^{-1}$	82.056	cm^3 atm $mole^{-1}$ deg^{-1}
22.4130	liter $mole^{-1}$	0.082054	liter atm $mole^{-1}$ deg^{-1}
		8.3143	J $mole^{-1}$ deg^{-1}
		8.3143×10^7	erg $mole^{-1}$ deg^{-1}
		1.9872	(Thermochem.) cal $mole^{-1}$ deg^{-1}

[a]The abbreviation deg means the Kelvin degree.

Equation (2.41) embodies the laws of Boyle and Gay-Lussac, eqns. (2.33) and (2.34), the definitions of the perfect gas Kelvin scale and the mole as a unit of mass for a gas, and also eqns. (2.36) and (2.39). Once the value of R has been determined from eqn. (2.39), we can employ the following relation, based on eqn. (2.41), to determine the value of a mole M for any real gaseous substance

$$M = RT \lim_{p \to 0} \frac{d}{p} \qquad (m, T = \text{constant}) \qquad (2.42)$$

where $d = m/V$ is the density of the gas at p and T. When interpreted in the light of Avogadro's law (the quantities of different perfect gases which have equal values of pV/T contain equal numbers of molecules), eqn. (2.42) is the basis of one experimental method for determining molecular weights of gases.

Every equation of thermodynamics which contains the universal gas constant R depends on eqn. (2.41), for it is through this relation that R enters thermodynamic formulae. The molar weight M of the gas used enters explicitly into eqn. (2.41). Hence, whenever R appears in an equation of thermodynamics, the molar weights of the substances composing the system must be known, or at least an assignment of molar weights must have been made. Of course these remarks do not apply if we choose to write eqn. (2.41) for unit mass (one gram or one pound) of gas. Then the quantity $p\bar{v}/T$, although constant for each perfect gas, would vary from one perfect gas to another. If we persist in using this unit of mass, we would find that many of the equations of thermodynamics would have an unfamiliar appearance.

2.11. Virial equations of state of a real gas. We shall present in this section two of the many equations of state which have been proposed for real gases. The pressure—volume product pV of a fixed mass of a non-dissociating pure real gas at a constant temperature T has been represented by a power series in the density n/V. From the discussion in the last two sections, we know that the first term of the series is nRT where n is the number of moles of the gas, R is the gas constant, and T is the perfect gas absolute temperature. The series written in closed form has been found empirically to represent the measured pV-products of gases quite satisfactorily over wide ranges of density and temperature, including the critical region. The infinite series can be rigorously derived from molecular (and hence not from thermodynamic) considerations. The following expression is called the virial equation of state, because it was first derived theoretically by the methods of kinetic theory from the virial theorem of Clausius

$$pV = nRT\left(1 + \frac{nB}{V} + \frac{n^2 C}{V^2} + \frac{n^3 D}{V^3} + \ldots\right) \qquad (2.43)$$

The parameters B, C, D, ... are called the second, third, fourth, ... virial coefficients of a gas. They are functions of temperature only.

Equation (2.43) can be converted into the following power series in pressure

$$pV = nRT + nB'p + nC'p^2 + nD'p^3 + \ldots \quad (2.44)$$

where B', C', D', ... are temperature functions. If we consider that this relation is derived from the virial equation (2.43) by mathematical manipulation, the following relations must hold:

$$B' = B;\ RTC' = C - B^2;\ (RT)^2 D' = D - 3BC + 2B^3;\ \ldots \quad (2.45)$$

Equation (2.44) is known as the pressure virial equation, and B', C', D', ... are called the second, third, fourth, ... pressure virial coefficients.

Investigators interested mainly in the representation of measured pV-products frequently omit the terms in odd powers of n/V and of p higher than the first in eqns. (2.43) and (2.44), since they have found that the experimental values can be represented quite satisfactorily with fewer terms by such a series. The number of terms retained in any of these series depends on the accuracy of representation of the measurements desired, the density (or pressure) range covered, and the relation of the temperature to the critical temperature of the gas under consideration. In general, three to six terms suffice, but as many as ten terms have been employed for isotherms in the critical region.

Whether we use all of the powers of n/V and p or omit the odd powers above the first, we find empirically the following results when the parameters of any of these equations of state are determined from measured pV-products:

1. The density series represents the experimental data better than the pressure series when the same number of terms is used in each.

2. A variation in the number of terms retained in each expansion has a small effect on the value found for B or B', but has, in general, a large effect on the values of all of the other coefficients.

3. The first relation of eqns. (2.45) holds quite satisfactorily for values of B and B' derived from experimental data; the remaining relations do not.

4. The temperature variation of the second virial coefficient determined from measured pV-products agrees fairly well with that predicted from theory over quite a wide temperature range for many gases. No such agreement is found for the higher virial coefficients, and hence their values when determined from experimental data have little physical significance.

The following expressions which contain only the first two terms of eqns. (2.43) and (2.44) give a satisfactory representation of the pV-product of real gases in the region of low pressures:

$$pV = nRT\left(1 + \frac{nB}{V}\right) \quad (2.46)$$

$$pV = nRT + nBp \quad (2.47)$$

Here B has the same value in each equation. The question naturally arises: what is the region of low pressures? The answer depends on the desired accuracy of representation of the data. In general, at temperatures higher than 100 deg. above the critical temperature of a gas, the errors introduced by neglecting the third and higher terms in eqns. (2.43) and (2.44) are less than 0.001% at 1 atmosphere and less than 0.1% at 10 atmospheres. The error of the two-term equations for the gases employed in gas thermometry at the pressures and temperatures used is several parts per million or less, much smaller than the uncertainties in any existing gas thermometric measurements.

The extension of the virial equation of state to gas mixtures will be presented in a later chapter.

2.12. Thermostats. In order to produce a given change in state by an isothermal (constant temperature) process, we enclose the system in a diathermal boundary and place it in thermal contact with a second system (the thermostat) whose temperature is maintained at the desired value within the required variation. The main mass of a thermostat may consist of a well-stirred fluid, of a thermally conducting solid surrounded by one or more enclosures of the same material (radiation shields), or of a liquid boiling under a constant pressure.

There are many kinds of thermostats. A typical instrument has a mass much larger than the mass of the system under investigation. Its outer enclosure is composed of materials of low thermal conductivity. The temperature of the thermostat is maintained constant by the action of a thermoregulator, while the temperature of its environment is less-carefully controlled at a somewhat lower value. One component of a thermoregulator circuit is a thermometer in thermal contact with the thermostat. For precise regulation the thermometer senses the direction and the amount of the variation of the thermostat temperature from the desired value. This information is used by the other components of the regulating circuit to vary the electrical current flowing through a resistance wire located within the thermostat.

REFERENCES

1 See H. F. Stimson, in C. M. Herzfeld (Ed.), Temperature, Vol. 3, Part 1, Reinhold, New York, 1962, pp. 59—66.
2 See F. K. Burgess, J. Res. Natl. Bur. Stand., 1 (1928) 635—640.
3 See H. F. Stimson, J. Res. Natl. Bur. Stand., 42 (1949) 209—217.
4 See H. F. Stimson, International Practical Temperature Scale of 1948 — Text Revision of 1960, J. Res. Natl. Bur. Stand., Sect. A, 65 (1961) 139—145.
5 J. A. Beattie, M. Benedict, E. B. Blaisdell and J. Kaye, J. Chem. Phys., 42 (1965) 2274—2282.

6 L. A. Guildner, H. F. Stimson, R. E. Edsinger and R. L. Anderson, Metrologia, 6 (1970) 1—18; L. A. Guildner and R. E. Edsinger, J. Res. Natl. Bur. Stand., Sect. A, 80 (1976) 703—738; T. J. Quinn, L. A. Guildner and W. Thomas, Metrologia, 13 (1977) 177—178; see also, The International Practical Temperature Scale of 1968, Amended Edition of 1975, Metrologia, 12 (1976) 7—17.
7 See H. J. Kostkowski and R. D. Lee, in C. M. Herzfeld (Ed.), Temperature, Vol. 3, Part 1, Reinhold, New York, 1962, pp. 449—481.
8 See, for example, D. de Klerk, M. J. Steenland and C. J. Gorter, Physica, 15 (1949) 649—666.
9 See F. G. Keyes, J. Phys. Chem., 17 (1949) 923—934, for the liquid water range; F. G. Brickwedde, H. van Dijk, M. Durieux, J. R. Clement and J. K. Logan, J. Res. Natl. Bur. Stand., Sect. A, 64 (1960) 1—17, for the liquid helium range; H. van Dijk, M. R. Moussa and R. Muijlwijk, Physica, 32 (1966) 805—822, 900—912, 945—953, for the liquid nitrogen and oxygen ranges.
10 J. A. Beattie, M. Benedict and J. Kaye, Proc. Am. Acad. Arts Sci., 74 (1941) 343—370.
11 J. A. Beattie, D. D. Jacobus, J. M. Gaines, Jr., M. Benedict and B. E. Blaisdell, Proc. Am. Acad. Arts Sci., 74 (1941) 327—342.
12 H. Moser, J. Otto and W. Thomas, Z. Phys., 147 (1957) 59—75; see also, J. A. Beattie, in G. A. Cook (Ed.), Argon, Helium and the Rare Gases, Vol. 1, Interscience, New York, 1961, p. 282.
13 J. Kistemaker and W. H. Keesom, Physica, 12 (1946) 227—240; W. E. Keller, Phys. Rev., 97 (1955) 1—8.
14 E. R. Cohen and W. M. DuMond, in W. H. Johnson, Jr. (Ed.), Nuclidic Masses, Springer, Vienna, 1964, pp. 152—186.

Chapter 3

WORK AND HEAT

The disturbances produced in the surroundings when a closed system undergoes a change in state by a specified process are classified in thermodynamics as either work or heat. These effects require careful definition, because work in thermodynamics is not defined in the same manner as work in mechanics, and heat in thermodynamics is not defined in the same way as heat in kinetic theory. We shall employ operational definitions for both work and heat. Each is expressed primarily in terms of measurements made in the surroundings, since the surroundings are readily accessible to the necessary experimentation, whereas the system is not. Thus, in thermodynamics, the terms work and heat mean primarily the work and heat effects produced in the surroundings, not the work and heat effects produced in or by the system. The latter interpretation can be given only for reversible processes where friction, impact, and other dissipative effects are absent.

In order to define the precise experiment to which work and heat pertain we must specify: (1) the complete change in state of the system; (2) the exact location and the nature of the boundary; (3) the external forces acting on the system throughout the change; and (4) the arbitrary constraints which we impose on the system. We shall use the expression "the process" to denote this description of what occurs in a system undergoing a change in state. The term process includes the specification of the change in state, but the expression "change in state" may be and frequently is used without reference to any particular process.

In principle we can identify and measure (or compute) the work and heat effects produced in the surroundings when a system undergoes a given quasistatic process, whether it be reversible or not. The reason for this lies in the fact that the system is, at all times, infinitesimally close to a state of equilibrium in such a change. Uncertainties in identifying work and heat may arise when the rate of the process is so large that the system departs far from an equilibrium state, as evidenced by large gradients, within the system, of one or more of its intensive properties*. However, as we shall see, we can distinguish and evaluate heat and work exactly for some processes occurring at finite rates, and we can determine these quantities well enough for practical purposes (although with some uncertainty) for many other such processes.

Frequently, we idealize the properties of the material of the boundary so as to make the identification of work and heat effects unambiguous; for

---

*In this connection, see ref. 1.

example, we suppose that the boundary is perfectly adiabatic or rigid. In addition, we may neglect the effects of gradients of pressure, temperature, and other intensive properties of the system resulting from a finite rate of a process. These simplifications permit us to obtain a definite answer for a problem which we could not otherwise solve. We realize that the idealization of the process introduces some error into the result, but in many cases we have reason to believe that the errors are not intolerably large.

Initially we restrict the discussion of work and heat to processes occurring in closed systems. However, we shall see that open systems can usually be brought into the category of closed systems for the purposes of determining work and heat effects by various devices. For example, if the system is a liquid flowing in a pipe, we may choose a boundary moving with the velocity of the fluid and always enclosing a fixed amount of matter.

We find experimentally that the work and heat effects produced in the surroundings of a system depend on how a specified change in state of the system is brought about: in general, each depends on the process, not solely on the change in state. Under these circumstances we cannot defend on any basis the statement that a system in a definite state contains a definite amount of work or a definite amount of heat. Thus work and heat are not properties of a system. They are not state variables, and cannot be expressed as functions of a set of independent state variables. We shall write đW and đQ for infinitesimal quantities of work and heat to denote that they are not exact differentials, and dX for a state variable X to denote that it is an exact differential.

The line integrals of đW and đQ from the initial to the final state of a system for a particular process are represented conventionally by writing W and Q (not ΔW and ΔQ), since in thermodynamics the symbol ΔX is reserved for the increment of a state property X of a system.

3.1. Work. After a system has undergone a change in state, certain disturbances are left in the surroundings resulting from the interactions of the system with its surroundings during the process employed to effect the change. In order to identify a particular disturbance as work and to determine the corresponding quantity of work produced in the surroundings, we must find the answer to the question first posed by Gibbs [2], which may be paraphrased for the present purpose as follows. How great a weight located in the surroundings does the disturbance in question enable us to raise through a given distance, no other permanent change being produced in the surroundings? Evidently, if more than one kind of disturbance is identified as work in a given process, the sum of all of these work effects constitutes the total work of the process.

It may happen that a particular process produces disturbances in the surroundings other than the raising of a weight. Any of these effects are classified as work if they can be converted completely into the elevation of a weight in the surroundings during or at the end of the process, without any other permanent change being produced exterior to the system. Some examples of such effects are the increase in velocity of a mass (in a vacuum), the charging of a reversible galvanic cell surrounded by an adiabatic boundary, and the reversible compression of a spring surrounded by an adiabatic boundary.

Friction, impact, fluid viscosity, and other such dissipative effects occurring entirely in the surroundings may result in a smaller amount of apparent work production in a given process than would have been performed if such effects were absent. In seeking the answer to Gibbs' question in such a case we must determine how much work would have been produced if the process were repeated in such a manner that all of the interactions of the system with its surroundings across the boundary and the change in state of the system were exactly the same as in the original process, but those dissipative effects located entirely in the surroundings were eliminated. Hence, in identifying and measuring the work of a process, we must place the boundary of the system in such a position that every dissipative effect is known to occur entirely inside or entirely outside the system, but not at the boundary or across the boundary. The position given the boundary is considered at some length in Sections 3.5—3.10, where examples of the work of reversible and irreversible expansions of a gas are treated.

3.2. Measure and units of work. The quantity of work W produced in the surroundings by a system undergoing a specified change in state by a given process is equal, by definition, to the net number of standard weights that the process enables us to raise through unit distance in the surroundings. Work is an algebraic quantity: we say that W has a positive value when a weight is raised in the surroundings; W has a negative value when a weight is lowered in the surroundings. In the first case, a positive quantity of work has been produced in the surroundings. In the second case, a negative quantity of work has been produced in the surroundings, that is, a positive quantity of work has been withdrawn from the surroundings. Sometimes we call W the work effect of a process, or simply the work of a process.

In the metric and English gravitational systems the units of force are, respectively, the weight at standard gravity of a mass of one kilogram, called the kilogram-force (kgf) and the weight under the same conditions of a mass of one pound, called the pound-force (lbf) (see the footnote to Table 2.1). The corresponding units of work are the kilogram-meter (kgf m) and foot-pound (ft lb).

In the MKS, CGS, and corresponding English systems the units of work are the newton-meter (N m) or joule (J), the dyne-centimeter (dyn cm) or erg, and the foot-poundal (ft pdl), respectively. Other units are also used, for example, the liter-atmosphere (l atm). Numerical relations between various units of work are given in Table 4.1.

In a particular process let a mass of m kg be raised through a vertical distance Δz meters at a place where the gravitational acceleration is g ms^{-2}, g being substantially constant over the interval Δz. The standard weight of the body lifted is mg/g_n kgf, where g_n is the standard acceleration of gravity (see Table 2.1), and the gravitational force acting on the mass m kg is mg N. The work W produced in the surroundings during the process is

$$W = mg\Delta z/g_n \text{ kgf m} \tag{3.1}$$

or

$$W = mg\Delta z \text{ J} \tag{3.2}$$

In general, we shall express work in MKS or CGS units rather than in a gravitational system of units.

3.3. Kinds of work. We frequently append a word or phrase to the term work to focus our attention on some aspect of the change produced in the system or of the interaction between the system and the surroundings during a process. Examples of these are: (1) the work to vary the position of a system or any finite or infinitesimal part thereof in a gravitational or in a centrifugal field; (2) the work to vary the velocity of a system or any part thereof relative to its surroundings; (3) the work to vary the volume of a system (work of expansion); (4) the work to vary the area or principal curvatures of a surface film (surface work); (5) the work to vary the linear extension or the degree of flexure of a system in the form of a rod, a coiled spring, or a leaf spring; (6) the work to vary the charge of a galvanic cell or of a capacitor (electrical work); (7) the work to vary the linear displacement of a piston rod or the angular displacement of a rotating shaft extending through the boundary of a system (shaft work); (8) the work to vary the polarization of a dielectric; (9) the work to vary the magnetization of a paramagnetic substance; and (10) the work to vary the state of strain in a crystalline solid.

The term work occurring in the above list denotes work produced in the surroundings as defined in Section 3.1, not the work done on or by the system, as the wording of each entry might lead one to believe. Moreover, each entry in the list simply identifies the work associated with a particular phenomenon; it does not define a process. We have not, for example, stated whether the boundary of the system is diathermal or adiabatic, or whether it is rigid or flexible.

The symbol W will be used to denote any one kind of work and also all kinds taken collectively. When we wish to distinguish work of expansion from all other kinds we shall write W_{EX}, and represent any other type, or all other types taken collectively, by the symbol W_X.

Since expansion work is produced in almost every process, we shall discuss it at length here. Other types of work will be considered when needed.

3.4. Expansion work. Let a closed system B undergo the change in state

$$B(\text{state 1}, V = V_1) = B(\text{state 2}, V = V_2) \tag{3.3}$$

by a specified process, whereby the volume enclosed by its boundary varies from an initial value V_1 to a final value V_2. The expansion work dW_{EX} produced in the surroundings in an infinitesimal part of the process is

$$dW_{EX} = \int_S dF_n \, dR_n \tag{3.4}$$

where dF_n is the normal externally-applied mechanical force acting over the element of area dA at a point P in the boundary, dR_n is the normal component of displacement of this same element of area, and S denotes that the integral extends over the entire boundary of the system. In order to be consistent with the convention we have adopted regarding the algebraic sign of work, we consider dF_n to be negative when it acts in the direction of the outwardly-directed normal to the boundary at P and positive when it acts in the opposite direction; and we consider dR_n to be positive when the displacement is in the direction of this normal and negative when the displacement is in the opposite direction. Thus dW_{EX} is positive when dF_n and dR_n are antiparallel, and dW_{EX} is negative when dF_n and dR_n are parallel.

When the ratio $\Delta F_n / \Delta A$ at each point P in the surface approaches a limit as ΔA becomes sufficiently small, eqn. (2.1) applies. Then we can write

$$dW_{EX} = \int_S \frac{dF_n}{dA} dA \, dR_n = \int_S p_{AP} \, dA \, dR_n \tag{3.5}$$

where p_{AP} is the externally-applied pressure at P. If, in addition, the applied pressure may be considered to be uniform over all parts of the boundary surface of the system which move, the work produced in the surroundings in an infinitesimal part of the given process is

$$dW_{EX} = p_{AP} \int_S dA \, dR_n = p_{AP} \, dV \tag{3.6}$$

where dV is the increase in the total volume V of the system. The work of the entire process is expressed by the line integral

$$W_{EX} = \int_{C \, V_1}^{V_2} p_{AP} \, dV, \, p_{AP} = p_{AP}(V) \tag{3.7}$$

The description of the process must be given in sufficient detail to determine the course of the curve C during the change. We may exhibit C in a number of ways: (1) by expressing p_{AP} as a known function of V; (2) by tabulating the function p_{AP} for various values of the argument V; and (3) by a graph of p_{AP} as ordinate against V as abscissa. Evidently work is the area under this curve from the initial to the final volume of the system.

If the process under consideration is reversible, the pressure of the system over each element of area which is displaced differs, at most, infinitesimally from the corresponding applied pressure p_{AP} throughout the entire process. Under these conditions, eqns. (3.6) and (3.7) become

$$dW_{EX} = p\,dV \tag{3.8}$$

$$W_{EX} = \int_{C}^{V_2}_{V_1} p\,dV, p = p(V) \tag{3.9}$$

Here the curve C is called the path of the change in state or, more appropriately, the path of the process. In general, the pressure of a system at equilibrium is a function of a set of independent state variables. The specification of the process must provide a sufficient number of independent relations among these variables to enable us to express p as a function of V.

Let us consider two examples. The equation of state of a gas mixture composed of n_1 moles of a gas 1 and n_2 moles of a gas 2 has the form

$$p = p(v, T, x_1) \equiv p\left(\frac{V}{n_1 + n_2}, T, \frac{n_1}{n_1 + n_2}\right) \tag{3.10}$$

where v is the molar volume of the gas and x_1 is the mole fraction of gas 1. In order to express p in terms of V, we need three additional constraints. Let us suppose that the gases do not react chemically and that the system is expanded reversibly while enclosed in an impermeable, diathermal boundary, through which it is in thermal equilibrium with a thermostat maintained at a fixed temperature. The equation of the curve C of the path of the change in state is found by substituting into the equation of state (3.10) of the gas mixture the following three condition relations which are determined entirely by the nature of the process

$$T = b, n_1 = a_1, n_2 = a_2 \tag{3.11}$$

Here the constants b, a_1, a_2 are the given temperature and mole numbers of the two gases, respectively. Thus the equation of the curve C in the pV-diagram is

$$p = p\left(\frac{V}{a_1 + a_2}, b, \frac{a_1}{a_1 + a_2}\right) \tag{3.12}$$

If the system just considered were expanded reversibly while enclosed in an impermeable, adiabatic boundary, the equation of the path is obtained by eliminating n_1, n_2, and T among the three condition relations

$$n_1 = a_1, n_2 = a_2, F\left(\frac{V}{a_1 + a_2}, T, \frac{a_1}{a_1 + a_2}\right) = 0 \tag{3.13}$$

and the equation of state of the gas mixture, eqn. (3.10). The last expression of (3.13) is the equation of a reversible adiabatic.

At a later point, we shall call the first of these two experiments a reversible, isothermal process; and we shall call the second a reversible, adiabatic process without further description.

Let us now turn our attention to the application of eqns. (3.7) and (3.9) to some specific processes for bringing about the change in state

$$n\,\text{He}(g, V_1) = n\,\text{He}(g, V_2)\;(T) \tag{3.14}$$

which may also be written in the form

$$n\,\text{He}(g, p_1, V_1) = n\,\text{He}(g, p_2, V_2);\, p_1 = p(V_1/n, T),\, p_2 = p(V_2/n, T) \tag{3.15}$$

Here the relation $p = p(V/n, T)$ is the equation of state of helium gas. This change in state may be brought about by infinitely many reversible and by infinitely many irreversible processes. Equations (3.14) and (3.15) give us only the initial and final states of the system. We are free to employ any process we desire to produce this change. Let us now consider the following processes from the standpoints of how they may conceivably be carried out and of where we should locate the boundary in order to facilitate the identification and determination of the work produced in the surroundings.

3.5. A reversible, isothermal process. Let the change in state (3.15) be brought about by a reversible, isothermal process. This ideal process may be carried out in the idealized piston and cylinder arrangement A shown in Fig. 3.1. The walls of the piston and cylinder are rigid, impermeable to helium, and good thermal conductors. Throughout the process the cylinder is held at a fixed level with its axis vertical, and it is in thermal contact with a thermostat B maintained at the fixed temperature T K. The piston confining the gas moves in the cylinder without friction. It is connected by a rod to a platform bearing the weight F, as shown in the sketch. We shall suppose that the piston, connecting rod, and platform are weightless. The weight F is composed of many infinitesimally small (in the thermodynamic sense) weights which can be transported to or from the platform by the action of a frictionless transfer device located in the surroundings of the system. The entire apparatus is enclosed in the evacuated vessel C. Let the inner surfaces of the piston and cylinder form the boundary of the system.

At the start of the expansion the upward force exerted on the piston due to the pressure of the helium in its initial state (p_1, V_1, n) is just balanced by the downward force of the weight F; and the piston is at rest. By means of the frictionless transfer device, we remove one of the small weights f from the platform placing it a short distance away and at its initial level, whereby the piston moves upward an infinitesimal distance dz. We repeat this operation, stacking the small weights in the surroundings of the system (Fig. 3.1b) until the helium has reached the final state (p_2, V_2, n). Throughout the process the temperature of the system has remained uniform and constant at T, since the system has been in thermal contact and hence,

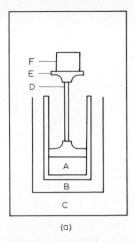 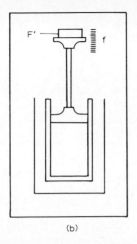

Fig. 3.1. Reversible, isothermal expansion of a gas, (a) at the start of the process, and (b) at the end of the process. The inner surfaces of the piston and cylinder form the boundary of the system. A, piston and cylinder; B, thermostat; C, evacuated jacket; D, connecting rod; E, platform; F, initial weight on piston; F′, final weight on piston; f, small weights.

under the present circumstances, in thermal equilibrium with the thermostat.

At any stage of the process we can reverse the portion of the direct process so far completed by transferring the small weights to their initial position in the reverse order of their removal.

In principle no net work is required to transfer one of the small weights from one position to another at the same level. Thus the net work produced in the surroundings in an infinitesimal step of the direct or reverse process is $mg\,dz$, by eqn. (3.2), where mg is the weight (downward force acting) on the platform at the start of the step and dz is the upward displacement of the piston. This may be written $p\,dV$, as in eqn. (3.8), where p is the pressure (mg/A) of the helium and dV is the volume ($A\,dz$) swept out by the piston of area A.

The work of the entire process is the integral of $mg\,dz$ from the initial to the final position of the platform. It is the work required to elevate the weight remaining on the platform at the end of the process through a distance Δz, equal to the total piston-travel, and to elevate the small weights f various distances ranging from zero to Δz (see Fig. 3.1b). Hence the work of the process is the integral of $p\,dV$ along the curve $p = p(V/n, T)$ with n and T fixed from the initial to the final state of the system, as indicated by eqn. (3.9).

A graph of the path of the process $p = p(V/n, T)$ on the pV-diagram is simply a plot of p against V for n moles of helium gas at the fixed temperature T. It is shown in Fig. 3.2, and the corresponding work is represented by the area W. This is the area of the portion of the pV-plane bounded by the curve of the path, the line $V = V_2$, the axis of zero pressure,

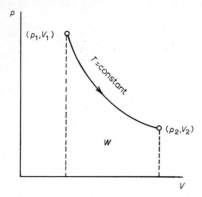

Fig. 3.2. Expansion work W of a reversible, isothermal process for producing the change in state (3.15).

and the line $V = V_1$. Correspondingly, the work of a reversible cycle is represented by the area on the pV-diagram bounded by the closed curve of the path of the cycle. Each such area is considered as positive when it is described in such a direction as to keep it upon the right, and it is considered as negative when described in such a direction as to keep it upon the left. The work shown in Fig. 3.2 is positive.

3.6. *A two-step, reversible process.* The change in state (3.15) is the sum of the two changes

$n \text{ He}(g, V_1) = n \text{ He}(g, V_2) \quad (p_1)$ \hfill (3.16)

$n \text{ He}(g, p_1) = n \text{ He}(g, p_2) \quad (V_2)$ \hfill (3.17)

Now each of these changes may be brought about by many processes. Let the first be produced by a reversible, isopiestic (constant pressure) process, and the second by a reversible, isometric (constant volume) process (see Fig. 3.3). Neither of these processes is isothermal.

The processes under consideration may be produced in the idealized piston and cylinder combination described earlier. The temperature of the system may be varied by allowing the system to come into thermal equilibrium successively with a series of thermostats whose temperatures increase or decrease monotonically by infinitesimal increments, while the applied pressure is correspondingly changed by the transfer of the small weights, one by one, to or from the platform. The inner surfaces of the piston and cylinder are chosen to be the boundary of the system.

The work of the present process for producing the change in state (3.15) is represented by the area W of Fig. 3.3, and is evidently equal to $p_1(V_2 - V_1)$. It is quite different from the work of the reversible, isothermal process shown in Fig. 3.2.

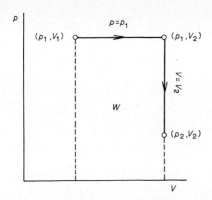

Fig. 3.3. Expansion work W of a two-step, reversible process (an isopiestic expansion followed by an isometric process) for producing the change in state (3.15).

3.7. A reversible cycle. Figure 3.4 shows the path of a reversible cycle composed of the one-step process of Fig. 3.2 for producing the change in state (3.15), followed by the reverse of the two-step process of Fig. 3.3 for restoring the system to its initial state. The work of the cycle is represented by the area W, and it is negative. The work would be positive if the cycle were described in the reverse direction.

3.8. Two fast, irreversible processes. Let us now make two modifications in the apparatus (Fig. 3.1) used for carrying out the change in state (3.15) by the reversible, isothermal process: (1) the cylinder is fitted with two sets of stops (see Fig. 3.5), and (2) the weight F is in one piece which may or may not be firmly bolted to the platform, as we choose. At the start of the first of the two fast, irreversible processes which we shall carry out in this apparatus, the helium is in its initial state (p_1, V_1, n), the piston is held by the lower stops, and the weight F is bolted to the platform. The magnitude

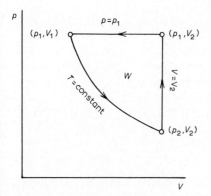

Fig. 3.4. Expansion work W of a reversible cycle composed of the process of Fig. 3.2 followed by the *reverse* of the process of Fig. 3.3.

of the weight has been chosen so that the piston, on being released from the lower stops, accelerates upward until it hits the upper stops, which have been so positioned that, after the establishment of equilibrium, the helium is in its final state (p_2, V_2, n). Thus the change in state (3.15) has been produced.

The finite upward velocity of the piston is attended by an upward flow of the gas composing the system, with a velocity ranging from zero at the inner surface of the cylinder head to the velocity of the piston at the inner surface of the piston.

In order to identify the work produced in the surroundings by the process, we must first determine the location with respect to the boundary of the dissipative effects resulting from: (1) the viscosity of the gas; (2) a portion of the change in the vertical component of momentum of the gas (produced by the arrest of the upward motion of the piston by the upper stops), which results in dissipative effects in the gas itself; and (3) the impulse of the force exerted on the upper stops by the impact of the piston caused by the remaining portion of the change in momentum of the gas mentioned above and the change in momentum of the weight F. (We recall that the piston, connecting rod, and platform were assumed to be weightless.) All of these effects occur inside the system if we place the boundary in the position indicated by the dashed lines in Fig. 3.5. As can be seen, this boundary encloses not only the system but a shell of the piston, cylinder,

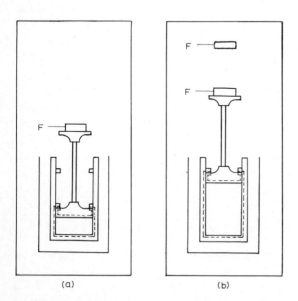

Fig. 3.5. Two fast, irreversible expansions of a gas, (a) at the start of the process, and (b) at the end of the process. In one the weight F is bolted to the platform, in the other F is not attached to the platform. The surface indicated by the dashed lines is the boundary of the system.

and stops which is infinitesimally thick (in the thermodynamic sense). The amount of material composing this shell, which has been added to the system under consideration, is so small that it need not be included in the change in state (3.15).

Evidently we cannot identify a path for this process because the system is too far from equilibrium during the expansion, as evidenced by large finite gradients of its intensive properties. When we place the boundary in the position indicated in Fig. 3.5, there has been no dissipation of work in the surroundings. Hence the work produced in the surroundings during the process can be identified without ambiguity. It is equal to the product $mg \Delta z$, where mg is the weight of the mass F bolted to the platform and Δz is the vertical distance travelled by the piston.

For the second fast, irreversible expansion carried out in the apparatus of Fig. 3.5, the weight F is not bolted to the platform. Now, when the piston hits the upper stops, the weight F continues to travel vertically upward through an additional distance (determined by the velocity of the piston just prior to impact and the local acceleration of gravity), where it could be caught on a platform (Fig. 3.5b). The work produced in the surroundings in this process can also be identified unambiguously. It is given by the product $mg \Delta z'$, where mg is the weight of F, and $\Delta z'$ is its total vertical travel. Thus the work attending the second process is larger than that attending the first, since $\Delta z'$ is greater than Δz, while mg is the same in both. This is to be expected, since the dissipative effect of the impact of the piston on the upper stops (which takes place inside the system) is smaller in the second process than in the first*.

3.9. A quasistatic, irreversible process. For our present purpose we make just one alteration in the specifications of the piston and cylinder combination shown in Fig. 3.1 and described in Section 3.7: we suppose that the motion of the piston in the cylinder is always opposed by a frictional force which is independent of the speed, position, and direction of motion of the piston. Since frictional effects now occur at the inner surfaces of the cylinder and piston, as the volume of the system is varied, we place the boundary of the system in the position shown by the dashed lines of Fig. 3.6. The boundary is so chosen that all of the surfaces of contact between the piston and cylinder are inside the system, except an infinitesimal portion at the top of the piston. With this location of the boundary, all dissipative effects arising from the motion of the piston in the cylinder occur inside the system, except for the negligible effect which takes place at the top of the piston where a small ring of contact between piston and cylinder lies outside the system.

Let us now bring about the change in state (3.15) in this apparatus by a

---

*The fast expansion of a gas enclosed in an adiabatic boundary is discussed by Kivelson and Oppenheim [3].

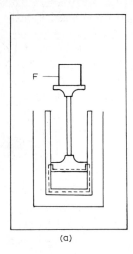 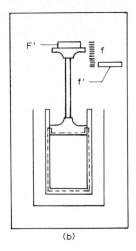

Fig. 3.6. Quasistatic, irreversible expansion of a gas, (a) at the start of the process, and (b) at the end of the process. The surface indicated by the dashed lines is the boundary of the system. F, initial weight on piston; F', final weight on piston; f, small weights; f', finite weight removed from piston before it starts its upward travel.

quasistatic, isothermal process. Initially the system is in the state (p_1, V_1, n); and we shall suppose that the weight F on the platform is of such a magnitude that the addition of one more of the small weights f would cause the piston to move downward an infinitesimal distance. By means of the frictionless transfer device we now move the small weights, one by one, from the platform to positions a short distance away and at their respective initial levels, waiting for the piston to assume its equilibrium position after each removal. In the present experiment the piston does not begin to move upward until we have removed from it a finite weight, composed of infinitely many small weights (see Fig. 3.6). Thereafter, the piston continues to move upward infinitely slowly until the system has reached its final state (p_2, V_2, n). Throughout the process the temperature of the system has remained constant because the expansion has been quasistatic.

We can reverse the direction of motion of the piston at any stage of the process by transferring the small weights f successively to their initial positions on the platform in the reverse order of their removal. In the present experiment the piston does not start to move downward until we have restored to the platform a finite weight equal in magnitude to the weight removed initially to start the expansion, and composed of infinitely many small weights. The latter have been elevated by various amounts ranging from an infinitesimal to a finite distance. Thereafter, the piston moves slowly downward as the small weights are each raised and placed on the platform. By the time the system has reached its initial state we have restored to the platform all of the small weights removed during the expansion. Evidently this requires a finite work consumption in the surroundings;

that is, a finite weight (composed of many small weights) must have been lowered a finite distance somewhere in the surroundings. Thus the process is irreversible.

We note that the path of the change in state of the system is exactly the same in this irreversible, isothermal process as in the reversible, isothermal expansion of Section 3.5. However, the area under the pressure vs. volume curve of the path in the pV-diagram has no significance with respect to the work produced in the surroundings in the irreversible process, since this is determined solely by the net number of standard weights that the process enables us to raise through unit distance in the surroundings (Fig. 3.6b). During the expansion process, the upward-directed force pA exerted on the piston by the gas is greater than the downward force mg of the weights on the platform by an amount equal to the downward-acting frictional force, that is $p > p_{AP}$. During the compression process, pA is less than mg by an amount equal to the upward-acting frictional force, that is, $p < p_{AP}$. We must compute the work produced in the irreversible expansion producing the change in state (3.15) by using eqn. (3.7) where p_{AP} is given by mg/A.

Usually the effects of both the finite speed of a process and of frictional forces enter into the determination of the work of an irreversible process. We must then apply a combination of the methods developed in Sections 3.8 and 3.9.

3.10. Work and the location of the boundary in reversible and irreversible processes. In any experiment, the location of the boundary of the system determines, in part, the work produced in the surroundings. The choice of a boundary location is made on the basis of the following considerations: (1) a precise distinction can be made between work and all other disturbances produced in the surroundings; (2) the work so identified pertains to the particular process we have in mind for bringing about the desired change in state of the system; and (3) this work can be evaluated by thermodynamic methods.

The first criterion requires that the boundary be so located that no frictional, impact, viscous, or other dissipative effects occur at or across it; the region in which these effects occur must be included in the system or in the surroundings.

In spite of the above restrictions on the location of the boundary, we may usually place it in any one of several positions. The work effect and the change in state of the system may vary from one location to another.

In Sections 3.5 and 3.6 above we discussed the computation of the work produced in each of two different reversible processes for bringing about the change in state (3.15). The inner surfaces of the piston and cylinder were taken to be the boundary of the system (Fig. 3.1). Evidently this boundary meets the three criteria given above. Since these processes are reversible, we could equally as well have moved the boundary outward (normally) an infinitesimal distance so that a thin shell of the piston and cylinder is inside the system, or inward (normally) a corresponding distance into the gas without appreciably affecting either the process (including the change in state) or the work produced in the surroundings. The location chosen was the simplest and the easiest to describe.

On the other hand, in each of these two experiments we may choose to enclose the piston, the cylinder, and all of the small weights composing the weight F (but not the thermostat) in a rigid, diathermal boundary. Then, the work attending each process would be zero, since no weights are elevated in the surroundings. These processes are entirely different from the corresponding ones treated in the last paragraph. Among other things, the change in state of each system must now include all of the small weights, together with their elevations in the initial and the final states of the system. The processes considered here differ from those of the last paragraph not because the apparatus or the experimental procedure has been varied, for neither has been altered; but simply because we have imagined the location of the boundary to have been varied, and this has changed the work produced in the surroundings.

In Section 3.8 we found that the boundary shown in Fig. 3.5 meets the three criteria stated above for two fast, irreversible expansions involving gas viscosity and impact effects; and in Section 3.9 we found that the boundary shown in Fig. 3.6 is suitable for a quasistatic, irreversible expansion where there is sliding friction between the piston and the cylinder.

Let us now choose a new location of the boundary for each of these two processes. We place it parallel to the inner surfaces of the piston and cylinder but displaced a small distance into the gas composing the system. The change in state (3.15) is not materially affected by this shift of the boundary. However, this position cannot be allowed for the fast processes (Section 3.8). One reason for this is the occurrence of a dissipative effect arising from the viscosity of the gas precisely at this boundary, which prevents the precise identification of the work produced in the surroundings.

On the other hand this position of the boundary is allowable for the quasistatic process (Section 3.9), because with this location all dissipative effects resulting from the sliding friction between the piston and cylinder occur in the surroundings. In order to determine the work of this process in accordance with the definition and discussion of Section 3.1, we must repeat the expansion in such a manner that no dissipation of work takes place in the surroundings. A little consideration will show that the work produced in the surroundings in this process is the same as that of the reversible, isothermal process of Section 3.5 (the area W of Fig. 3.2), since the path of the change in state and all interactions of the system with its surroundings are essentially the same in the two processes, but the dissipative effect of friction in the first process is absent in the second.

When several different locations of the boundary, each leading to essentially the same change in state are equally allowable in a given experiment, we choose the position that yields the process in which we are interested.

3.11. Equations for the expansion work of several processes. Equation (3.7) is the general expression for the expansion work of any process which conforms to the conditions stated in its derivation from eqn. (3.4). This relation reduces to eqn. (3.9) for reversible processes. In Sections 3.5—3.9 we indicated how several expansion processes may be brought about in a manner which permits the direct application of Gibbs' definition of work (Section 3.1). We shall now apply the methods developed there to the derivation of equations for the expansion work of several processes.

3.11A. A constant-volume process. When the change in state

$$n \, \text{He}(g, p_1, T_1) = n \, \text{He}(g, p_2, T_2) \tag{3.18}$$

is produced inside a boundary of fixed volume, the expansion work produced exterior to this boundary is zero:

$$W_{EX} = 0 \tag{3.19}$$

This relation holds whether the process is reversible or irreversible, and whether the boundary is diathermal or adiabatic. It holds no matter how complicated the process may be; for example, chemical reactions may occur. It also holds for the expansion work when other types of work (Section 3.3) are also produced in the surroundings.

As an application of eqn. (3.19) we may consider the Joule, or free expansion, experiment. A gas, for example helium, is confined in an inner vessel fitted with a virtually frictionless valve. This vessel is enclosed in an outer vessel whose walls are rigid and are thermal insulators, and the intervening space is evacuated. When the valve is opened, the gas rushes out and fills the available space, whereby the change in state (3.18) occurs. The expansion work is zero if we place the boundary of the system parallel to the inside surface of the outer vessel but displaced an infinitesimal distance outward. The expansion work is not zero if the boundary is similarly positioned with respect to the inside surface of the inner vessel. In the latter instance we are dealing with an open system (Section 1.2), and the determination of the work effect has been solved only for special cases (see ref. 4).

The process occurring in the system may be much more complicated than the free expansion of a gas. The space between the two vessels may contain another gas, for example nitrogen, whereupon the gases mix and fill the available space when the stopcock is opened. The inner vessel may contain a combustible gas and the space between the vessels may be filled with oxygen, whereupon a chemical reaction accompanied by an explosion may occur when the stopcock is opened. The expansion work is zero for all of these processes if the boundary is placed just outside the inner surface of the outer vessel.

3.11B. A quasistatic process under a constant applied pressure. In the two changes in state

$$n \, \text{He}(g, V_1, T_1) = n \, \text{He}(g, V_2, T_2) \tag{3.20}$$

$$n \, \text{H}_2\text{O}(l, V_1, T_1) = n \, \text{H}_2\text{O}(g, V_2, T_1) \tag{3.21}$$

let the pressure of each system in its final state be the same as the pressure in its initial state. We propose to produce each change by a quasistatic process in which the applied pressure p_{AP} is always constant. If the process is also reversible, p_{AP} is equal to the pressure p of the system.

In the first process, the temperature of the gas is increased or decreased slowly under a constant applied pressure; in the second, a liquid is evaporated slowly and isothermally under a constant applied pressure. Application of eqn. (3.7) gives the following equation for the work of each of these two processes:

$$W_{EX} = p_{AP}(V_2 - V_1) \tag{3.22}$$

If the process is reversible, this relation becomes

$$W_{EX} = p(V_2 - V_1) \text{ (reversible process)} \tag{3.23}$$

where p is the constant pressure of the system. Equation (3.23) may be written in the following form for the reversible heating or cooling of n moles of a perfect gas from T_1 to T_2

$$W_{EX} = nR(T_2 - T_1) \tag{3.24}$$

3.11C. A reversible, isothermal expansion of a gas. Let the change in state

$$n\,\text{He}(g, V_1) = n\,\text{He}(g, V_2)\ (T = \text{constant}) \tag{3.25}$$

be produced by a reversible, isothermal process.

Let us suppose that helium gas may be considered to be perfect. Substitution from the perfect gas equation (2.41) into the relations (3.08) and (3.09) gives

$$đW_{EX} = nRT\,d\ln V \tag{3.26}$$

$$W_{EX} = nRT\ln\frac{V_2}{V_1} \quad (T = \text{constant}) \tag{3.27}$$

$$W_{EX} = nRT\ln\frac{p_1}{p_2} \quad (T = \text{constant}) \tag{3.28}$$

Here p_1 and p_2 are the initial and final pressures of the system.

We may employ the virial equations (2.43) and (2.44) for a more accurate value of the reversible, isothermal work of expanding a real gas. Substitution of the values of p from the first of these relations and of V from the second into the equations

$$đW_{EX} = p\,dV \equiv d(pV) - V\,dp \tag{3.29}$$

and integration of the resulting expressions from V_1 to V_2 at constant temperature gives the following relations:

$$W_{EX} = nRT\ln\frac{V_2}{V_1} - n^2\,RTB\left(\frac{1}{V_2} - \frac{1}{V_1}\right) - \frac{n^3\,RTC}{2}\left(\frac{1}{V_2^2} - \frac{1}{V_1^2}\right) - \cdots$$
$$(T = \text{constant}) \tag{3.30}$$

$$W_{EX} = nRT\ln\frac{p_1}{p_2} + \frac{nC'}{2}(p_2^2 - p_1^2) - \ldots \quad (T = \text{constant}) \tag{3.31}$$

Note that the perfect gas equation for the work is the leading term in each of the above expressions.

3.11D. A reversible, polytropic expansion of a gas. In many expansion processes the pressure—volume product of a gaseous system is adequately represented by the equation

$$pV^k = K \tag{3.32}$$

The constant k may have any value from a large positive number to a large negative number. The constant K can be computed from the given value of k, and the pressure and volume of the system in its initial state. Polytropic processes include the reversible expansion of a gas at constant pressure ($k = 0$); the reversible expansion of a perfect gas at constant temperature ($k = 1$); and the reversible expansion of a gas under adiabatic conditions ($k = C_p/C_V$, where C_p and C_V are the heat capacities of the gas at constant pressure and constant volume, respectively).

Let the change in state

$$n\,\text{He}(V_1, T_1) = n\,\text{He}(V_2, T_2) \qquad (C \text{ given by } pV^k = K) \tag{3.33}$$

be produced by a reversible, polytropic process. Substitution from eqn. (3.32) into eqns. (3.08) and (3.09) gives

$$dW_{EX} = KV^{-k}\,dV$$

$$\left.\begin{array}{l} W_{EX} = \dfrac{K}{k-1}\left(\dfrac{1}{V_1^{k-1}} - \dfrac{1}{V_2^{k-1}}\right) \qquad (k \neq 1) \\[1em] \phantom{W_{EX}} = \dfrac{p_1 V_1 - p_2 V_2}{k-1} \end{array}\right\} \tag{3.34}$$

If we suppose that the perfect gas law gives an adequate representation of the pressure—volume—temperature behavior of helium, we may replace pV by RT in the last equation above. Thus we find

$$W_{EX} = \frac{nR(T_1 - T_2)}{k-1} \tag{3.35}$$

3.11E. A reversible, isothermal expansion of a condensed phase. Let the change in state

$$n\,\text{H}_2\text{O}(l, V_1, T) = n\,\text{H}_2\text{O}(l, V_2, T) \tag{3.36}$$

be produced by a reversible, isothermal process. The equation of state of the liquid is

$$V = V_0(1 - \bar{\beta}_T p) \tag{3.37}$$

where $\bar{\beta}_T$ is the mean coefficient of isothermal compressibility of the liquid from zero pressure to p at the temperature T, and is a function of p and T;

while V_0 is its (computed) volume at zero pressure and T, and is a function of temperature alone. Under the assumption that $\bar{\beta}_T$ is independent of pressure we find

$$dW_{EX} = -V_0 \bar{\beta}_T p \, dp \tag{3.38}$$

$$W_{EX} = \tfrac{1}{2} V_0 \bar{\beta}_T (p_1^2 - p_2^2) \qquad (T = \text{constant}) \tag{3.39}$$

For greater accuracy we can express $\bar{\beta}_T$ as a power series in p in eqn. (3.37).

3.11F. A steady-flow process. Let a fluid flow into a vessel A enclosing a fixed volume through an inlet tube and flow out through an outlet tube (see Fig. 3.7). Various processes may occur in the vessel: the walls may be thermal conductors or thermal insulators, shaft or other types of work may be produced in the surroundings, and chemical reactions may occur. A steady-flow process is one in which the following conditions hold: (1) each intensive property of the fluid is uniform and independent of time in the inlet stream and in the outlet stream; (2) the mass rate of flow of fluid into the vessel equals the mass rate of flow out; and (3) the rate of heat flow from the surroundings and the rate of work production in the surroundings are each constant. Hence in the neighborhood of every point in the vessel each of the intensive properties of the fluid (within the accuracy of its determination) is independent of time, or at least the states of the fluid in the neighborhood of all points in the vessel pass periodically and simultaneously through states which have existed earlier at the respective points [5].

In order to treat the fluid under consideration as a closed system, we choose a boundary composed, in part, of surfaces parallel to the inner surfaces of the walls of the vessel A and the inlet and outlet tubes but displaced an infinitesimal distance outward, indicated by the dashed lines in Fig. 3.7, and, in part, of two imaginary planes perpendicular to the axes of the tubes, indicated by the broken lines 1 and 1', each plane moving with the velocity of fluid flow in its respective tube. During the time Δt, the plane in the inlet tube sweeps out the volume V_1 in moving from position 1 to position 2, and the plane in the outlet tube sweeps out the volume V_2 in moving from position 1' to position 2'. These planes act as imaginary pistons always confining a constant mass of fluid, which is not, however, the system of present interest.

Fig. 3.7. A steady-flow process. The surface indicated by the dashed and broken lines is the boundary of the system. A, a vessel of constant volume.

Let us suppose that the flow of the fluid through the inlet and outlet tubes takes place with negligible shear. Then the pressure in each tube is essentially independent of position. We define a system for the present experiment and determine its change in state in the following manner. After steady-flow conditions have been established, the system, in its initial state, is the fluid occupying the volume V_1 under the constant pressure p_1 at any instant of time, and the system, in its final state, is the fluid in the volume V_2 under the constant pressure p_2 at the same instant. If the experiment involves the flow of n moles of gaseous water through a turbine during the time Δt, the change in state of the system of interest is

$$n\, H_2O(g, p_1, V_1) = n\, H_2O(g, p_2, V_2) \tag{3.40}$$

Under the supposition of fluid flow with negligible shear, the externally-applied pressure acting on the imaginary piston 1 is essentially p_1, and in the time Δt the piston sweeps out a volume V_1 moving in the direction in which this applied pressure acts, thereby producing in the surroundings the expansion work $-p_1 V_1$ (see Section 3.4). During the same time, the corresponding piston 1' in the outlet tube sweeps out the volume V_2 moving in the direction opposite to that in which the externally-applied pressure p_2 acts, producing the expansion work $+p_2 V_2$. The total expansion work W_{EX} produced in the time Δt is

$$W_{EX} = p_2 V_2 - p_1 V_1 \tag{3.41}$$

A flow process may also be attended by other types of work, for example, shaft work in a turbine, electrical work in a voltaic cell, etc.

In the adiabatic Joule—Thomson experiment a fluid flows under steady-flow conditions through a thermally insulated tube containing a plug of a porous material, such as unglazed porcelain, where it experiences a finite drop in pressure. If the fluid were gaseous water, the change in state (3.40) occurs and the work of the process is given by eqn. (3.41).

If we chose a boundary which did not move with respect to the apparatus, the expansion work would evidently be zero, but other types of work could be produced. We shall treat such an open system in a later chapter.

In some flow processes there are several entrance tubes and several exit tubes. Under these conditions each term in eqn. (3.41) is replaced by a sum of pV-products over the given number of tubes.

3.12. Heat. In the course of defining thermal contact and thermal equilibrium (Section 2.3), we described some experiments made with two simple, homogeneous systems A and B, each composed of a vessel of strictly constant volume containing a fixed mass of helium gas at the same density ρ, but at different initial pressures, and hence at different initial temperatures. Let us suppose that the vessels are made of the same thermally conducting material. When systems A and B are placed in thermal contact under conditions of complete isolation from the rest of the universe, their pressures, and hence their temperatures, are observed to approach each other and finally attain essentially the same value.

Many careful experiments have shown that the changes in state of the systems A and B in this type of experiment, and hence their common equilibrium temperature, are uniquely determined by their initial states. The rate of attainment of equilibrium depends on the thermal conductivity of the walls separating one system from the other, and on whether the two systems are kept in thermal contact continuously or intermittently, but their final equilibrium states do not.

For our present purpose we take the outer surface of the vessel enclosing the gas as the boundary of the system, which thus includes both the gas and the vessel.

In thermodynamics we explain the attainment of thermal equilibrium of systems A and B by the following thermal postulate. A physical entity, heat, flows through the boundary of a system enclosed in diathermal walls, by virtue of a temperature difference between the system and its surroundings, the flow of heat at each point in the boundary being in the direction of the negative temperature gradient.

Many careful quantitative experiments have shown that this definition of heat can be applied to all types of systems undergoing all types of processes. We note that heat is identified and defined only as it flows across the boundary of a system.

3.13. Measure and units of heat. The number of units of heat Q withdrawn from the surroundings by a system undergoing a specified process was originally equated to the number of units of mass of liquid water, under a pressure of one standard atmosphere, that this quantity of heat would decrease one degree in temperature on a designated scale, starting from a specified temperature. Heat is an algebraic quantity. We say that Q has a positive value when the temperature of the water is decreased in the above experiment; Q has a negative value when the temperature of the water is increased in this experiment. In the first case a positive quantity of heat has flowed from the surroundings to the system; in the second case a negative quantity of heat has flowed from the surroundings to the system, that is, a positive quantity of heat has flowed to the surroundings from the system. Sometimes we speak of Q as the heat effect attending a process, or simply as the heat of a process.

It follows directly from the above definition of quantity of heat that, if the heat effect attending a process 1 is Q_1 and the heat effect attending a second process 2 is Q_2, then the heat effect attending both processes is the algebraic sum of Q_1 and Q_2. This is evidently true for successive processes occurring in the same system, and also for processes taking place in different systems. Thus the total heat effect of any number of processes is simply the algebraic sum of the heat effects of the individual processes.

The heat effect attending a process depends on the location of the boundary of a system, because heat is only identified as it passes the boundary and, like work, it is primarily measured by the disturbances it produces in the surroundings. We have found as a result of experiment that the heat withdrawn from the surroundings by a system undergoing any change is not determined solely by the change in state, but by the complete process. Thus, quantity of heat is not a state property. This is the reason we speak of the heat of a process, not the heat attending a change in state.

Originally, quantities of heat were expressed in terms of calories (cal) and British thermal units (Btu), defined as follows:

The 15° calorie (cal_{15}) is the quantity of heat which, when absorbed by one gram of liquid water under a pressure of one standard atmosphere, will increase its temperature from 14.5—15.5°C. For the 20° calorie (cal_{20}) the temperature interval is 19.5—20.5°C. The corresponding 15° kilocalorie ($kcal_{15}$) and 20° kilocalorie ($kcal_{20}$) refer to one kilogram of liquid water.

For the Btu the mass of water involved is one pound (mass) and the temperature interval (on the Fahrenheit scale) is 59.5—60.5°F.

Other temperature ranges have been used in defining both the calorie and the Btu.

There are a number of experimental difficulties in measuring quantities of heat accurately in terms of these units. Clearly, if the system is at temperatures greatly different from the standard temperature interval during the process of interest, we may have to employ several intermediate systems in order to relate the heat effect of the given process to the desired heat unit. For this reason, most of the vast number of thermal investigations of the nineteenth century, particularly those of the heat effects attending processes involving chemical reactions at constant volume or constant pressure, were made at temperatures near 20°C, which was close to room temperature in the laboratories where these measurements were made.

With the advent of methods for the accurate measurement of quantities of electrical work, experimental investigators found that a convenient, flexible, and precise method of measuring quantities of heat without the use of a series of intermediate systems was to employ an "electric heater". The flow of an electric current through a resistor causes the resistor's temperature to rise, which, in accordance with the thermal postulate, will result in a flow of heat to any body at a lower temperature in thermal contact with it. An electric heater consists of a resistor, from each end of which run wires for the introduction and measurement of the electrical power imput. The resistor and short lengths of the wires may be enclosed in a metal case for protection.

In the present chapter we regard the electric heater simply as a transfer instrument which permits us to compare the heat effect Q attending a given process at any temperature with the heat effect attending the process defining the unit of heat. Apparatus and experimental methods for accomplishing this result are described in detail in Sections 3.18—3.24, where we consider calorimeters for measuring heat effects attending processes taking place at temperatures ranging from many hundred degrees Celsius down to the lowest attainable values. Whenever measurements of a given heat effect by the electric method and by the older method employing water have been made, the values agree within the accuracies of the two methods.

Actually, the electric method expresses quantities of heat directly in electrical work units: joules in the MKS system. In fact, the Ninth General Conference on Weights and Measures in 1948 adopted the joule as the unit of quantity of heat, and requested that the results of calorimetric experiments be expressed in joules when possible (see ref. 6). However, most experimental investigators continue to express the results of thermal measurements in calories, although quantities of heat are in fact first measured by the electrical method and then expressed in calories by use of a conversion factor.

Each of two international groups has proposed a defined conversion factor [6] for changing joules to calories. Each factor has a fixed value, without reference to the thermal properties of water. These two defined or "dry" calories are: the thermochemical calorie (cal_{th}) which is used by scientists, and the International Steam Table calorie (cal_{IT}) used by engineers. A defined conversion factor for Btu is so chosen that 1.8(=9/5) Btu per lb is equal to 1 cal_{IT} per g, i.e. 1 Btu = (453.5 9237/1.8) cal_{IT}. Factors for converting various heat units to joules are listed in Table 3.1. We note that the two defined calories differ in absolute magnitude by about 7 parts in 10,000.

3.14. Definition of heat capacity. Earlier we pointed out that the heat effect attending a process is primarily determined from the disturbances it produces in the surroundings, except for reversible processes where it can be computed from properties of the system. Let a simple, homogeneous system A, consisting of a fixed mass of a pure substance or of a solution of constant composition, undergo the following change in state

$$A(T_1) = A(T_2) \tag{3.42}$$

by a reversible process along some definite path C, in which the quantity of heat Q_C is withdrawn from the surroundings. We define the mean heat capacity $\bar{C}_C$ of the system along the path C over the temperature range T_1 to T_2 by the equation

TABLE 3.1

CONVERSION FACTORS FOR HEAT UNITS

1 cal_{15}	=	4.1855 J
1 cal_{20}	=	4.1816 J
1 cal_{th}	=	4.1840 J[a]
1 cal_{IT}	=	4.1868 J[a]
1 Btu	=	1.055 0559 × 10^3 J = 2.519 9576 × 10^2 cal_{IT}^b

[a] Exact by definition.
[b] 1 Btu lb^{-1} = $\frac{5}{9}$ $cal_{IT} g^{-1}$, by definition.

$$\bar{C}_C = \frac{Q_C}{T_2 - T_1} \tag{3.43}$$

Thus the heat effect attending the process is

$$Q_C = \bar{C}_C (T_2 - T_1) \tag{3.44}$$

Now let us allow the temperature interval ΔT to approach zero. The heat capacity C_C along the path C of the system at T_1 is defined by the relation

$$C_C = \lim_{\Delta T \to 0} \frac{Q_C}{\Delta T} = \left(\frac{đQ}{dT}\right)_C \tag{3.45}$$

provided that this limit exists. Whence

$$đQ_C = C_C \, dT \tag{3.46}$$

and

$$Q_C = \int_{C\,T_1}^{T_2} đQ_C = \int_{C\,T_1}^{T_2} C_C \, dT \tag{3.47}$$

We can relax the restrictions placed on the system and the process somewhat. The process need not be strictly reversible, as long as we can identify a path well enough to determine the work with reasonable accuracy. Thus, in the determination of the constant-pressure heat capacity of a gas (Section 3.24) in a flow calorimeter, the gas is allowed to flow through the calorimeter at a finite velocity and thus experiences a finite drop in pressure. A constant-volume process in a simple, homogeneous system of constant composition need not be reversible, because the total work is zero, independently of what takes place inside the system. The system need not be homogeneous; but we shall find in Section 3.15 that, when the process includes a change in aggregation state (see Fig. 3.8) or a phase change of the second order (see Fig. 3.10), the limit of eqn. (3.45) may not exist. Finally, we shall define in a later chapter a heat capacity at constant pressure in a system in which chemical reaction is occurring under essentially reversible conditions.

Experimentally we find that C_C is a function of any set of independent state variables of a system, for example a function of $p, T, n_1, n_2, \ldots$ for a simple, homogeneous system. Also, it is extensive, since the heat effect attending λ identical processes is λ times the heat effect of each process.

The notation $(đQ/dT)_C$ in eqn. (3.45) calls attention to the fact that this quantity is not a partial derivative of any function of independent state variables. Nevertheless, when the state of a given homogeneous system is constrained to move along a fixed path C, from a fixed initial temperature T_0 to a variable temperature T, Q_C measured along the curve C is a function of T containing T_0 as a parameter. The quantity $(đQ/dT)_C$ is the derivative of this function at T. Thus, as long as the state of the system remains on a fixed path C, the corresponding heat capacity C_C depends only on T, and the mean heat capacity $\bar{C}_C$ from T_0 to T depends not only on T but also on the value of T_0.

From eqns. (3.43) and (3.47) we find the following relations for the heat effect Q_C attending a process in which the state of a system moves along the path C from a fixed temperature T_0 to a variable temperature T:

$$Q_C = \bar{C}_C(T-T_0) = \int_{C\,T_0}^{T} C_C\,dT \qquad (3.48)$$

Whence

$$\bar{C}_C = \frac{1}{T-T_0} \int_{C\,T_0}^{T} C_C\,dT \qquad (3.49)$$

$$C_C = \frac{d}{dT}[(T-T_0)\bar{C}_C] \qquad (3.50)$$

A system in a given state has as many heat capacities as there are identifiable paths passing through this state. In the next two sections we consider three of these heat capacities.

3.15. Constant-pressure and constant-volume heat capacities. Let a simple, homogeneous system A, consisting of a fixed mass of a pure substance or of a solution of constant composition, undergo a reversible, constant-pressure process:

$$A(p_1, T_1) = A(p_1, T_2) \qquad \text{(C given by } p = \text{constant)} \qquad (3.51)$$

where p is the pressure of the system. In a second experiment let the system A undergo a constant-volume process (not necessarily reversible):

$$A(V_1, T_1) = A(V_1, T_2) \qquad \text{(C given by } V = \text{constant)} \qquad (3.52)$$

In accordance with the relations derived in the last section, we define the constant-pressure heat capacity C_p and the constant-volume heat capacity C_V of the system in any given state by the following relations:

$$C_p = \left(\frac{dQ}{dT}\right)_p \text{ and } C_V = \left(\frac{dQ}{dT}\right)_V \qquad (3.53)$$

Each has been found by experiment to be an extensive state property of a given system.

The heat effect Q_p attending the constant-pressure process (3.51) and the heat effect Q_V attending the constant-volume process (3.52) are given by the relations:

$$Q_p = \int_{p\,T_1}^{T_2} C_p\,dT = \bar{C}_p(T_2 - T_1) \qquad (3.54)$$

and

$$Q_V = \int_{VT_1}^{T_2} C_V \, dT = \bar{C}_V (T_2 - T_1) \tag{3.55}$$

Here $\bar{C}_p$ is the mean constant-pressure heat capacity of the system A from T_1 to T_2 at a constant pressure p; and $\bar{C}_V$ is its mean constant-volume heat capacity over the same temperature range at a constant volume V.

We usually measure C_p calorimetrically by methods discussed in Sections 3.21, 3.22, and 3.24. We then compute C_V from C_p and quantities that can be determined from an equation of state. We shall derive the thermodynamic relation connecting these quantities in a later chapter.

In general, the molar heat capacities C_p and C_V are functions of p and T for a pure substance, and of p, T, and the composition for a solution.

Usually, the constant-pressure and constant-volume specific or molar heat capacities of a substance in a given aggregation state are continuous functions of temperature along a given isobar. Discontinuities do occur: (1) at the temperature where a given isobar intersects any saturation curve of the substance and a change in aggregation state (solid to solid, solid to vapor, solid to liquid, liquid to vapor) takes place at constant temperature and pressure (Fig. 3.8); and (2) at the temperature (called a critical temperature T_c) where a solid conductor, on being cooled, undergoes a transition from a normally conducting to a superconducting state (see Fig. 3.9). This critical temperature is little affected by pressure, but it is decreased when the solid is placed in a magnetic field H. We can make H sufficiently large to suppress entirely the appearance of the superconducting state; and under these conditions the heat capacity of the substance in the normal state can be followed below T_c, shown as dashed line in Fig. 3.9.

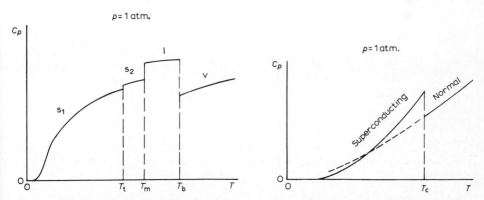

Fig. 3.8. Molar constant-pressure heat capacity C_p of a substance in several different aggregation states, as a function of temperature. T_t, transition point of crystalline solid s_1 to crystalline solid s_2; T_m melting point of s_2 to liquid l; T_b, normal boiling point of l to vapor v.

Fig. 3.9. Molar constant-pressure heat capacity C_p of a solid in the superconducting and normal states, as a function of temperature. T_c is the critical temperature.

Changes in state of the type described in (1) above are almost always accompanied by a finite heat effect; changes described in (2) are accompanied by zero heat effect.

The C_p vs. T and C_V vs. T curves of some substances along an isobar exhibit abnormal behavior (see ref. 7) in certain temperature ranges. They are of two types: the λ anomaly and the Schottky anomaly. In each, the system undergoes a phase change from a more ordered to a less ordered arrangement, and the transition occurs throughout a range of temperature rather than at a sharply defined point.

The λ anomaly is characterized by a specific heat curve which rises to a rather sharp peak over a fairly narrow temperature range (Fig. 3.10). Nonthermodynamic theory indicates that the functions $C_p(T)$ and $C_V(T)$ along an isobar may have a logarithmic or other type of singularity at the temperature of the peak. The small rate of attainment of equilibrium by systems in the neighborhood of the peak and the effects of errors of measurement make the experimental confirmation of this prediction difficult, if not impossible. The experimental curves are always more or less rounded at the top. Some examples of processes leading to λ anomalies in specific heat curves are: (1) the transition from a ferromagnetic, antiferromagnetic, or ferrimagnetic state to a paramagnetic state; (2) the transition from a ferroelectric, antiferroelectric, or ferrielectric state to a paraelectric state; (3) the transition from an ordered arrangement or orientation of atoms, molecules, or ions in a crystal lattice to a disordered arrangement; (4) the transition from a superfluid to a normal fluid in ^{4}He; (5) the liquid–vapor critical point; and (6) the transition from an ordered to a disordered arrangement of the rotational axis in solid parahydrogen.

In the Schottky anomaly the C_p vs. T and C_V vs. T curves along an isobar have camel-like humps (Fig. 3.11). Schottky anomalies are associated

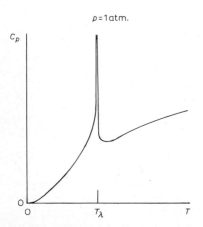

Fig. 3.10. Molar constant-pressure heat capacity C_p of a substance having a λ anomaly, as a function of temperature.

with a transition from an ordered to a disordered arrangement of the magnetic moments in: (1) paramagnetic salts where the paramagnetic ions are well separated by other ions or molecules, for example $NiSO_4 \cdot 6H_2O$ at about 2 K and $KCr(SO_4)_2 \cdot 12H_2O$ at about 0.1 K; and (2) solids containing atoms whose nuclei have permanent nuclear magnetic moments. The latter effect is observed at temperatures well below 0.01 K.

Other types of behavior of heat capacity vs. temperature curves are encountered which do not clearly belong to either of these anomalies. A transition [8] from torsional oscillation or hindered rotation to free rotation of an ion in a crystal lattice can give rise to a broad, low hump in the heat capacity curve which may not rise to a maximum, for example, the NH_4^+ ion in NH_4BF_4 and the ND_4^+ ion in ND_4BF_4. In many supercooled liquids, for example glycerol, the heat capacity curve of the supercooled liquid, which lies markedly above the curve for the crystalline solid at the melting point, falls rather abruptly at a lower temperature, so that it lies only slightly above the curve for the crystalline phase.

3.16. Saturation heat capacities and latent heat of phase change. We know from experiment that all of the intensive properties of each phase of a simple system, consisting of two aggregation states of a pure substance in equilibrium, are functions of one intensive property of one of the phases, for example, temperature, pressure, or density. Such a system is called univariant. The intensive properties of each phase of the system are called orthobaric or saturation properties. Under equilibrium conditions, both phases have the same pressure and temperature but not necessarily the same density, index of refraction, etc. The plot of pressure against temperature for a univariant one-component system is called the saturation curve of the two phases in equilibrium, and the equation of this curve is

$$F(p, T) = 0, \text{ or } p_{SAT} = p_{SAT}(T) \tag{3.56}$$

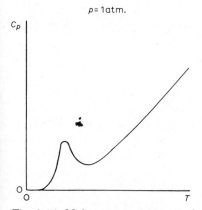

Fig. 3.11. Molar constant pressure heat capacity C_p of a solid having a Schottky anomaly, as a function of temperature.

The changes in aggregation state that occur under conditions of constant temperature and pressure on a saturation curve are given special names: transition (one crystalline form to a second crystalline form), sublimation (solid to vapor), fusion or melting (solid to liquid), vaporization or boiling (liquid to vapor). As an illustration of the procedure applicable to all orthobaric one-component systems, let us now consider certain saturation properties along a liquid—vapor curve.

In separate experiments let the temperature of one mole of liquid water and one mole of water vapor be increased by a reversible process along the liquid—vapor saturation curve C from T_1 to T_2 without any vaporization or condensation occurring. This may be accomplished, for example, in a piston and cylinder combination (see Fig. 3.1). As the temperature of each system is increased, the applied pressure is varied by the addition of weights to the platform in such amount that the pressure of the system has the value given by eqn. (3.56), and hence the system is always at its saturation pressure. Thus the changes in state and processes are:

(a) $H_2O(l, T_1) = H_2O(l, T_2)$ (C given by $p = p_{SAT}$)

(b) $H_2O(g, T_1) = H_2O(g, T_2)$ (C given by $p = p_{SAT}$)   (3.57)

We can now define two molar saturation heat capacities, one for the liquid phase and one for the vapor phase:

$$C^l_{SAT} = \left(\frac{dQ^l}{dT}\right)_{SAT} \text{ and } C^g_{SAT} = \left(\frac{dQ^g}{dT}\right)_{SAT} \qquad (3.58)$$

Here dQ^l is the heat effect attending the increase in temperature of one mole of liquid water from T to $T + dT$ along the liquid—vapor saturation vapor curve, and dQ^g is the heat effect attending the corresponding process for one mole of water vapor.

The saturation heat capacities are extensive state properties. They are related to the constant-pressure and constant-volume heat capacities of the corresponding phase by thermodynamic equations which we shall derive later.

The heat effects attending the finite processes (3.57) are

$$Q^l_{SAT} = \int_{p_{SAT} T_1}^{T_2} C^l_{SAT}\, dT \text{ and } Q^g_{SAT} = \int_{p_{SAT} T_1}^{T_2} C^g_{SAT}\, dT \qquad (3.59)$$

Let a simple system A, whether homogeneous or not, consisting of a fixed mass of a pure substance or a mixture of constant gross composition, undergo a reversible, isothermal expansion. The heat effect $(dQ/dV)_T$ attending this process is called the latent heat of dilation l_T of the system (see Section 3.28). Examples of the type of changes under consideration here are (a) the isothermal variation in volume of a homogeneous system; and (b) an isothermal phase change from one aggregation state to another of a pure substance under saturation conditions:

$$\text{He}(g, V_1) = \text{He}(g, V_2) \qquad (T = \text{constant}) \qquad (3.60\text{a})$$

$$\text{H}_2\text{O}(l) = \text{H}_2\text{O}(g) \qquad (p, T = \text{constant}) \qquad (3.60\text{b})$$

The latent heat of dilation l_T of a homogeneous system, for example the system of eqn. (3.60a), is an intensive state property of the phase.

The latent heat of dilation l_{pT} attending a change in aggregation state of a pure substance under orthobaric conditions is a function of a single independent variable, for example p or T. It is usually replaced by $(đQ/dn)_{pT}$ where dn is the mass of the substance taking part in the reaction; and is sometimes written λ_{pT}. If n is expressed in moles, λ_{pT} is called the molar heat of phase change (e.g. fusion, vaporization, etc.)

3.17. Five thermal experiments. The existence of the physical entity, heat, and the state property, heat capacity along a path, are confirmed by the results of many careful, quantitative experiments which can be consistently interpreted by the use of these concepts. Among these are the results of the five somewhat idealized experiments which we shall now consider. They deal with two simple, homogeneous systems A and B, each composed of a vessel of strictly constant volume containing a fixed mass of helium gas at the same density ρ but at different initial temperatures T_A and T_B (Section 3.12); and in addition three simple, homogeneous systems S$'$, S$''$, and S$'_2$, each composed of a vessel of strictly constant volume containing a fixed mass of a standard substance, for example liquid water, at the same density ρ_S. Systems S$'$ and S$''$ are identical, but the mass of S$'_2$ is exactly twice that of S$'$ or S$''$. The vessels enclosing each of the five systems are made of the same thermally conducting material, and the volume of each system is strictly constant throughout each experiment. The outer surface of the vessel is the boundary of each system.

Experiment 1. System A initially at T_A and system B initially at T_B are placed in thermal contact while isolated from the rest of the universe. The final equilibrium temperature is found to be T. The observed changes in state of the individual systems, each at strictly constant volume, are:

$$\text{A}(\rho, T_A) = \text{A}(\rho, T) \qquad (3.61\text{a})$$

$$\text{B}(\rho, T_B) = \text{B}(\rho, T) \qquad (T_A < T < T_B) \qquad (3.61\text{b})$$

Experiment 2. This procedure is repeated with A initially at T_A and S$'$ initially at $T + \Delta'T$, where $\Delta'T$ is so chosen that the final equilibrium temperature has the value T found in Experiment 1. The observed changes in state are:

$$\text{A}(\rho, T_A) = \text{A}(\rho, T) \qquad (3.62\text{a})$$

$$\text{S}'(\rho_S, T + \Delta'T) = \text{S}'(\rho_S, T) \qquad (3.62\text{b})$$

Experiment 3. We again repeat the procedure of the first two experiments with B initially at T_B and S$''$ initially at $T - \Delta''T$, where $\Delta''T$ is so chosen that the final equilibrium temperature is again T.

$$B(\rho, T_B) = B(\rho, T) \tag{3.63a}$$

$$S''(\rho_S, T - \Delta''T) = S''(\rho_S, T) \tag{3.63b}$$

Experiment 4. Now we place S' initially at $T + \Delta'T$, and its twin S'' initially at $T - \Delta''T$ in thermal contact, while isolated from the rest of the universe. Here $\Delta'T$ and $\Delta''T$ have the values found in Experiments 2 and 3, respectively. The final equilibrium temperature is again found to be T. The observed changes in state are thus:

$$S'(\rho_S, T + \Delta'T) = S'(\rho_S, T) \tag{3.64a}$$

$$S''(\rho_S, T - \Delta''T) = S''(\rho_S, T) \tag{3.64b}$$

Experiment 5. Finally we repeat the procedure employed above with S_2' initially at $T + \Delta'T$ and two identical systems S'' each initially at $T - \Delta''T$. After the three systems have reached thermal equilibrium, they are found to have the same temperature T as that observed in Experiment 1. Thus the changes in state are:

$$S_2'(\rho_S, T + \Delta'T) = S_2'(\rho_S, T) \tag{3.65a}$$

$$2S''(\rho_S, T - \Delta''T) = 2S''(\rho_S, T) \tag{3.65b}$$

We note that the final equilibrium temperature T has the same value in each of these five experiments. Also $\Delta'T$ has the same value wherever it appears, and so does $\Delta''T$.

In Table 3.2 we list for each experiment the quantity of heat withdrawn from the surroundings Q_{Vj} by each system j, computed from its change in

TABLE 3.2

HEAT EFFECTS PRODUCED IN THE FIVE EXPERIMENTS

Experiment	Change in state (equation)	Heat withdrawn from the surroundings
1	(3.61a)	$Q_{VA} = \bar{C}_{VA}(T - T_A) = -Q_{VB}$
	(3.61b)	$Q_{VB} = \bar{C}_{VB}(T - T_B) = -Q_{VA}$
	(3.61a and b)	$Q_{VA} + Q_{VB} = 0$
2	(3.62a)	$Q_{VA} = \bar{C}_{VA}(T - T_A) = -Q_{VS'}$
	(3.62b)	$Q_{VS'} = \bar{C}_{VS'}(-\Delta'T) = -Q_{VA}$
	(3.62a and b)	$Q_{VA} + Q_{VS'} = 0$
3	(3.63a)	$Q_{VB} = \bar{C}_{VB}(T - T_B) = -Q_{VS''}$
	(3.63b)	$Q_{VS''} = \bar{C}_{VS''}\Delta''T = -Q_{VB}$
	(3.63a and b)	$Q_{VB} + Q_{VS''} = 0$
4	(3.64a)	$Q_{VS'} = \bar{C}_{VS'}(-\Delta'T) = -Q_{VS''}$
	(3.64b)	$Q_{VS''} = \bar{C}_{VS''}\Delta''T = -Q_{VS'}$
	(3.64a and b)	$Q_{VS'} + Q_{VS''} = 0$
5	(3.65a)	$Q_{VS_2'} = \bar{C}_{VS_2'}(-\Delta'T) = 2(-Q_{VS''})$
	(3.65b)	$2Q_{VS''} = 2\bar{C}_{VS''}\Delta''T = -Q_{VS_2'}$
	(3.65a and b)	$Q_{VS_2'} + 2Q_{VS''} = 0$

$Q_{VA} = -Q_{VB} = -Q_{VS'} = Q_{VS''} = -\tfrac{1}{2}Q_{VS_2'}$ = a positive number temperature and its mean constant volume heat capacity $\bar{C}_{Vj}$ [see eqn. (3.55)]. The third line for each experiment gives the heat effect for the overall process undergone by the composite system composed of the two systems inside the isolating boundary. This heat effect must be zero. At the bottom of Table 3.2, we give the algebraic sign of each quantity of heat Q_{Vj}. A little consideration will show that the equations given in Table 3.2 offer a consistent explanation of the results of the five experiments.

If initially we choose the mass of S' in Experiment 2, and hence the mass of S'' in Experiment 3, so that $\Delta'T$ is about one degree, we find that the ratio $\Delta''T/\Delta'T$ is approximately unity. As we increase the masses of S' and S'' used in these two experiments, we find that the ratio $\Delta''T/\Delta'T$ rapidly approaches unity as $\Delta'T$ becomes smaller, unless the standard substance has a phase change or a λ-point in the vicinity of T. Thus the constant volume heat capacity C_V of the substance composing the systems S' and S'' is a continuous function of T, except at isolated points. This is a general property of C_V.

From eqns. (3.64a) and (3.65a) of Table 3.2, we find that

$$\bar{C}_{VS_2'} = 2\bar{C}_{VS}, \text{ whence } C_{VS_2'} = 2C_{VS}' \tag{3.66}$$

which shows that C_V is extensive.

With some modification in procedure, we could carry out the above experiments under conditions of constant pressure rather than constant volume. Then C_V in Table 3.2 would be replaced by C_p.

Experiments 2 and 3 are examples of the calorimetric method known as the "method of mixtures", which has been used to determine the heat capacity of a substance in terms of the properties of water. In the usual procedure, a known mass of the substance under investigation at a uniform temperature T_A is immersed in a known mass of water at a uniform temperature T_W contained in a thermally insulated vessel, and the final equilibrium temperature T measured. The experiment is carried out under atmospheric pressure. By application of relations similar to those derived for Experiment 2 in Table 3.2, we can express the mean heat capacity of the given substance from T to T_A in terms of the mean heat capacity of water from T_W to T. A number of corrections to the observations are necessary, particularly for the vessel enclosing the substance A if such a vessel is used.

3.18. Calorimetry. The determination of the quantity of heat withdrawn from the surroundings by a system undergoing a given process is known as calorimetry, and the apparatus employed for making this measurement is a calorimeter [9]*. Formerly, the heat effect attending a process was computed from the temperature change it caused in a known mass of liquid water. In the hands of careful investigators the method gave surprisingly accurate results with a relatively simple apparatus, owing, in part, to the

---

*For a description of early types of calorimeter see ref. 9a; some modern calorimeters are described by Kubaschewski et al. [9b].

fact that the specific heat capacity of liquid water has a minimum at about 35°C and varies by a maximum of only ~0.2% over the entire temperature range 14—65°C. In a second method, the given heat effect was computed from the mass of a pure substance it caused to undergo a change in aggregation state, for example, the mass of ice melted.

Both of these experimental procedures are employed in modern calorimetry. The given heat effect is not, however, computed from the thermal properties of water, but is determined electrically in a "blank run". Hence, heat quantities are expressed primarily in joules. They are then converted to calories by use of a defined conversion factor. The values so determined have been found to agree with those measured in terms of the thermal properties of water, within the accuracy of the latter results.

We shall now describe several types of modern calorimeter.

3.19. Bomb calorimeter. The bomb calorimeter [10] is used to determine the heat effect attending a constant-volume process in which a chemical reaction (particularly combustion) takes place under pressures in excess of one atmosphere. For the present, let us suppose that the heat effect of this process is negative. Known amounts of the reactants are confined in a sealed, thick-walled metal bomb B (see Fig. 3.12), which is immersed in a calorimeter C filled with a liquid (usually water) and fitted with a stirrer S, a heater H with provision for measuring the electric power input, and a precision thermometer T. The stirrer S runs continuously at constant speed throughout all measurements, but electrical current flows through the heater H only during the blank run. The calorimeter is separated by an air space (for partial thermal insulation) from a liquid-filled jacket J, fitted with a stirrer S', a heater H', and a thermoregulator R. By proper control of the electrical current passing through the resistor H', we can either maintain the temperature of the jacket J constant throughout a measurement, or (for adiabatic calorimetry) vary it so that it follows closely the temperature of the calorimeter C.

After the initial temperature of the calorimeter has been measured periodically for a time, the given chemical reaction is initiated electrically, and temperature—time measurements are continued until the experiment is completed. If the operation is not adiabatic, the temperature rise $T_2 - T_1$ of the system, corrected to adiabatic conditions for thermal leakage and the work dissipated in the calorimeter by the stirrer S, must be determined from a temperature—time plot. We shall call this procedure the main run.

In a blank run, a corrected temperature rise $T_2' - T_1'$, substantially the same as $T_2 - T_1$ and made, as nearly as possible, under the same conditions as those in the main run, is produced in the calorimeter and empty bomb by the dissipation of a measured quantity of electrical work W_{EL} in the heater H. This is equal to the quantity of heat Q^{CB} which would produce the same corrected rise in temperature $T_2' - T_1'$ in the calorimeter-bomb CB during a blank run made without the dissipation of electrical work in the system (see

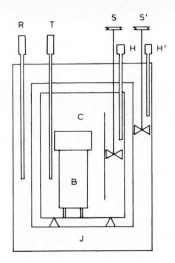

Fig. 3.12. Bomb calorimeter (after Jessop and Green [10]). B, bomb; C, calorimeter; J, jacket; H,H', heaters; R, thermoregulator; S,S', stirrers; T, thermometer.

Section 3.13). Thus the mean heat capacity $\bar{C}^{CB}$ of the system CB under the conditions of the blank run, sometimes called the energy equivalent of the calorimeter, is:

$$\bar{C}^{CB} = Q^{CB}/(T'_2 - T'_1) \text{ J deg}^{-1} \qquad (3.67)$$

For work of the highest accuracy we must take into account any change in the mean heat capacity of the system CB produced by the difference between the pressure in the bomb during the main run and the pressure during the blank run.

The mean heat capacity of the calorimeter and empty bomb may also be determined by burning in the bomb a substance of known heat of combustion, for example benzoic acid. The amount of the substance used is adjusted to produce approximately the corrected temperature rise $T_2 - T_1$ of the main run.

Let us place the boundary of the system so that it includes the substances inside the bomb parallel to the inner surface of the bomb but displaced slightly outward. The exterior surface of the calorimeter vessel C is taken to be the outer boundary of the system CB. The portions of the heater H, stirrer S, and thermometer T inside this latter boundary are considered to be parts of CB.

In the main run, a composite system consisting of the calorimeter, bomb, and the substances in the bomb has experienced a corrected rise in temperature from T_1 to T_2 in a constant-pressure process which is adiabatic except for the very small heat effect $Q_1 = Q_I$ arising from the electrical work dissipated inside the bomb in order to activate the fuse igniting the reaction. Let us consider separately the two parts constituting this composite system.

The part consisting of the calorimeter and bomb has undergone an increase in temperature from T_1 to T_2 by a process in which the inner boundary is essentially unchanged in volume and the outer boundary is under a constant pressure. The attending heat effect Q_2 is

$$Q_2 = \bar{C}^{CB}(T_2 - T_1) \qquad (3.68)$$

where $\bar{C}^{CB}$ is given by eqn. (3.67). The part of the system composed of the substances in the bomb has undergone a chemical reaction by an essentially constant-volume process, whereby the temperature has increased from T_1 to T_2. The attending heat effect Q_3 is evidently $Q_1 - Q_2$ (see Section 3.13).

The above information is best presented in the form of thermochemical equations:

Reactants (T_1) + CB(T_1) = Products (T_2) + CB(T_2) $\quad Q_1 = Q_I \qquad$ (3.69a)

CB(T_1) = CB(T_2) $\qquad\qquad\qquad\qquad\qquad Q_2 = \bar{C}^{CB}(T_2 - T_1) \qquad$ (3.69b)

Reactants (T_1) = Products (T_2) $\qquad\qquad Q_3 = Q_I - \bar{C}^{CB}(T_2 - T_1) \quad$ (3.69c)

Here the notation Reactants (T_1) stands for a complete description of the initial state of the system enclosed in the bomb, which consists of the substance under consideration, a large excess of the gas with which it is to react, the fuse to initiate the reaction, and perhaps some auxiliary substances such as water; Products (T_2) stands for a corresponding description of the final state of this system; and CB(T_i) represents the state of the calorimeter and empty bomb at T_i.

We wish to determine the heat effect attending an isothermal, constant-volume process in which the chemical reaction (3.69c) takes place either at T_1 or at T_2. We can determine this heat quantity at the initial temperature T_1, if we know the mean constant volume heat capacity $\bar{C}_V^p$ of the Products over the range T_1 to T_2, by the following method, where all changes in state are carried out by constant-volume processes.

Reactants (T_1) = Products (T_2) $\qquad Q_a = Q_I - \bar{C}^{CB}(T_2 - T_1) \qquad$ (3.70a)

Products (T_1) = Products (T_2) $\qquad Q_b = \bar{C}_V^p(T_2 - T_1) \qquad\qquad$ (3.70b)

Reactants (T_1) = Products (T_1) $\qquad Q_c = Q_I - (\bar{C}^{CB} + \bar{C}_V^p)(T_2 - T_1) \quad$ (3.70c)

Here the change in state (3.70c) is evidently the result of the change (3.70a) followed by the reverse of the change (3.70b), and hence Q_c is equal to Q_a minus Q_b (see Section 3.13).

In order to obtain the heat effect for the corresponding isothermal, constant-volume process at the final temperature T_2 we need to know the mean constant-volume heat capacity $\bar{C}_V^R$ for the Reactants. Then, instead of eqns. (3.70), we have

Reactants (T_1) = Products (T_2) $\qquad Q_a = Q_I - \bar{C}^{CB}(T_2 - T_1) \qquad$ (3.71a)

Reactants (T_1) = Reactants (T_2) $\qquad Q_b = \bar{C}_V^R(T_2 - T_1) \qquad\qquad$ (3.71b)

Reactants (T_2) = Products (T_2) $\qquad Q_c = Q_I - (\bar{C}^{CB} + \bar{C}_V^R)(T_2 - T_1) \quad$ (3.71c)

As we have indicated above, the changes in state (3.70c) and (3.71c) involve complex mixtures of a number of substances distributed among two or more phases. We desire to know the heat effect attending a particular isothermal standard change in state, in which a stoichiometric set of pure reactants, each in its standard state at T, is transformed into a set of products, each in its standard state at T, by an isothermal, constant-volume process (see Section 6.4). This may involve a laborious, painstaking calculation requiring the consideration of more than 20 steps and upwards of 100 items [11].

Now let us suppose that the heat effect attending the process under consideration is positive. Then, in the main run, the final temperature T_2 of the calorimeter is lower than the initial temperature T_1. The method we have just described requires no revision. The mean heat capacity of the system CB and of the Reactants or the Products must now be known over the temperature range T_2 to T_1. In general, a bomb is not required for an endothermic reaction. Such a process would be carried out in a vessel open to the atmosphere (Section 3.20); and it would ordinarily be initiated by mixing or by bringing the reactants together, rather than electrically, so that Q_I in eqns. (3.69) would be zero.

A direct method for determining the heat effect attending an endothermic process is to maintain the temperature of the calorimeter C constant during the process by dissipating a measured quantity of electrical work in the heater H.

3.20. Open calorimeter. The heat effects accompanying a number of processes occurring at atmospheric pressure, including chemical reaction, the solution of a solid in a liquid [12], the mixing of two liquids, and the dilution of a liquid solution by the addition of solvent [13], have been measured in an apparatus similar to a bomb calorimeter (see Fig. 3.12) with the following modifications. A sealed bomb B is not used. The process under investigation is carried out either in a vessel open to the atmosphere immersed in the calorimeter vessel C, or in C itself, which may be made of metal or glass. The substances taking part in the process may be introduced into the reaction vessel through one or more tubes extending through the top of the jacket J. The calorimeter C may be separated from the jacket J by an air space, or it may be made in the form of a Dewar vessel. For the measurement of very small changes in temperature, such as may be encountered in heats of dilution, the resistance thermometer T is replaced by a multijunction thermocouple.

The procedure and the reduction of the observations are similar to those used in Section 3.19, except that all processes take place at constant pressure rather than at constant volume, and C_V must be replaced by C_p in eqns. (3.69) and (3.70).

3.21. Low temperature heat capacity calorimeter. There are many variations in the design details of calorimeters [14] built for measuring constant-pressure heat capacities of solids and liquids from temperatures close to 0 K to room temperature or somewhat above. A sketch embodying the basic elements of an adiabatic, vacuum calorimeter is shown in Fig. 3.13. The substance under investigation is enclosed in a sealed copper calorimeter C which contains a number of radial copper vanes to promote uniformity of temperature throughout the sample. A small amount of helium gas may be introduced into C for the same purpose. A combination thermometer–heater T–H is inserted into C through the tube E_1. Surrounding the calorimeter is a radiation shield S which is not gas-tight. For adiabatic calorimetry, this thermal shield is maintained, as nearly as possible, at the temperature of the calorimeter, by controlling the electrical currents passing through three resistance wires (not shown): one on the top, one on the side, and one on the bottom. Outside the adiabatic shield is a gas-tight jacket J. The tube E_3 is used for evacuating the space between the calorimeter and the jacket during a measurement, and for admitting helium gas whenever the calorimeter is to be cooled. The calorimeter, shield, and case are placed in a Dewar vessel D which in turn is enclosed in a gas-tight metal case B fitted with a tube E_2 for introducing the refrigerant into D and for controlling the pressure in D.

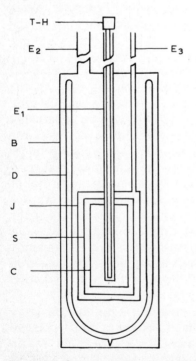

Fig. 3.13. Low temperature calorimeter. C, calorimeter shell; S, radiation shield; J, vacuum jacket; D, Dewar vessel; B, metal case; E_1, tube for thermometer–heater T–H; E_2, tube for refrigerant and pressure control; E_3, tube for evacuation.

A known mass of the substance under investigation is placed in C and helium gas introduced (if used). The calorimeter and its contents are brought to the chosen initial temperature T_1. After T_1 has remained constant for a time as indicated by T—H acting as a thermometer (with a very small current flowing through it), a much larger current is passed through T—H, now acting as a heater, for a sufficient time to increase the calorimeter temperature by the desired amount. The elapsed time and the electrical power imput during this period are measured. Then the final temperature T_2 is determined. In a blank run on the empty calorimeter (or the calorimeter partly filled with the substance under investigation) the electrical power imput and the exact time required to duplicate the same temperature increase under as nearly as possible the same conditions as in the original run are determined.

From these measurements we can compute the number of joules of heat Q_p withdrawn from the surroundings when the temperature of a mass m of the solid is increased from T_1 to T_2 by a process essentially at constant pressure (taken to be one atmosphere). The heat capacity C_p of unit mass is given by the relation

$$C_p = Q_p/[m(T_2 - T_1)] \tag{3.72}$$

The temperature interval $T_2 - T_1$ is chosen small enough so that we may consider C_p to be the true constant-pressure heat capacity of the substance at the mean temperature $(T_2 - T_1)/2$ without appreciable error.

3.22. High temperature heat capacity calorimeter. The mean constant-pressure heat capacity of a solid from 0°C to temperatures as high as 1500°C has been measured by the "drop method" [15]. A weighed sample of the substance under investigation encased in a thin metal capsule, which may also contain helium at a low pressure, is heated in an electric furnace to a given temperature and then dropped through an evacuated tube into a hole drilled in a block of copper surrounded by an adiabatic shield, or into a Bunsen ice calorimeter (Fig. 3.14) immersed in a mixture of ice and water. A blank run is made duplicating the conditions of the original experiment as closely as possible, but with the capsule empty, except for the helium. If we know the constant-pressure heat capacity of the copper calorimeter or the quantity of heat required to melt unit mass of ice in the ice calorimeter, we can compute the mean constant-pressure heat capacity $\bar{C}_p$ of the given substance over the temperature interval of the measurement.

A sketch of an ice calorimeter is shown in Fig. 3.14. The calorimeter D is a glass Dewar vessel closed gas-tight by a metal top through which passes a metal tube A for introduction of the system under investigation, and a steel capillary B which connects a pool of mercury M in the bottom of D to a pool M in a weighing bottle W. The calorimeter and a coiled portion C of the capillary B are immersed in a mixture of ice and water held in the jacket J. The space above the mercury in D is completely filled with air-free water. Before the start of an experiment, a thick layer of ice is frozen around the

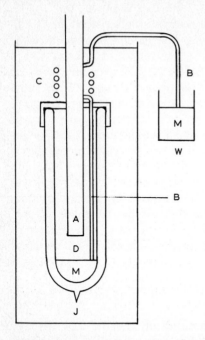

Fig. 3.14. Ice calorimeter (after Furukawa et al. [15]). A, sample tube; B, steel capillary; C, coil in B; D, Dewar vessel; J, jacket containing an ice-water mixture; M, mercury; W, weighing bottle.

tube A. Since the specific volume of ice is about 9% greater than that of water, the melting of ice produced by a flow of heat to the calorimeter through the walls of A causes mercury to be withdrawn from the weighing bottle W.

In a given run let w_1 be the number of grams of mercury withdrawn from the weighing bottle when the capsule containing m grams of the substance under investigation is heated to a temperature T_2 in the electric furnace, and then dropped into the ice calorimeter at the temperature T_1. And let a be the number of units of heat flowing through the walls of A to the calorimeter per gram of mercury withdrawn from W, as determined electrically in a calibration run. Then the quantity of heat withdrawn from its surroundings by the system composed of the capsule and its contents is $-aw_1$; and the heat effect Q_1 attending the *increase* in temperature of this system from T_1 to T_2 in a constant-pressure process is $+aw_1$.

In a blank run made with the capsule empty (except for the helium), the procedure of the original run is duplicated as nearly as possible; and w_2 grams of mercury are withdrawn from W. The heat effect Q_2 attending the increase in temperature of the capsule (and helium) from T_1 to T_2 is $+aw_2$. The mean constant-pressure heat capacity $\bar{C}_p$ of unit mass of the substance is given by the relation

$$\bar{C}_p = a(w_2 - w_1)/[m(T_2 - T_1)] \tag{3.73}$$

Since $\bar{C}_p$ for a solid changes so little with pressure, we may consider that all measurements were made at one atmosphere.

From a set of values of $\bar{C}_p$ for a series of temperature intervals starting from various upper temperatures and all ending at the same lower temperature, we can compute the true value of C_p as a function of temperature from eqn. (3.50).

3.23. Calorimeters for measurement of saturation thermal properties. The heat effect attending a change in aggregation state and the heat capacity of a condensed phase along a saturation curve of a substance have been measured [16] below room temperature in a calorimeter similar to the one sketched in Fig. 3.13, except that the calorimeter shell C is connected by a tube to the outside for the removal of vapor when heats of sublimation and vaporization are measured.

Calorimeters for determination of the saturation properties of liquids above room temperature have the same general features (calorimeter shell, thermometer, heater, radiation shields, jacket, thermostat, vapor outlet) as the low temperature model, and may also have an inlet tube to permit the apparatus to be used as a flow calorimeter [17]. Because of the high temperatures and pressures involved, such calorimeters are of quite complicated construction.

3.24. Flow calorimeter. The constant-pressure heat capacities of gases and vapors from about -180 to $+200°C$ have been measured in a flow calorimeter [18]. In the apparatus shown in Fig. 3.15, the gas passes through a coiled tube immersed in a thermostat (not shown in the drawing) which surrounds the calorimeter and is maintained at the initial temperature of the run. It then enters the calorimeter at I, passes successively through the thermometer windings T_1, a heating coil H, the thermometer windings T_2, and leaves the apparatus at O. The various directions of flow of the gas in the tubes and in the thermometer and heater cases are indicated by arrows. A number of thermal shields are provided to assure that: (1) the thermometers come into thermal equilibrium with the gas; (2) the thermometers are protected from the radiation of the heater; and (3) the calorimeter system is protected from thermal leakage. Provision is made to reduce to a minimum heat conduction along all wires (not shown in the drawing) leading from the thermometer and heating coils to the outside, and along tubes (also not shown) leading from the thermometer cells to pressure gauges.

At the start of a run, the rate of flow of gas through the calorimeter is adjusted to the required value and maintained constant by: (1) boiling the liquified gas electrically at a controlled rate if feasible; or (2) maintaining a constant pressure by means of a gas holder or the liquid boiling under pressure on the upstream side of a throttle valve in the entering gas

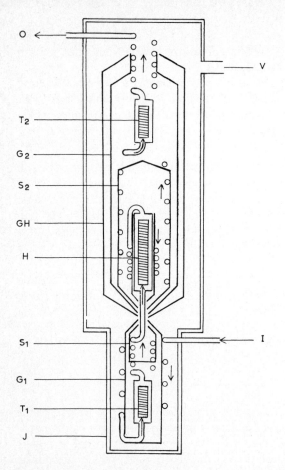

Fig. 3.15. Flow calorimeter (after Osborne et al. [18]). Electrical and pressure connections are not shown. V, vacuum line; I, gas inlet; O, gas outlet; T_1, T_2, thermometers; G_1, G_2, isothermal guards; S_1, S_2, isothermal shields; GH, electrically heated shield for G_2; H, heater; J, vacuum jacket.

stream. A constant electrical current large enough to produce the desired temperature rise in the gas is passed through the heater. After steady-state conditions have been reached, as indicated by the thermometers, the electrical power imput and the rate of flow of gas through the calorimeter are measured.

The mass of gas which has flowed through the calorimeter during a given time interval is determined by: (1) condensing it and weighing the liquid; or (2) collecting it in a gas holder and computing the mass by the equation of state from the measured pressure, volume, and temperature. The constant-pressure heat capacity of unit mass of the gas at the average temperature of the thermometers T_1 and T_2 and at the mean pressure in the calorimeter can be computed from eqn. (3.72).

Heat capacities of gases have been determined from measurements of the velocity of sound in the gas. This has been related by non-thermodynamic methods to the ratio $\gamma = C_p/C_V$. We can calculate the difference $C_p - C_V$ by thermodynamic methods if we are given the equation of state of the gas. From these two equations the values of C_p and C_V can be found.

The molar heat capacities C_p and C_V of many gaseous substances at low pressure have been computed, from the lowest temperatures at which they remain gaseous to temperatures far above those reached in experimental calorimetric investigations, by the methods of statistical mechanics. Analyses of the spectra of diatomic and polyatomic gases are required.

3.25. Heat, work, and the location of the boundary. We must exercise care in distinguishing between the heat and work effects attending a given process, since ambiguity may arise owing to the fact that temperature changes in bodies in the neighborhood of a system may be caused either by the withdrawal of heat from the surroundings, or by the dissipation of part or all of the work produced in the surroundings by frictional, fluid viscosity, impact, or other such processes. An appropriate location of the boundary can usually, but not always, resolve the difficulty. The essential criterion is to place the boundary in such a position that every dissipative effect is definitely known to occur entirely inside the system or entirely in the surroundings, but not across the boundary (see Section 3.10).

In Sections 3.5, 3.8, and 3.9 we found acceptable positions for the boundary of a system for the identification and calculation of the work of any reversible process and of two irreversible processes producing the change in state (3.15). The corresponding heat effect attending each process is the quantity of heat which has flowed from the surroundings across the same boundary as that chosen for the identification of work. This quantity of heat could be measured by calorimetric methods. Clearly, the location of the boundary of the system in the above examples permits the separate identification of work and heat, because all effects of friction, fluid viscosity, and impact definitely occur inside the boundary of the system. An example of an experiment in which a frictional effect occurs entirely in the surroundings is given below (Fig. 3.16).

Let us now turn our attention to a process which does not involve the expansion of a gas. In the apparatus (Fig. 3.16) used by Joule [19] for his famous paddle-wheel experiment, a metal framework C carrying a paddle-wheel fitted with eight sets of revolving blades AA working between four sets of stationary vanes BB (only one set of each is shown in the sketch) is enclosed in a cylindrical, copper vessel D filled with water. The wheel is rotated by the action of the weights F,F which are connected to the drum G by long lines passing over the pulleys E,E. The rotation of the paddle wheel tends to produce a corresponding rotation of the fluid, but this is prevented by the stationary vanes. Joule's experiment consisted in allowing the weights to fall through a known distance, and measuring the initial and final

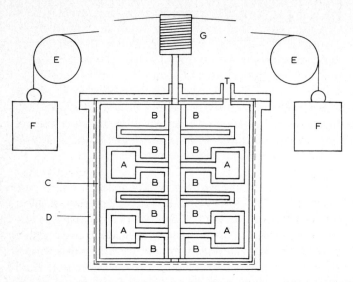

Fig. 3.16. Joule's paddle-wheel apparatus. AA, revolving blades; BB, stationary vanes; C, metal framework; D, copper jacket; E,E, pulleys; F,F, weights; G, drum; T, hole for thermometer.

temperature of the water, and the duration of the run. The quantity of heat a which has flowed to the surroundings through the outer surface of the copper vessel during the experiment was determined in a blank run; and the frictional force in each of the pulleys E,E was also determined in a separate run. The experiment was carried out at atmospheric pressure (1 atm).

Let us place the boundary of the system in the outer surface of the copper container, as Joule did. Then the change in state of the system, and the work and heat effects attending the process are:

$$n\,H_2O(l, T_1) + S(T_1) + D(T_1) = n\,H_2O(l, T_2) + S(T_2) + D(T_2) \qquad (p = 1\text{ atm}) \tag{3.74a}$$

$$W = -(w_1 d_1 + w_2 d_2) + p\Delta V; Q = -a \tag{3.74b}$$

Here n is the mass of water in the system; S denotes all of the metal parts forming the stirring mechanism and baffles; D is the copper container (with cover); w_1 is the force exerted on the drum G by one of the weights F (the weight of F reduced by the frictional force of the bearing of its pulley E), and d_1 is the distance it falls; w_2 and d_2 are the corresponding quantities for the other weight F; $p\Delta V$ is the (very small) expansion work produced by the system; and $-a$ is the quantity of heat withdrawn from the surroundings during the experiment.

Let us now place the boundary parallel to the inner surface of the copper container but displaced a slight distance outward. With this location of the boundary the copper vessel is not part of the system. The change in state of the system, and the work and heat effects attending the process are:

$$n\,\text{H}_2\text{O}(l, T_1) + \text{S}(T_1) = n\,\text{H}_2\text{O}(l, T_2) + \text{S}(T_2) \qquad (p = 1 \text{ atm}) \qquad (3.75a)$$

$$W = -(w_1 d_1 + w_2 d_2) + p\Delta V',\ Q = -a - \bar{C}_p(\text{D})(T_2 - T_1) \qquad (3.75b)$$

Here $\bar{C}_p(\text{D})$ is the mean constant-pressure heat capacity of the copper container D, which was supposed by Joule to undergo the same change in temperature as the water, and $p\Delta V'$ is the expansion work produced by the system.

With either of these two positions of the boundary there is a clear operational distinction between the work and the heat of the process.

On the other hand it would be extremely difficult, if not impossible, to evaluate the work and heat effects separately from thermodynamic considerations alone, if we chose to place the boundary parallel to the inner surface of the container but displaced slightly into the liquid water where viscous effects are occurring.

There are experiments in which an operational distinction between work and heat effects cannot be made within the framework of thermodynamics. In Fig. 3.17 a closed, rigid cylinder made of thermally insulating material is divided into two chambers by a piston C which is made of thermally conducting material. The piston moves in the cylinder without friction; and its upward travel is limited by two sets of stops. The upper chamber of the cylinder contains a real gas A and the lower chamber is filled with a real gas B (which may be the same as the gas in the upper chamber if we so choose). Initially the two gases have the same temperature T_1, the pressure of B is considerably higher than the pressure of A, and the piston is held in the lower set of stops (Fig. 3.17a). When the piston is released, it accelerates upward until it hits and is held by the upper set of stops (Fig. 3.17b). After equilibrium has been established, the gases A and B are found to have the same temperature T_2, which is different from T_1.

We wish to determine the work and heat effects attending the expansion

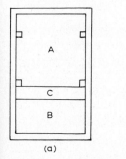 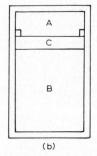

Fig. 3.17. Simultaneous expansion of one gas and compression of another in a closed cylinder, (a) at the start of the experiment, and (b) at the end of the experiment. A, gas being compressed; B, expanding gas; C, piston.

of gas B. In order to eliminate difficulties arising from the impact of the piston on the upper set of stops, we place the boundary of B in the location indicated in Fig. 3.5. Then for all practical purposes, the surroundings of B are composed of the gas A and the remaining part of the piston and cylinder. The work and heat effects attending the expansion of B must be identified and measured by the disturbances they produce in these surroundings. The temperature change $T_2 - T_1$ of the gas A is produced, in part, by a flow of heat across the piston and, in part, by a dissipation through viscous effects in A of some of the work produced by B in its surroundings. The determination of the amount of work so dissipated is beyond the field of thermodynamics. Thus there is no thermodynamic operational method for determining separately the work and heat effects attending the expansion of B in the experiment under consideration.

A number of writers claim that the heat withdrawn from the surroundings of a system undergoing a given process is determined operationally from a knowledge of the change in state of the system and the measurement of the work produced in the surroundings during the process. It would seem however that, if the work effect attending a process can be measured, the heat effect can be measured as well. Therefore we prefer to treat both heat and work as measureable quantities.

3.26. Review of the properties of work and heat. We have defined work and heat in Sections 3.1 and 3.12, respectively; and we have discussed the measurement and units of these quantities in Sections 3.2 and 3.13.

We find as a result of experiment that the net quantity of work W produced in the surroundings and the net quantity of heat Q withdrawn from the surroundings by a system in undergoing a cyclical process are not, in general, zero:

$$\oint đW \neq 0 \text{ and } \oint đQ \neq 0 \text{ (in general)} \tag{3.76}$$

The following statements are consequences of these experimentally established facts.

1. The quantities $đW$ and $đQ$ are not exact differentials. They simply represent small amounts of work and heat, respectively.

2. The line integrals of $đW$ and $đQ$ along the path C of a reversible process from an initial state 1 to a final state 2 of a system are conventionally represented by W and Q, respectively, and not by ΔW and ΔQ. In thermodynamics the latter notation is reserved for the increment of a state variable, for example, ΔV or Δp. Thus for a reversible process we may write

$$W = \int_{C\,1}^{2} đW \text{ and } Q = \int_{C\,1}^{2} đQ \tag{3.77}$$

In order to evaluate these integrals for a given reversible process we must know the path C followed by the state of the system. Hence the specification of a change in state alone does not determine W and Q. If the process is reversible the path must be specified. For an irreversible, quasistatic process, W and Q, in general, will depend on well-defined variables which may not be specified by the path of the process (Section 3.9). For processes which proceed at a finite rate, the situation may be so complicated that we cannot make any general statement regarding the evaluation of W and Q.

In any case, W and Q are not functions of any set of independent state variables of a system; they are not state properties. We speak of the work and heat of a process, but we do not speak of the work and heat attending a change in state. We cannot defend the statement that a system in a specified state has a definite quantity of work or heat in it, either in the absolute sense (as we speak of the volume of a system) or relative to any fixed state of reference (as we speak, later on, of the energy or the entropy of a system).

3. As long as we confine our attention to changes in state produced by a definite, reversible process whereby the state of a system moves along a given path C starting from a fixed fiducial state, W and Q are functions of a single independent state variable, and $đW$ and $đQ$ are the differentials of these functions. For a simple, homogeneous system, quantities such as $(đW/dT)_p$ and $(đQ/dT)_p$ are the derivatives of the functions $W(T)$ and $Q(T)$. The notation used here calls attention to the fact that the above two symbols are not the partial derivatives of a function of a set of independent variables of the system; but, along the path C given by p = constant, W and Q are functions of a single independent variable T, and the above expressions are the derivatives of these functions.

Let n moles of gaseous oxygen (considered to be a perfect gas) undergo a change from the fiducial state (p_0, T_0) to the variable state (p_0, T) by a reversible, constant-pressure process:

$$n\,O_2(g, p_0, T_0) = n\,O_2(g, p_0, T) \qquad (C \text{ given by } p = p_0) \qquad (3.78)$$

Then

$$W = nR(T - T_0), \quad đW = nR\,dT, \quad \left(\frac{đW}{dT}\right)_p = nR \qquad (3.79a)$$

$$Q = nC_p(T - T_0), \quad đQ = nC_p\,dT, \quad \left(\frac{đQ}{dT}\right)_p = nC_p \qquad (3.79b)$$

Here we have assumed that C_p is independent of temperature in the temperature interval T_0 to T.

3.27. *Quantities that determine the behavior of a simple system.* The behavior of a simple, homogeneous system consisting of a fixed mass of a pure substance or of a solution of constant composition under the action of various external influences is described by a number of differential coefficients. Some are state properties of a system; others depend on the given state of the system and also on the properties of the system in a fiducial state. The coefficients are functions of an independent set of state variables. A property of the system in a fiducial state occurs as a parameter in some of these functions; and the composition variables occur as parameters in all of the functions for a solution phase. The coefficients are continuous and bounded except at particular points, for example, the liquid—vapor critical point and λ-points. In general, they have a discontinuity at a point where a change in aggregation state occurs.

These quantities are called the calorimetric coefficients, the thermometric coefficients, and the elastic coefficients.

3.28. *Calorimetric coefficients.* Let a simple, homogeneous system A of fixed mass and composition undergo the change in state

$$A(p, V, T) = A(p + dp, V + dV, T + dT) \tag{3.80}$$

Only two of these three variables are independent. Let this change in state be brought about by any one of the following three two-step reversible processes.

1. A constant-temperature process followed by one at constant pressure, the changes in state being

$$A(p, T) = A(p + dp, T) = A(p + dp, T + dT) \tag{3.81}$$

2. A constant-temperature process followed by one at constant volume

$$A(V, T) = A(V + dV, T) = A(V + dV, T + dT) \tag{3.82}$$

3. A constant-pressure process followed by one at constant volume

$$A(p, V) = A(p, V + dV) = A(p + dp, V + dV) \tag{3.83}$$

The differential quantities of heat đQ withdrawn from the surroundings in these three two-step reversible processes are given by the following expressions:

$$đQ = C_p dT + \Lambda_T dp \tag{3.84}$$

$$đQ = C_V dT + l_T dV \tag{3.85}$$

$$đQ = \Gamma_V dp + \eta_p dV \tag{3.86}$$

where the calorimetric coefficients are:

$$C_p = \left(\frac{đQ}{dT}\right)_p = \text{heat capacity at constant pressure} \tag{3.87}$$

$$\Lambda_T = \left(\frac{đQ}{dp}\right)_T = \text{latent heat of pressure increase} \tag{3.88}$$

$$C_V = \left(\frac{đQ}{dT}\right)_V = \text{heat capacity at constant volume} \tag{3.89}$$

$$l_T = \left(\frac{đQ}{dV}\right)_T = \text{latent heat of expansion} \tag{3.90}$$

$$\Gamma_V = \left(\frac{\mathrm{d}Q}{\mathrm{d}p}\right)_V = \text{heat of pressure increase at constant volume} \tag{3.91}$$

$$\eta_p = \left(\frac{\mathrm{d}Q}{\mathrm{d}V}\right)_p = \text{heat of expansion at constant pressure} \tag{3.92}$$

Of these quantities, $C_p, \Lambda_T, C_V,$ and Γ_V are extensive state properties of a system, while l_T and η_p are intensive state properties. Since đQ is not an exact differential, we find in general that

$$\left(\frac{\partial C_p}{\partial p}\right)_T \neq \left(\frac{\partial \Lambda_T}{\partial T}\right)_p, \left(\frac{\partial C_V}{\partial V}\right)_T \neq \left(\frac{\partial l_T}{\partial T}\right)_V, \left(\frac{\partial \Gamma_V}{\partial V}\right)_p \neq \left(\frac{\partial \eta_p}{\partial p}\right)_V \tag{3.93}$$

The calorimetric coefficients tell us how a simple, homogeneous system A of constant mass and composition behaves with respect to the withdrawal of heat from the surroundings when subject to certain variations in state under fixed conditions of restraint. For example Λ_T is the ratio of the heat đQ withdrawn from the surroundings to the increase in pressure dp of a system during an infinitesimal reversible, isothermal expansion. For this infinitesimal process we can write

$$\text{đ}Q = \Lambda_T(p, T)\,\mathrm{d}p \ (T = \text{constant}) \tag{3.94}$$

The heat effect Q attending a corresponding finite process is given by the expression

$$Q = \int_{T\,p_1}^{p_2} \Lambda_T(p, T)\,\mathrm{d}p \tag{3.95}$$

Let the system A undergo a finite change in state by a reversible process in which the state follows the path C given by the equation $p = p(T)$:

$$A(p_1, T_1) = A(p_2, T_2) \ [C \text{ given by } p = p(T)] \tag{3.96}$$

The heat effect attending the process can be found by integrating eqn. (3.84) along the curve $p = p(T)$:

$$Q = \int_{C\,p_1,T_1}^{p_2,T_2} [C_p\,\mathrm{d}T + \Lambda_T\,\mathrm{d}p] \tag{3.97}$$

If we know C_p and Λ_T as functions of p and T, we can express them, as well as dp, in terms of T by use of the relation $p = p(T)$.

Except for C_p and C_V the calorimetric coefficients do not play an important part in the development of thermodynamics. Their role is taken by partial derivatives of the entropy, which is a state property of a system.

3.29. Thermometric coefficients. The various coefficients of thermal expansion at constant pressure of a simple, homogeneous system of fixed mass and composition are defined by the relations:

$$\alpha_p = \frac{1}{V}\left(\frac{\partial V}{\partial T}\right)_p \quad = \text{true coefficient of thermal expansion at constant pressure} \tag{3.98}$$

$$\alpha'_p = \frac{1}{V_0}\left(\frac{\partial V}{\partial T}\right)_p \quad = \text{true coefficient of thermal expansion at constant pressure referred to } T_0 \tag{3.99}$$

$$\bar{\alpha}_p = \frac{1}{V_0}\left(\frac{V-V_0}{T-T_0}\right)_p \quad = \text{mean coefficient of thermal expansion at constant pressure from } T \text{ to } T_0 \tag{3.100}$$

Here V_0 is the volume $V(T_0, p)$ of the system at the fiducial temperature T_0 and the fixed pressure p.

The corresponding coefficients of thermal linear expansion are defined through the relations:

$$\alpha_l = \frac{1}{l}\left(\frac{\partial l}{\partial T}\right)_p, \; \alpha'_l = \frac{1}{l_0}\left(\frac{\partial l}{\partial T}\right)_p, \; \bar{\alpha}_l = \frac{1}{l_0}\left(\frac{l-l_0}{T-T_0}\right)_p \tag{3.101}$$

The various coefficients of thermal pressure increase at constant volume of a simple, homogeneous system of fixed mass and composition are defined by the relations:

$$\alpha_V = \frac{1}{p}\left(\frac{\partial p}{\partial T}\right)_V \quad = \text{true coefficient of thermal pressure increase at constant volume} \tag{3.102}$$

$$\alpha'_V = \frac{1}{p_0}\left(\frac{\partial p}{\partial T}\right)_V \quad = \text{true coefficient of thermal pressure increase at constant volume referred to } T_0 \tag{3.103}$$

$$\bar{\alpha}_V = \frac{1}{p_0}\left(\frac{p-p_0}{T-T_0}\right)_V \quad = \text{mean coefficient of thermal pressure increase at constant volume from } T_0 \text{ to } T \tag{3.104}$$

Here p_0 is the pressure $p(T_0, V)$ of the system at the fiducial temperature T_0 and the fixed volume V.

Each of the above coefficients has the units of $\deg^{-1}$. The first coefficient in each group of three is an intensive state property of a system; the last two in each group depend not only on the given state of the system, but also on a fiducial state.

Equations (3.100) and (3.104) are the basic relations on which determinations of temperatures by means of constant-pressure and constant-volume gas thermometers rest [see eqns. (2.25) and (2.21) in Section 2.7]. For this purpose they are written in the following manner:

$$t_p = T - T_0 = \frac{1}{\bar{\alpha}_p}\left(\frac{V}{V_0} - 1\right) \quad (p = \text{constant}) \tag{3.105a}$$

$$t_V = T - T_0 = \frac{1}{\bar{\alpha}_V}\left(\frac{p}{p_0} - 1\right) \quad (V = \text{constant}) \tag{3.105b}$$

3.30. Elastic coefficients. The various coefficients of isothermal compressibility of a simple, homogeneous system of fixed mass and composition are defined by the relations:

$$\beta_T = -\frac{1}{V}\left(\frac{\partial V}{\partial p}\right)_T = \text{true coefficient of isothermal compressibility} \quad (3.106)$$

$$\beta'_T = -\frac{1}{V_0}\left(\frac{\partial V}{\partial p}\right)_T = \text{true coefficient of isothermal compressibility referred to } p_0 \quad (3.107)$$

$$\bar{\beta}_T = -\frac{1}{V_0}\left(\frac{V-V_0}{p-p_0}\right)_T = \text{mean coefficient of isothermal compressibility from } p_0 \text{ to } p \quad (3.108)$$

Here V_0 is the volume $V(p_0, T)$ of the system at the fiducial pressure p_0 and the fixed temperature T.

The corresponding coefficients of adiabatic compressibility are

$$\beta_S = -\frac{1}{V}\left(\frac{\partial V}{\partial p}\right)_S, \beta'_S = -\frac{1}{V_0}\left(\frac{\partial V}{\partial p}\right)_S, \bar{\beta}_S = -\frac{1}{V_0}\left(\frac{V-V_0}{p-p_0}\right)_S$$

where the subscript S denotes a reversible, adiabatic process; and V_0 is the volume $V(p_0, S)$ of the system at the fiducial pressure p_0 on the fixed reversible, adiabatic curve denoted by S. The symbol S stands for entropy which is defined by the second law of thermodynamics, and β_S can be called the true coefficient of isentropic compressibility.

Each of the above coefficients has the units p^{-1}. The first coefficient in each group of three is an intensive state property of a system; the last two in each group depend not only on the given state of the system, but also on a fiducial state.

The reciprocal of a coefficient of compressibility is the corresponding bulk modulus of a system.

REFERENCES

1 G. N. Hatsopoulos and J. H. Keenan, Principles of General Thermodynamics, Wiley, New York, 1965, p. 43.
2 J. W. Gibbs, Scientific Papers, Vol. I, Thermodynamics, Longmans, Green and Co., New York, 1906, footnote p. 51.
3 D. Kivelson and I. Oppenheim, J. Chem. Educ., 43 (1966) 233—235.
4 L. J. Gillespie and J. R. Coe, Jr., J. Chem. Phys., 1 (1933) 102—113.
5 J. H. Keenan, Thermodynamics, Wiley, New York, 1941, p. 35.
6 H. F. Stimson, Am. J. Phys., 23 (1945) 614—622.
7 E. S. R. Gopal, Specific Heats at Low Temperatures, Plenum Press, New York, 1966.
8 C. C. Stephenson, R. S. Orehotsky and D. Smith, personal communication.
9 (a) J. R. Partington and W. G. Shilling, The Specific Heat of Gases, Van Nostrand, New York, 1924, Chapters II—IV; A. Eucken, in W. Wien and F. Harms (Eds.), Handbuch der Experimental-Physik, Akademische Verlagsgesellschaft M. B. H., Leipzig, 1929, Vol. 8, Part 1, Energie und Wärmeinhalt. (b) O. Kubaschewski, E. Ll. Evans and C. B. Alcock, Metallurgical Thermochemistry, Pergamon, Oxford, 1967, Chapter II.
10 H. C. Dickinson, Bull. Bur. Stand., 11 (1915) 189—257; R. S. Jessup and C. B. Green, J. Res. Natl. Bur. Stand., 13 (1934) 469—495.
11 E. W. Washburn, J. Res. Natl. Bur. Stand., 10 (1933) 525—558; J. Coops, R. S. Jessup and K. van Nes, in F. D. Rossini (Ed.), Experimental Thermochemistry, Vol. I, Interscience, New York, 1956, Chapter 3; W. N. Hubbard, D. W. Scott and G. Waddington, ibid., Chapter 5; note also other chapters in this publication, and in Vol. II [1962, H. A. Skinner (Ed.)] of the same series.

12 E. S. Newman and L. S. Wells, J. Res. Natl. Bur. Stand., 20 (1938) 825—836, for heats of solution of solids in water; B. W. Howlett, J. S. Ll. Leach, L. B. Ticknor and M. B. Bever, Rev. Sci. Instrum., 33 (1962) 619—624, for heats of solution of solids in liquid tin.
13 E. Lange and A. L. Robinson, Chem. Rev., 9 (1931) 89—115.
14 J. C. Southard and F. G. Brickwedde, J. Am. Chem. Soc., 55 (1933) 4378—4384; W. F. Giauque and C. J. Egan, J. Chem. Phys., 5 (1937) 45—54; H. M. Huffman, Chem. Rev., 40 (1947) 1—14; E. F. Westrum, Jr., J. B. Hatcher and D. W. Osborne, J. Chem. Phys., 21 (1953) 419—423; W. B. Hadley and J. W. Stout, J. Chem. Phys., 39 (1963) 2205—2210.
15 J. C. Southard, J. Am. Chem. Soc., 63 (1941) 3142—3146, for the copper calorimeter; G. T. Furukawa, T. B. Douglas, R. E. McCoskey and D. C. Ginnings, J. Res. Natl. Bur. Stand., 57 (1956) 67—82, for the Bunsen ice calorimeter.
16 R. B. Scott, C. H. Meyers, R. D. Rands, Jr., F. G. Brickwedde and N. Bekkedahl, J. Res. Natl. Bur. Stand., 35 (1945) 39—85.
17 N. S. Osborne, H. F. Stimson and D. C. Ginnings, J. Res. Natl. Bur. Stand., 18 (1937) 389—447; ibid., 23 (1939) 197—260, the classic work on the thermal properties of water from 0—374°C.
18 K. Scheel and W. Heuse, Ann. Phys. (Leipzig), 37 (1912) 79—95; ibid., 40 (1913) 473—492; N. S. Osborne, H. F. Stimson and T. S. Sligh, Jr., Sci. Pap. Bur. Stand., 20 (1924—1926) 119—151; G. Waddington, S. S. Todd and H. M. Huffman, J. Am. Chem. Soc., 69 (1947) 22—30.
19 J. P. Joule, Philos. Trans. R. Soc. London, 140 (1850) 61—82; The Scientific Papers of James Prescott Joule, The Physical Society, London, 1884, pp. 298—328.

Chapter 4

THE FIRST LAW OF THERMODYNAMICS

In the last chapter we found that the work and heat effects produced in the surroundings when a system undergoes a change in state are not determined solely by the initial and final states of the system, but that they depend on the process employed to produce this change. Thus, these two quantities are not necessarily conserved. They are not state properties. The first law of thermodynamics asserts the existence of a new state property which we call the energy (or energy content) E of a system. By the first law the increment ΔE in the energy of a system, defined as the difference between the heat and work effects attending any process producing a specified change in state, depends only on the end states of the system.

4.1. Historical introduction. The concept of work did not present any difficulty to the scientists of the nineteenth century; but the understanding of the nature of heat did.

From a consideration of the thermal effects produced by the friction and impact of two bodies, Bacon (1561—1626) concluded that "heat is motion". The notion that heat is a manifestation of the vibration or perhaps the rotation of the particles constituting bodies was held by other seventeenth-century natural philosophers, notably Boyle (1627—1691), Hooke (1635—1703), Locke (1632—1704), and Newton (1642—1727). This view was eclipsed in the eighteenth and first half of the nineteenth centuries by the caloric theory, in which heat was considered to be an indestructible fluid. However, some scientists continued to uphold the older non-conservative concept which, in the last half of the nineteenth century, evolved into the mechanical (or dynamical) theory of heat. This concept was of fundamental importance in the development of the laws and equations of thermodynamics by Clausius and Kelvin, the kinetic theory of gases by Clausius and Maxwell, and statistical mechanics by Boltzmann and Gibbs.

As a result of his studies of the pendulum, Huygens (1629—1695) concluded that perpetual motion, in which force (energy) is generated without any corresponding expenditure, is impossible in purely mechanical systems. Building on this foundation Johann Bernoulli (1667—1748) arrived at the law of the conservation of *vis viva* (the sum of the products of the mass multiplied by the square of the velocity of the particles constituting a system). The failure of this law for systems in which friction is present puzzled scientists until the middle of the nineteenth century.

Black [1] was apparently the first to distinguish clearly between heat and temperature. Shortly after 1760 he began a series of careful calorimetric investigations, using the method of mixtures. He introduced the concept of specific heat to explain the changes in temperature produced when equal masses or equal volumes of two liquids initially at different temperatures are mixed. He applied the term latent to heats of fusion and vaporization. He explained his experimental results on the basis of the transfer from one body to another of a subtle, invisible, weightless fluid (caloric) which passes freely between the particles of a body. His conclusion that heat is conserved applies under the particular conditions of his experiments, which were carried out at constant pressure or constant volume. Unfortunately this conservation principle was applied by his followers to processes for which it did not hold.

Rumford [2] (1798) vigorously attacked the caloric theory. Impressed by the very considerable quantity of heat generated in the boring of brass cannons, he designed his famous blunt-borer equipment. An iron rod was inserted in a hole bored in the upper section of a cannon, which was connected to the main portion by a slender neck of metal and was surrounded by water. In one experiment he found that the amount of metal and water having a heat capacity equivalent to 26.58 lbs. of water was increased in temperature by 180 deg. F in 2.5 hours by the rotation of the cannon against the blunt-borer. He noted that the lathe could easily be turned by the power of one horse. He found that 4,125 grains of metallic powder were produced. (Joule, in 1850, calculated from these observations a value of the mechanical equivalent of heat about 30% larger than his own.) Rumford showed that the powder produced had the same heat capacity as an equal weight of the original metal. He wrote "Is it possible that the very considerable quantity of Heat that was produced in this experiment could have been furnished by so inconsiderable a quantity of metallic dust? and this merely in consequence of *a change* in its capacity for Heat?" And he concluded: "...anything which any *insulated* body... can continue to furnish *without limitation*, cannot possibly be a *material substance*; and it appears to be extremely difficult, if not quite impossible, to form any distinct idea of anything capable of being excited and communicated in the manner of Heat in these experiments, except it be MOTION". The calorists answer to Rumford's argument was that the powder had less "capacity for heat", that is, it contained less caloric than the original metal, because some had been squeezed out by friction; and this "capacity for heat" was not measured by "heat capacity" as determined by Rumford.

Davy [3] (1799) found that two pieces of ice could be melted simply by rubbing them together, although the ambient temperature was below the melting point of ice. He concluded that the immediate cause of the phenomenon of heat is motion, and that caloric does not exist. The objection raised by the calorists to Rumford's contention does not apply to Davy's argument, since Black had shown that ice absorbs heat when it melts and hence water must contain more caloric than the same weight of ice. Davy's work should have been conclusive, but it received little attention at the time, and the caloric theory continued to be used for another fifty years.

Carnot [4] thought in terms of the caloric theory in his investigations of the efficiency of heat engines which led to his discovery of the principle named after him, and now known as the second law of thermodynamics. But, in a footnote, he wrote with regard to the caloric theory that several experimental facts seem to be almost inexplicable in the actual state of that theory.

Clapeyron [5] gave analytical form to the results found by Carnot, but he continued to use the ideas of the caloric theory. Fortunately, his adherence to this theory did not invalidate his derivation of the relation known as the Clapeyron equation, because this equation applies to a change in state produced by a reversible process in which the state of the system moves along a fixed path, a saturation curve (see Section 3.16 and Section 3.26, Paragraph 3).

Mayer [6], in a paper not free from obscurity, postulated that "forces are causes". Hence, he reasoned, we can apply to them the principle *causa aequat effectum*, from which he concluded that forces, like causes, are indestructible and convertible from one form into another. On the basis of these considerations he deduced the law of the conservation of *vis viva*, and the equivalence of both "falling force" and "motion" (two forms of force) to heat. Mayer considered that, when a falling body strikes the earth, or when two metal plates are rubbed together, the falling force (equal to the weight of the body times its distance of fall) on the one hand and the motion producing the rubbing on the other hand have not been annihilated, but have been converted into heat, another form of force. On the hypothesis that "the sinking of a mercury column by which a gas is compressed is equivalent to the quantity of heat set free by the compression" he calculated a value of the mechanical equivalent of heat (365 kgf m kcal^{-1}) which is about 15% too small. No details of the calculation are given. Apparently he did not consider a direct experimental determination of this equivalent, but he equated the value of the gas constant in mechanical units computed from the perfect gas law $R = pV/T$ to its value in heat units derived from the perfect gas relation $R = C_p - C_V$ (see eqn. 5.39). He carried out one experiment in which he observed that the temperature of water was raised from 12 to 13°C by violent shaking, a fact referred to by Joule.

Helmholtz [7] based his discussion of the conservation of energy on the proposition: "it is not possible by any combination whatever of natural bodies to derive an unlimited amount of mechanical force [energy]". He related it to the premise that all actions in nature can ultimately be referred to attractive and repulsive forces whose magnitudes depend on the distance between their points of application. He applied the above proposition to all known transformations of energy. He concluded that "Nature as a whole possesses a store of energy which cannot in any wise be added to or subtracted from: the quantity of energy in inorganic nature is as eternal and unalterable as the quantity of matter". He did not attempt to calculate the mechanical equivalent of heat or to formulate the first law of thermodynamics, and he did no experimental work. This paper and that of Mayer were refused publication in the Annalen der Physik und Chemie by Poggendorff on the basis that they were not sufficiently experimental.

During the period 1840—1850, Joule [8] conclusively overthrew the caloric theory and placed the mechanical theory of heat on a firm experimental basis by means of a series of diverse and painstaking experimental investigations designed to determine the value of the mechanical equivalent of heat $\mathscr{J}$ (Section 4.2). He found values of $\mathscr{J}$ ranging from about 770 to 840 ft lb per Btu by: (1) the conversion of mechanical work into electrical work and dissipation of the latter in water (1843); (2) the flow of water under pressure through narrow tubes (1843); (3) the expansion and compression of air (1845); and (4) his paddle-wheel experiment using water, sperm oil, and mercury (1847). He then concentrated on the paddle-wheel experiment using water (Fig. 3.16). At the end of the paper describing this work (1850), he concluded: "That the quantity of heat produced by the friction of bodies, whether solid or liquid, is always proportional to the work expended". At that time he found $\mathscr{J}$ = 772 ft lb per Btu. In a later paper describing a new series of measurements (1878), he reported $\mathscr{J}$ = 772.55 ft lb per Btu (the heat required to increase the temperature of 1 lb of water from 60 to 61°F). The modern value corresponding to the International Steam Table conversion factor is 778.17 ft lb per Btu.

At first, W. Thomson (Lord Kelvin) [9] found himself unable to reconcile Joule's conclusion with the theory of Carnot. He wrote (in 1848) "...the conversion of heat (or *caloric*) into mechanical effect is probably impossible, certainly undiscovered", with a footnote stating "A contrary opinion has been advocated by Mr. Joule of Manchester". A year later (1849) Thomson was still puzzled. He again mentioned Joule's work, called it "extremely important", but brushed it under the rug for the time being with the statement that "The fundamental axiom adopted by Carnot may be considered as still the most probable basis for an investigation of the motive power of heat; although this... may ultimately require to be constructed upon another function, when our experimental data are more complete". Thomson was loath to give up Carnot's theory because he had been able to correlate Regnault's [10] measurements of the saturation properties of water from 0 to 230°C by means of an equation derived by Clapeyron [5] from the Carnot principle.

Clausius [11] (1850) realized that Joule's principle "...does not stand in contradiction to the essential principle of Carnot, but only to the subsidiary statement *that no heat is lost*, since in the production of work it may very well be the case that at the same time a certain quantity of heat is consumed... another quantity [is] transferred from a hotter to a colder body, and both quantities of heat stand in definite relation to the work that is done". Thus, a portion of the heat withdrawn from the surroundings at temperature T_1 by a reversible engine operating on a Carnot cycle is converted into an equivalent quantity of work produced in the surroundings, and the remaining portion of heat is rejected to the surroundings at a lower temperature T_2.

Clausius then proposed an axiom now called the first law of thermodynamics: "In all cases in which work is produced by the agency of heat [that is to say, in a cyclical process] a quantity of heat is consumed which is proportional to the work done; and, conversely, by the expenditure of an equal quantity of work an equal quantity of heat is produced". He proceeded to apply Joule's principle and Carnot's principle to a perfect-gas heat engine and to a steam engine, each operating on a (reversible) Carnot cycle between the temperatures t and $t-dt$. Clausius made a clear distinction between heat, which is not a function of a set of independent state variables, and energy, which is a function of these variables.

Finally [9] (1851) the confusion in Thomson's mind was dissipated, and he saw his way clearly. He based his discussion of the dynamic theory of heat on two propositions or axioms which are now known as the first and second laws of thermodynamics, attributing the first to Joule and the second to Carnot and Clausius. His statement of the first law is: "When equal quantities of mechanical effect are produced by any means whatever from purely thermal sources, or lost in purely thermal effects, equal quantities of heat are put out of existence or are generated".

4.2. Mechanical equivalent of heat.

Carnot [4] (1824) was the first to realize the importance of the cyclical process in the investigation of the work and heat effects produced in the surroundings by a system undergoing change. Thomson [9] (1851) wrote: "According to an obvious principle, first introduced, however, into the theory of the motive power of heat by Carnot, mechanical effect produced in any process cannot be said to have been derived from a purely thermal source, unless at the end of the process all the materials used are in precisely the same physical and mechanical circumstances as they were in the beginning". These same considerations also hold for the conversion of work into heat.

Actually, Joule and the investigators who followed him did not use cyclical processes in their determinations of the mechanical equivalent of heat; but they operated in such a manner that we can readily complete a cycle which consists in part of the process employed by them. For example, in the paddle-wheel experiment (Section 3.25) Joule measured the work effect W (in ft lbs) attending a constant-pressure process in which the change in state (3.74) was produced by the rotation of the paddle-wheel (Fig. 3.16). From the results of a blank run with the paddle-wheel stationary, he corrected the measured temperature rise to what it would have been if the process had been adiabatic as well as isopiestic. From the known masses of the water and the metal parts of the apparatus and the mean specific heat of the latter, he computed the quantity of heat Q (in Btu) required to produce this corrected increase in the temperature of the system in a constant-pressure process with the paddle-wheel stationary. He then calculated the mechanical equivalent of heat $\mathscr{J}$ from the relation

$$W = \mathscr{J}Q \tag{4.1}$$

treating W and Q as arithmetic quantities. We would prefer to suppose that the second step of the process was reversed, so that the system was restored to its initial state, and then treat W and Q as algebraic quantities, both being negative for the complete cycle.

In spite of the clear statements of Carnot and Thomson regarding the conversion of work into heat, it has commonly been believed that the work withdrawn from the surroundings in the paddle-wheel experiment was converted into heat inside the system, and that this heat produced the observed temperature rise. This confusion of heat and temperature, which was resolved by Black [1] in the eighteenth century, is still common today. Heat cannot be converted into work and work cannot be converted into heat *by any process occurring inside a system*, since neither heat nor work exists as such in a system (Section 3.26).

We can, however, consider that heat has been converted into work, or work converted into heat, *by a system which has undergone a cyclical process* at the end of which the system has returned to its initial state, and the work W has been produced in and the heat Q withdrawn from the surroundings where both are measured. Both W and Q may be positive, or both may be negative.

Let us suppose that the constant-pressure process studied by Joule [see eqn. (3.74)] was followed by a second constant-pressure process in which the paddle-wheel was stationary and heat flowed to the surroundings until the system A (water, stirring mechanism S, and copper container D) had regained its initial temperature:

$$A(T_1) = A(T_2) \quad W_1 = -\Sigma_i w_i d_i + p\Delta V \quad Q_1 = -a \quad (p = 1 \text{ atm}) \tag{4.2a}$$

$$A(T_2) = A(T_1) \quad W_2 = -p\Delta V \quad Q_2 = -\bar{C}_p \Delta T \quad (p = 1 \text{ atm}) \tag{4.2b}$$

$$\Sigma W = -\Sigma_i w_i d_i \quad \Sigma Q = -a - \bar{C}_p \Delta T \tag{4.2c}$$

Here $-\Sigma_i w_i d_i$ is the shaft work and $-a$ is the heat effect (calculated from the results of the blank run) in the first step of the cycle, ΔT and ΔV are the corresponding increases in the temperature and volume of the system A, and $\bar{C}_p$ is the mean constant-pressure heat capacity of A in the range T_1 to T_2. The quantities ΣW and ΣQ are the work and heat effects attending the cycle. Substitution of these values into equation (4.1) gives $\mathscr{J}$.

Many careful measurements have shown that $\mathscr{J}$ has the same value for all the cyclical processes studied, no matter what kinds of work are involved.

The mechanical equivalent of heat allows us to express work and heat in the same units. The unit of work in the MKS system is the joule (Section 3.2); and the recommended unit of heat is also the joule, although thermochemical and IT calories (Section 3.13) are in common use.

4.3. First law of thermodynamics. We shall state the first law in the following form: Whenever a system undergoes a cyclical process, the net quantity of work produced in the surroundings is proportional to the net quantity of heat withdrawn from the surroundings, and the constant of proportionality depends only on the units in which work and heat are expressed.

This law is the expression of the results of many careful investigations of the kind discussed in the last section. Its correctness is confirmed by the agreement of the deductions made from it with well-established principles and facts. Among these is the principle of the impossibility of perpetual motion of the first kind, which may be stated as follows:

It is impossible for a system operating in a cycle and withdrawing no net quantity of heat from the surroundings to produce net work in the surroundings perpetually.

Let $đW$ be the work produced in the surroundings and $đQ$ be the heat withdrawn therefrom in an infinitesimal part of any cyclical process undergone by any system. Then, according to the first law

$$\oint đW = \mathscr{J} \oint đQ \text{ (all cycles)} \tag{4.3}$$

where $\mathscr{J}$ depends only on the units in which W and Q are expressed. If we agree to measure W and Q in the same units, $\mathscr{J}$ is unity. Equation (4.3) may then be written in the following forms:

$$\oint đW = \oint đQ \text{ or } \oint (đQ - đW) = 0 \text{ (all cycles)} \tag{4.4}$$

Equations (4.4) say in mathematical terms that the difference $(đQ - đW)$ is an exact differential; that is, this expression is the total differential of a function which, in the present instance, is a function of a set of independent state variables of the system under consideration. We shall discuss this function, which we call the energy of the system, in the following section.

4.4. Definition of the energy of a closed system. Let any closed system A undergo the change in state

$$A(\text{state 1}) = A(\text{state 2}) \tag{4.5}$$

by any process whatever. The work and heat effects attending an infinitesimal part of this process are $đW$ and $đQ$. We define the corresponding change dE of the energy E of the system by the relation

$$dE \equiv đQ - đW \text{ (definition of } E\text{)} \tag{4.6}$$

Then, by eqns. (4.4) and (4.6), the following relation must hold for every cyclical process occurring in any closed system A:

$$\oint dE = 0 \text{ (property of } E\text{)} \tag{4.7}$$

Application of eqns. (4.6) and (4.7) to the finite change in state (4.5) gives

$$\int_1^2 dE = \int_1^2 (đQ - đW) = Q - W \qquad \text{(all processes)} \qquad (4.8)$$

$$\int_1^2 dE = E_2 - E_1 = \Delta E \qquad \text{(all processes)} \qquad (4.9)$$

Therefore

$$\Delta E = Q - W \qquad \text{(all processes)} \qquad (4.10)$$

Equations (4.6) and (4.7) are the complete formulation of the first law for a closed system in differential form, and eqns. (4.8) and (4.9) are the corresponding formulation of the law in integral form. The relations (4.6) and (4.8) simply define the differential dE and its line integral. The relations (4.7) and (4.9) assert in mathematical terms the existence of an energy function. The energy E of a closed system is a single-valued function of a set of independent state variables of the system and dE is the total differential of this function.

The notation used in eqns. (4.6)–(4.10) is designed to indicate the distinction between energy, which is a state function, and work and heat, which are not. Thus dE is an exact differential, and $đW$ and $đQ$ are not. The definite integral of dE is written ΔE, the increment of a function; and the definite integrals of $đW$ and $đQ$ are represented conventionally by the symbols W and Q.

The first law defines only the difference between the energies of two states of a system, not the energy content of each state. In order to complete the definition of energy content, we assign an arbitrary value to the energy of the system when in some selected state, called the standard state (see Section 6.2). In the change in state (4.5) let State (1) be selected as the standard state of the system and let State (2) be a variable state. Then, eqns. (4.8) and (4.9) express the energy of the system as a function of a set of independent state variables, which contains the assigned value of the energy in the standard state as an integration constant.

Fortunately the assignable integration constant in the energy function does not present any real difficulty in thermodynamic computations, because we are interested only in computing the *difference* between the energies of two states of a system, as given by eqn. (4.6) or (4.10).

From the foregoing considerations and the discussion of the measure of work and heat in Sections 3.2 and 3.13, we find that energy is an extensive state property of a system. When the effects of long-range forces and shape (capillarity) may be neglected, we may consider that a heterogeneous system is separated by mathematical surfaces into phases, each of which is unaffected by the presence of the others and hence is homogeneous right up to the dividing surfaces. Under these conditions the energy of the system is equal to the sum of the energies of its individual phases.

It may happen that we cannot calculate the increment of energy ΔE produced in a given system by a particular process because we cannot measure or compute separately the work and heat effects attending the process. Since, according to the first law, ΔE depends only on the change in state, we can replace the given process by any other producing the same change for which W and Q can be evaluated separately. Substitution of these values into eqn. (4.10) gives the required change in energy.

If there should be a state of a system which cannot be reached from a standard state by at least one process for which W and Q can be determined separately, we could not determine its energy.

4.5. *Law of the conservation of energy.* One of the consequences of the first law of thermodynamics is the law of the conservation of energy: Energy cannot be created or destroyed. An equivalent form of the law is: The energy of an isolated system is constant. Thus the increase or decrease in the energy of a system in a process is exactly equal to the corresponding decrease or increase in the energy of its environment. Furthermore, when one kind of energy is withdrawn from the surroundings by a system in a cyclical process, an exactly equivalent amount of another kind (or kinds) of energy is produced in the surroundings. This is called the conversion of energy from one form into another form, and it always refers to effects produced in the surroundings of a system which has undergone a cyclical process.

At the end of a paper on thermodynamics Clausius [12] wrote "Die Energie der Welt ist konstant".

Let a system A and that portion of its surroundings M (called the medium) which is appreciably disturbed by changes occurring in A be enclosed in a boundary which isolates the composite system A + M from the rest of the universe. When A undergoes the change in state (4.5) by any given process, the work W is produced in M and the quantity of heat Q is withdrawn from M. Application of the first-law equation (4.10) first to the system A and then to the isolated composite system A + M gives

$$\Delta E_A = Q - W, \Delta E_A + \Delta E_M = 0 \therefore \Delta E_M = -(Q - W) \tag{4.11}$$

These equations which are based on the first law are evidently consistent with the two versions of the law of the conservation of energy given above.

We cannot properly speak of the conversion of one kind of energy into another kind in the above experiment unless the process is cyclical, so that the system A has returned to its initial state. Let us suppose that this is the case. Then the first law requires that $Q = W$ in each of the three equations (4.11), and hence ΔE_A and ΔE_M are each zero. From the standpoint of M (the surroundings), which has not returned to its initial state, the system A has produced the conversion of a quantity of heat into an equivalent quantity of work if Q and W are both positive, or work into an equivalent quantity of heat if both are negative.

Among the many conservation principles which were announced in the seventeenth, eighteenth, and nineteenth centuries, some, such as the law of the conservation of heat, were found to be subject to such broad restrictions that we no longer consider them to be laws. Others have been found useful in spite of the limitations on their use. Among these are the laws of the conservation of *vis viva* (limited to conservative systems), of entropy (limited to systems undergoing reversible processes), of the chemical elements (limited to systems with negligible nuclear reactions), and of mass (limited to systems with negligible relativity effects).

Finally, a small number of these conservation principles, such as the laws of the conservation of momentum, angular momentum, and energy, have such widespread application and freedom from restrictive conditions that an apparent exception to any one of them would cause a thorough investigation designed to uncover a heretofore unknown phenomenon or to lay a basis for a reformulation of the pertinent laws of physics. For example, the existence of the neutrino was postulated because of the apparent failure of the laws of the conservation of momentum, angular momentum, and energy.

According to the special theory of relativity, the increment of the energy ΔE of a system at rest attending an increment of its rest mass ΔM_0 is given by the expression

$$\Delta E = c^2 \Delta M_0 \tag{4.12}$$

where c is the velocity of light. Hence, mass is one form of energy which is interconvertable with other forms only under special conditions, the "mechanical equivalent of mass" being approximately 9×10^{16} J kg^{-1} in the MKS system.

4.6. Measure and units of energy. Energy may be expressed in either work or heat units. Conversion factors from one energy unit to another may be readily calculated from the values listed in Table 4.1.

4.7. Review of the properties of energy. The following properties of energy are consequences of eqns. 4.6 and 4.7 and the definitions of work and heat we discussed in Chapter 3 (see particularly Section 3.26).

TABLE 4.1

CONVERSION FACTORS FOR ENERGY UNITS

1 cal$_{15}$	= 4.1855 J	1 kwh	= 3.6 × 10^6 J[a]
1 cal$_{20}$	= 4.1816 J	1 gf cm	= 9.806 65 × 10^{-5} J[a]
1 cal$_{th}$	= 4.1840 J[a]	1 kgf m	= 9.806 65 J[a]
1 cal$_{IT}$	= 4.1868 J[a]	1 cm^3 atm	= 1.013 250 × 10^{-1} J[a]
1 Btu	= 1.055 0559 × 10^3 J[b]	1 liter atm	= 1.013 2784 × 10^2 J
1 Btu	= 2.519 9576 × 10^2 cal$_{IT}$[b]	1 ft pdl	= 4.214 0110 × 10^{-2} J
1 Btu	= 7.781 6926 × 10^2 ft lbf	1 ft lbf	= 1.355 8179 J
1 erg	= 1 × 10^{-7} J[a]	1 ft^3 psi	= 1.952 3778 × 10^2 J
1 W sec	= 1 J[a]	1 ft^3 atm	= 2.869 2045 × 10^3 J

[a] Exact by definition.
[b] 1 Btu lb^{-1} = $\frac{5}{9}$ cal$_{IT}$ g^{-1} by definition.

1. The quantity dE is an exact differential.

2. The increment ΔE of the energy attending any finite change in the state of a system is independent of the process producing this change. It depends only on the end states of the process. Thus we may properly call ΔE the increment of energy attending a change in state.

When the work and heat effects attending a given process for producing a particular change in state cannot be evaluated separately, we can replace the given process by one for which Q and W can be determined separately. Substitution of these values into eqn. (4.10) gives ΔE for the change in state in question.

3. The energy E of a closed system is an extensive state property, determined except for an additive constant of integration whose value is fixed by selecting some one state of the system (called the standard state) and specifying its energy. The energies of all other states are then completely determined. Hence we can defend the statement that a system in a given state has a definite energy (or energy content). Thus E is a single-valued function of a set of independent thermodynamic state variables, dE is the total differential of this function, and ΔE is the difference $E_1 - E_2$ between the values of this function for two finitely different states of a system.

4. When the effects of long-range forces and shape may be neglected, the energy of a heterogeneous system is the sum of the energies of its individual phases.

5. Energy cannot be created or destroyed. The increase or decrease in the energy of a system attending any process is equal to the corresponding decrease or increase of the energy of its environment produced in the same process. Also, whenever one kind of energy is produced in the surroundings by a system in a cyclical process, an equal quantity of another kind (or kinds) of energy is withdrawn from the surroundings.

4.8. Definition of the enthalpy of a closed system. The enthalpy, also called the heat content, of a closed system having a volume V and a uniform pressure p is defined by the equation

$$H = E + pV \qquad (4.13)$$

It bears the same relation to the non-expansion work and heat of a reversible, constant-pressure process* that the energy bears to the corresponding quantities of a constant-volume process. Evidently the enthalpy, like the energy and volume, is an extensive state property of a system. When the effects of long-range forces and shape may be neglected, the enthalpy of a

---

*All aspects of the process need not be reversible or even quasistatic, provided the expansion work is essentially reversible. This provision is met if throughout the process the system has an identifiable pressure p and if the expansion work is represented satisfactorily by the integral of pdV along the path of the process. For example, in Joule's paddle-wheel experiment (Fig. 3.16) the shaft work is not reversible, but the expansion work is essentially reversible.

heterogeneous system is the sum of the enthalpies of the individual phases.

For an infinitesimal change in state we find

$$dH = dE + d(pV) \tag{4.14}$$

and for a finite change we have

$$\Delta H = \Delta E + \Delta(pV) \tag{4.15}$$

The enthalpy of a system in a given state contains an additive constant of integration. Strict application of eqn. (4.13) to the standard state would require this constant to be the sum of the energy and the pressure—volume product of the system in its standard state. Since the constant always disappears in thermodynamic calculations, see eqns. (4.14) and (4.15), we do not follow this procedure but use instead the methods outlined in Section 6.2.

Substitution of the values of dE and ΔE from eqns. (4.6) and (4.10) into eqns. (4.14) and (4.15), respectively, gives

$$dH = đQ - đW + d(pV) \tag{4.16}$$

$$\Delta H = Q - W + \Delta(pV) \tag{4.17}$$

If a system is divided by rigid walls into two or more parts, each under a uniform pressure, eqns. (4.13)—(4.15) become

$$H \equiv E + \Sigma_i(p_iV_i) \tag{4.18a}$$

$$dH = dE + \Sigma_i d(p_iV_i) \tag{4.18b}$$

$$\Delta H = \Delta E + \Sigma_i \Delta(p_iV_i) \tag{4.18c}$$

Here p_i is the uniform pressure and V_i the volume of the i-th part of the system, and the summation extends over all parts of the system.

4.9. Some first-law equations for closed systems. Equations (4.6) and (4.10) take the following forms for a quasistatic process* producing the change in state (4.5) in a closed system A:

$$dE = đQ - p_{AP}dV - đW_X \tag{4.19}$$

$$\Delta E = Q - \int_{C\,V_1}^{V_2} p_{AP}\,dV - W_X \tag{4.20}$$

Here p_{AP} is the applied pressure acting on the system, the integral is evaluated along the curve C given as $p_{AP} = p_{AP}(V)$ (Section 3.4), and W_X represents all of the non-expansion work produced in the surroundings. The corresponding equations for the enthalpy are:

---

*All aspects of the process need not be quasistatic, provided the expansion work is essentially quasistatic.

$$dH = đQ - p_{AP}dV + d(pV) - đW_X \tag{4.21}$$

$$\Delta H = Q - \int_{C\,V_1}^{V_2} p_{AP}\,dV + (p_2V_2 - p_1V_1) - W_X \tag{4.22}$$

In these equations, p is the pressure of the system.

If the process is reversible*, eqns. (4.19) and (4.20) become:

$$dE = đQ - pdV - đW_X \tag{4.23}$$

$$\Delta E = Q - \int_{C\,V_1}^{V_2} pdV - W_X \tag{4.24}$$

Here p is the pressure of the system, and C is the curve of the path of the process given by the relation $p = p(V)$. The corresponding equations, (4.21) and (4.22), for the enthalpy are:

$$dH = đQ + Vdp - đW_X \tag{4.25}$$

$$\Delta H = Q + \int_{C\,p_1}^{p_2} Vdp - W_X \tag{4.26}$$

The curve C is now given by the equation $V = V(p)$.

The following two sets of experiments illustrate the use of the energy and enthalpy functions for some constant volume and some constant pressure processes.

Experiment 1. Let a system composed of 1 mole of gaseous nitrogen undergo the change in state

$$N_2\,(g, T_1, V_1) = N_2\,(g, T_2, V_1) \qquad (T_2 > T_1) \tag{4.27}$$

which is brought about by each of the following two processes: (a) an adiabatic, constant-volume process in which the shaft work W'_X is produced in the surroundings by the rotation in the system of a shaft carrying several sets of stationary blades, (such as in Fig. 3.16); and (b) a constant-volume process in which the quantity of heat Q_V is withdrawn from the surroundings with the fan-blades stationary. Application of eqns. (4.20) and (4.22) gives:

$$Q = 0 \quad W = W'_X \quad \Delta E = -W'_X \quad \Delta H = -W'_X + V_1(p_2 - p_1) \tag{4.28a}$$

$$Q = Q_V \quad W = 0 \quad \Delta E = Q_V \quad \Delta H = Q_V + V_1(p_2 - p_1) \tag{4.28b}$$

Here p_1 and p_2 are the initial and final pressures of the system. From eqns. (4.28) we find

$$-W'_X\,(\text{process a}) = Q_V\,(\text{process b}). \tag{4.29}$$

---

*See footnote, p. 109.

In the present case W'_X is negative and Q_V is positive.

Experiment 2. In the same apparatus let the change in state

$$N_2(g, T_1, p_1) = N_2(g, T_2, p_1) \qquad (T_2 > T_1) \qquad (4.30)$$

be brought about by each of the following two processes: (a) an adiabatic, constant-pressure process in which the expansion work $p_1(V_2 - V_1)$ and the shaft work W''_X are produced in the surroundings; and (b) a constant-pressure process in which the expansion work $p_1(V_2 - V_1)$ is produced in the surroundings, the heat Q_p withdrawn therefrom, and the fan-blades are stationary. We have supposed that the expansion work is essentially reversible*.

Application of eqns (4.24) and (4.26) gives:

$$Q = 0, \quad W = p_1(V_2 - V_1) + W''_X, \quad \Delta E = -p_1(V_2 - V_1) - W''_X, \quad \Delta H = -W''_X \qquad (4.31a)$$

$$Q = Q_p, \quad W = p_1(V_2 - V_1) \qquad \Delta E = Q_p - p_1(V_2 - V_1) \qquad \Delta H = Q_p \qquad (4.31b)$$

Here V_1 and V_2 are the initial and final volumes of the system in eqn. (4.30). From eqns. (4.31) we find

$$-W''_X \text{ (process a)} = Q_p \text{ (process b)} \qquad (4.32)$$

where W''_X is negative and Q_p is positive.

We note from eqns. (4.28) and (4.31) that $\Delta E = -W_X$ for an adiabatic, constant-volume process and $\Delta H = -W_X$ for an adiabatic, constant-pressure process, in which the expansion work is essentially reversible. Also, when no non-expansion work is produced in a process, $\Delta E = Q_V$ for a constant-volume process, and $\Delta H = Q_p$ for a constant-pressure process, in which the expansion work is essentially reversible.

4.10. Tabulation of some first-law equations for closed systems. In Table 4.2 are presented some first-law equations of general validity for closed systems.

The equations of Table 4.3 apply to quasistatic processes in closed systems under an applied pressure p_{AP}. They also hold for the larger class of processes mentioned in the footnote on p. 110.

TABLE 4.2

SOME GENERAL FIRST-LAW EQUATIONS FOR CLOSED SYSTEMS

E	$H = E + pV$	(1)
dE	$dH = dE + d(pV)$	(2)
$dE = đQ - đW$	$dH = đQ - đW + d(pV)$	(3)
ΔE	$\Delta H = \Delta E + \Delta(pV)$	(4)
$\Delta E = Q - W$	$\Delta H = Q - W + \Delta(pV)$	(5)

*See footnote, p. 109.

TABLE 4.3

SOME FIRST-LAW EQUATIONS FOR QUASISTATIC PROCESSES IN CLOSED SYSTEMS[a]

Process	E	H	
(a) $đW_{EX} = p_{AP}dV$			
	$dE = đQ - p_{AP}dV - đW_X$	$dH = đQ - p_{AP}dV + d(pV) - đW_X$	(1)
Constant V	$\Delta E = Q_V - W_X$	$\Delta H = Q_V + V\Delta p - W_X$	(2)
Constant p_{AP}	$\Delta E = Q_p - p_{AP}\Delta V - W_X$	$\Delta H = Q_p - p_{AP}\Delta V + \Delta(pV) - W_X$	(3)
(b) $đW_{EX} = p_{AP}dV$, $W_X = 0$			
	$dE = đQ - p_{AP}dV$	$dH = đQ - p_{AP}dV + d(pV)$	(4)
Constant V	$\Delta E = Q_V$	$\Delta H = Q_V + V\Delta p$	(5)
Constant p_{AP}	$\Delta E = Q_p - p_{AP}\Delta V$	$\Delta H = Q_p - p_{AP}\Delta V + \Delta(pV)$	(6)
(c) $đW_{EX} = p_{AP}dV$, $Q = 0$			
	$dE = -p_{AP}dV - đW_X$	$dH = -p_{AP}dV + d(pV) - đW_X$	(7)
Constant V	$\Delta E = -W_X$	$\Delta H = V\Delta p - W_X$	(8)
Constant p_{AP}	$\Delta E = -p_{AP}\Delta V - W_X$	$\Delta H = -p_{AP}\Delta V + \Delta(pV) - W_X$	(9)
(d) $đW_{EX} = p_{AP}dV$, $Q = 0$, $W_X = 0$			
	$dE = -p_{AP}dV$	$dH = -p_{AP}dV + d(pV)$	(10)
Constant V	$\Delta E = 0$	$\Delta H = V\Delta p$	(11)
Constant p_{AP}	$\Delta E = -p_{AP}\Delta V$	$\Delta H = -p_{AP}dV + \Delta(pV)$	(12)

[a] All aspects of the process need not be quasistatic provided the expansion work is essentially quasistatic, see footnote on p. 110.

TABLE 4.4

SOME FIRST-LAW EQUATIONS FOR REVERSIBLE PROCESSES IN CLOSED SYSTEMS[a]

Process	E	H	
(a) $đW_{EX} = pdV$			
	$dE = đQ - pdV - đW_X$	$dH = đQ + Vdp - đW_X$	(1)
Constant V	$\Delta E = Q_V - W_X$	$\Delta H = Q_V + V\Delta p - W_X$	(2)
Constant p	$\Delta E = Q_p - p\Delta V - W_X$	$\Delta H = Q_p - W_X$	(3)
(b) $đW_{EX} = pdV$, $W_X = 0$			
	$dE = đQ - pdV$	$dH = đQ + Vdp$	(4)
Constant V	$\Delta E = Q_V$	$\Delta H = Q_V + V\Delta p$	(5)
Constant p	$\Delta E = Q_p - p\Delta V$	$\Delta H = Q_p$	(6)
(c) $đW_{EX} = pdV$, $Q = 0$			
	$dE = -pdV - đW_X$	$dH = Vdp - đW_X$	(7)
Constant V	$\Delta E = -W_X$	$\Delta H = V\Delta p - W_X$	(8)
Constant p	$\Delta E = -p\Delta V - W_X$	$\Delta H = -W_X$	(9)
(d) $đW_{EX} = pdV$, $Q = 0$, $W_X = 0$			
	$dE = -pdV$	$dH = Vdp$	(10)
Constant V	$\Delta E = 0$	$\Delta H = V\Delta p$	(11)
Constant p	$\Delta E = -p\Delta V$	$\Delta H = 0$	(12)

[a] All aspects of the process need not be reversible provided the expansion work is essentially reversible, see footnote on p. 109.

The relations of Table 4.4 hold for reversible processes in closed systems having a uniform pressure p. They may also be applied to the larger class of processes mentioned in the footnote on p. 109.

When the different phases or parts of a system are under different pressures so that H is defined by eqn. (4.18), we must make the following changes in the relations given in Tables 4.2—4.4.

Replace $d(pV)$ by $\sum_i d(p_i V_i)$ Replace $\Delta(pV)$ by $\sum_i \Delta(p_i V_i)$ (4.33a)

Replace pdV by $\sum_i p_i dV_i$ Replace $p\Delta V$ by $\sum_i p_i \Delta V_i$ (4.33b)

Replace Vdp by $\sum_i V_i dp_i$ Replace $V\Delta p$ by $\sum_i V_i \Delta p_i$ (4.33c)

When the different parts of the system are under different applied pressures, we use eqns. (4.33b) with p_i replaced by p_{APi}.

REFERENCES

1 J. Black, Lectures on the Elements of Chemistry, edited by J. Robison, M. Carey, Philadelphia, 1807, Vol. 1.
2 Count Rumford, The Complete Works of Count Rumford, American Academy of Arts and Sciences, Boston, 1870, Vol. 1, pp. 469—493.
3 H. Davy, The Collected Works of Sir Humphry Davy, bart., edited by J. Davy, Smith, Elder and Co., London, 1839—1840, Vol. II, p. 11.
4 S. Carnot, Refléxions sur la Puissance Motrice du Feu et sur les Machines Propres à Développer cette Puissance, Bachelier, Paris, 1824; see The Second Law of Thermodynamics, translated and edited by W. F. Magie, Harper and Brothers, New York, 1899, pp. 3—61.
5 E. Clapeyron, Ann. Phys. Chem., 59 (1843) 446—451, 566—592, translated from J. Ecole Polytech., 14 (1834) 70.
6 J. R. Mayer, Ann. Chem. Pharm., 42 (1842) 233—240; see also H. S. Smith, A History of Science, Harper and Brothers, New York, 1904, pp. 259—267.
7 H. Helmholtz, Über die Erhaltung der Kraft, G. Reimer, Berlin, 1847; see also L. Koenigsberger, Hermann von Helmholtz, translated by F. A. Welby, Dover Publications, New York, 1965, pp. 38—51.
8 J. P. Joule, The Scientific Papers of James Prescott Joule, The Physical Society, London, 1884, (1843) pp. 123—157, (1843) 157—159; (1845) 172—189; (1847) 277—281, 283—286; (1850) 298—328; (1878) 632—657.
9 W. Thomson, Mathematical and Physical Papers, University Press, Cambridge, 1882, (1848) pp. 100—106; (1849) 113—155; (1851—1854, 1878) 174—332; (with J. P. Joule), (1852—1862) 333—455.
10 H. V. Regnault, Relations des Expériences, F. Didot Frères, Paris, 1847, Vol. 1, pp. 635—748; Ann. Phys. Chem., 78 (1849) 196—216, 523—563.
11 R. Clausius, Ann. Phys. Chem., 79 (1850) 368—397, 500—524; ibid., 93 (1854) 481—506; ibid., 125 (1865) 353—400; see The Mechanical Theory of Heat, translated by W. R. Browne, Macmillan and Co., London, 1879; see also W. F. Magie, ref. 4, pp. 65—108.
12 R. Clausius, Ann. Phys. Chem., 125 (1865) 353—400.

Chapter 5

APPLICATIONS OF THE FIRST LAW TO PHYSICAL CHANGES

The term temperature does not occur in the formal statement of the first law of thermodynamics; and temperature itself does not occur explicitly in a theoretical first-law equation, although its differential does [see for example eqns. (5.7) below]. For the integration of first-law relations we represent the properties of substances by functions of a set of independent state variables, one of which is usually temperature. These relations are either purely empirical, or they are based on principles other than the first law, for example $p = f(V, T)$ or $C_p = F(p, T)$. Evidently we may use any temperature scale we please in these relations. We choose to use the perfect gas absolute temperature scale (Section 2.9), called the Kelvin scale and written T. The fact that this temperature scale is given a deeper significance by the second law of thermodynamics than we attribute to it here need not concern us at this point.

In the present chapter we shall apply the first-law equations to systems without chemical reaction; in Chapter 6 we shall apply them to systems where chemical reactions occur. Unless otherwise stated, we shall be interested here in closed, simple systems of constant composition and at rest with respect to their environments or moving so slowly that their properties are not appreciably affected. When such a system undergoes a reversible process without the production of non-expansion work W_X, the following first-law equations (Table 4.4) apply

$$dE = đQ - pdV \tag{5.1}$$

$$dH = đQ + Vdp \tag{5.2}$$

We also have the defining relations

$$dH = dE + d(pV) \qquad \Delta H = \Delta E + \Delta(pV) \tag{5.3}$$

connecting changes in enthalpy with changes in energy and the pV-product of the system.

5.1. Effect of volume or pressure and temperature on energy and enthalpy. The energy E and enthalpy H of a closed, simple, homogeneous system of constant composition are functions of any two of the three variables (p, V, T). In general, equations for E are simplest when expressed in terms of the independent variables (V, T); while the relations for H are simplest when expressed in terms of the set (p, T). We shall therefore at this point use only these sets of independent variables.

The effects of variations of V and T on E and of p and T on H are given by the following equations:

V, T-independent variables $\qquad\qquad p, T$-independent variables

$$dE = \left(\frac{\partial E}{\partial T}\right)_V dT + \left(\frac{\partial E}{\partial V}\right)_T dV \qquad dH = \left(\frac{\partial H}{\partial T}\right)_p dT + \left(\frac{\partial H}{\partial p}\right)_T dp \qquad (5.4)$$

$$\left(\frac{\partial E}{\partial T}\right)_V = \left(\frac{dQ}{dT}\right)_V = C_V \qquad \left(\frac{\partial H}{\partial T}\right)_p = \left(\frac{dQ}{dT}\right)_p = C_p \qquad (5.5)$$

$$\left(\frac{\partial E}{\partial V}\right)_T = \left(\frac{dQ}{dV}\right)_T - p = l_T - p \qquad \left(\frac{\partial H}{\partial p}\right)_T = \left(\frac{dQ}{dp}\right)_T + V = \Lambda_T + V \qquad (5.6)$$

$$dE = C_V dT + (l_T - p)dV \qquad\qquad dH = C_p dT + (\Lambda_T + V)dp \qquad (5.7)$$

The relations for E are derived from eqns. (5.1), (3.53), and (3.90); while those for H come from eqns. (5.2), (3.53), and (3.88).

Let us apply eqns. (5.7) to the determination of ΔE and ΔH for a system undergoing the following change in state:

$$NH_3(g, p_1, V_1, T_1) = NH_3(g, p_2, V_2, T_2) \qquad (5.8)$$

in which only two of the three variables are independent. Since ΔE and ΔH depend only on the change in state, we are free to employ the most convenient paths for the integration of eqns. (5.7). We use the following two-step paths:

1. Integrate the expression for dE from T_1 to T_2 along the isometric $V = V_1$, and then from V_1 to V_2 along the isothermal $T = T_2$.
2. Integrate the expression for dH from T_1 to T_2 along the isopiestic $p = p_1$, and then from p_1 to p_2 along the isothermal $T = T_2$.

Thus we find the following relations:

$$\Delta E = \int_{V_1, T_1}^{T_2} C_V dT + \int_{T_2, V_1}^{V_2} (l_T - p)dV \qquad (5.9)$$

$$\Delta H = \int_{p_1, T_1}^{T_2} C_p dT + \int_{T_2, p_1}^{p_2} (\Lambda_T + V)dp \qquad (5.10)$$

We can derive from the second law of thermodynamics (but not from the first law) relations expressing the following properties of the system in terms of quantities that can be evaluated from its equation of state: (1) l_T and Λ_T; (2) $(\partial C_V/\partial V)_T$ and $(\partial C_p/\partial p)_T$; and (3) the difference $C_p - C_V$. The integrals appearing in eqns. (5.9) and (5.10) can therefore be evaluated for the system if we are given: (1) its equation of state $F(p, V, T) = 0$; and (2) the temperature variation of its constant-volume heat capacity $C_V = f_1(T)$ along some one isometric, or of its constant-pressure heat capacity $C_p = f_2(T)$ along some one isopiestic. The values of the constants appearing in these functions must be determined from experimental measurements.

When V is constant in eqns. (5.8) and (5.9), and when p is constant in eqns. (5.8) and (5.10), we find:

$$\Delta E = \int_{T_1}^{T_2} C_V dT \qquad (V = \text{constant}) \qquad (5.11)$$

$$\Delta H = \int_{T_1}^{T_2} C_p dT \qquad (p = \text{constant}) \qquad (5.12)$$

When T is constant in eqns. (5.8)–(5.10), we have:

$$\Delta E = \int_{V_1}^{V_2} (l_T - p) dV \qquad (T = \text{constant}) \qquad (5.13)$$

$$\Delta H = \int_{p_1}^{p_2} (\Lambda_T + V) dp \qquad (T = \text{constant}) \qquad (5.14)$$

5.2. Heat capacities. If we solve each of the eqns. (5.1) and (5.2) for dQ and substitute from eqns. (5.4) and (5.5), we find the following relations for a closed, simple, homogeneous system of constant composition, undergoing a reversible process:

$$dQ = C_V dT + \left[\left(\frac{\partial E}{\partial V}\right)_T + p\right] dV \qquad (5.15)$$

$$dQ = C_p dT + \left[\left(\frac{\partial H}{\partial p}\right)_T - V\right] dp \qquad (5.16)$$

Let the change in state (5.8) be carried out by a reversible process in which the state of the system moves along the curve C. From the standpoint of the variables (V, T) we say that the state of the system moves from (V_1, T_1) to (V_2, T_2) along the curve C given by $V = V(T)$. From the standpoint of the set (p, T) we consider that the state of the system moves from (p_1, T_1) to (p_2, T_2) along the same path C now given by $p = p(T)$. If we divide eqns. (5.15) and (5.16) by dT subject to these conditions, we find the following relations for the heat capacity C_C of the system along the curve C:

$$\left(\frac{dQ}{dT}\right)_C = C_C = C_V + \left[\left(\frac{\partial E}{\partial V}\right)_T + p\right] \left(\frac{dV}{dT}\right)_C \qquad (5.17)$$

$$\left(\frac{dQ}{dT}\right)_C = C_C = C_p + \left[\left(\frac{\partial H}{\partial p}\right)_T - V\right] \left(\frac{dp}{dT}\right)_C \qquad (5.18)$$

If the curve C is given by $p = $ constant in eqn. (5.17) and by $V = $ constant in eqn. (5.18), we obtain:

$$C_p - C_V = \left[\left(\frac{\partial E}{\partial V}\right)_T + p\right] \left(\frac{\partial V}{\partial T}\right)_p \qquad (5.19)$$

$$C_p - C_V = \left[V - \left(\frac{\partial H}{\partial p}\right)_T\right]\left(\frac{\partial p}{\partial T}\right)_V \qquad (5.20)$$

Here $(\partial V/\partial T)_p$ and $(\partial p/\partial T)_V$ are to be evaluated from the equation of state $F(p, V, T) = 0$ of the system.

Now let the curve C in eqns. (5.17) and (5.18) be the saturation curve $p = p(T)$ along which two different aggregation states of a pure substance are in equilibrium. Then the saturation heat capacity C_{SAT} of each phase is related to the constant-volume and constant-pressure heat capacities of that phase by the equations:

$$C_{SAT} = C_V + \left[\left(\frac{\partial E}{\partial V}\right)_T + p\right]\left(\frac{dV}{dT}\right)_{SAT} \qquad (5.21)$$

$$C_{SAT} = C_p + \left[\left(\frac{\partial H}{\partial p}\right)_T - V\right]\left(\frac{dp}{dT}\right)_{SAT} \qquad (5.22)$$

The relations of $(\partial E/\partial V)_T$ to l_T and of $(\partial H/\partial p)_T$ to Λ_T are given in eqns. (5.6). Hence, like l_T and Λ_T, they can be related to quantities that can be evaluated from an equation of state by use of a second-law equation.

5.3. Latent heats. The latent heat of expansion l_T and the latent heat of pressure increase Λ_T of a closed, simple, homogeneous system of constant composition have been defined by eqns. (3.90) and (3.88), respectively. Their relations to $(\partial E/\partial V)_T$ and $(\partial H/\partial p)_T$ are given by eqns. (5.6) which may be written in the forms:

$$l_T = \left(\frac{dQ}{dV}\right)_T = \left(\frac{\partial E}{\partial V}\right)_T + p \qquad \Lambda_T = \left(\frac{dQ}{dp}\right)_T = \left(\frac{\partial H}{\partial p}\right)_T - V \qquad (5.23)$$

We now turn to the consideration of latent heat effects in simple, univariant, heterogeneous systems. In Section 3.16 we noted that in such systems all of the intensive properties of each phase are functions of one intensive property of one of the phases, for example temperature or pressure. A simple system composed of two aggregation states of a pure substance in equilibrium is one example of a univariant system. Let the change in state

$$n\ H_2O(l) = n\ H_2O(g) \qquad (p, T = \text{constant}) \qquad (5.24)$$

be brought about in a univariant system consisting of liquid water and water vapor in equilibrium by a reversible, isothermal, isopiestic process at the fixed temperature T and the corresponding saturation pressure p. In general, such a phase transformation is accompanied by both a heat effect and a change in volume.

The latent heat of expansion l_{pT} and the latent heat of phase transformation λ_{pT} (heat of vaporization in the present case) of the system of eqn. (5.24) are defined in the following manner:

$$l_{pT} = \left(\frac{dQ}{dV}\right)_{pT} = \left(\frac{\partial E}{\partial V}\right)_{pT} + p \qquad l_{pT} = \left(\frac{\partial H}{\partial V}\right)_{pT} \qquad (5.25)$$

$$l_{pT} = \frac{\Delta E + p\Delta V}{\Delta V} \qquad\qquad l_{pT} = \frac{\Delta H}{\Delta V} \qquad (5.26)$$

$$\lambda_{pT} = \left(\frac{\mathrm{d}Q}{\mathrm{d}n}\right)_{pT} = \left(\frac{\partial E}{\partial n}\right)_{pT} + p\left(\frac{\partial V}{\partial n}\right)_{pT} \qquad \lambda_{pT} = \left(\frac{\partial H}{\partial n}\right)_{pT} \qquad (5.27)$$

$$\lambda_{pT} = \frac{\Delta E + p\Delta V}{n} \qquad\qquad \lambda_{pT} = \frac{\Delta H}{n} \qquad (5.28)$$

$$\lambda_{pT} = \left(\frac{\Delta V}{n}\right) l_{pT} \qquad (5.29)$$

In the above equations $\mathrm{d}E$, $\mathrm{d}H$, and $\mathrm{d}V$ are the changes in the energy, enthalpy, and volume of the system attending the transfer of $\mathrm{d}n$ units of mass from one phase to the other in the change in state (5.24); while ΔE, ΔH, and ΔV are the increments of the corresponding quantities accompanying the transfer of n units. Both l_{pT} and λ_{pT} are independent of the amount of phase transformation occurring. For a particular phase transformation, each depends only on the temperature or on the pressure.

5.4. Free expansion experiment, Joule effect. During his work on the mechanical equivalent of heat, Joule investigated the changes in temperature produced by the expansion and compression of gases [1]. As a part of this investigation he made a few measurements on the temperature changes produced in air by free expansion, that is, by expansion into a vacuum. His apparatus (Fig. 5.1) consisted of two copper vessels A and B, each having a

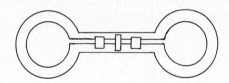

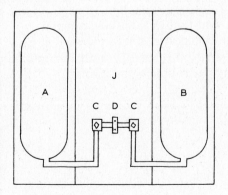

Fig. 5.1. Joule's free expansion apparatus. A, bulb initially filled with gas; B, bulb initially evacuated; C, C, stopcocks; D, coupling; J, jacket.

volume of approximately 135 in³ and each fitted with a tube leading to a stopcock C and thence to a coupling D. After A was filled with dry air to 22 atm (2.0 moles of air) and B was evacuated, they were joined and placed in a bath containing 16.5 pounds of water. The temperature of the water was read, the stopcock was opened allowing the gas to flow into B, and the water temperature again noted. The average increase in temperature of the system in 6 runs was 0.0062 ± 0.0066°F, and the average increase in 5 blank runs (without gas expansion) was 0.0068 ± 0.0056°F. Joule concluded that no change in temperature occurs when air undergoes an adiabatic expansion without doing work.

The free expansion experiment is not sufficiently sensitive to yield a quantitative result. In Joule's apparatus the ratio of the constant-pressure heat capacity of the water and metal container to the constant-volume heat capacity of the air was of the order of 1,000. We can estimate from the data given in Table 5.1 below that the actual decrease in temperature produced in the water jacket J by the gas expansion in this experiment would amount only to several thousandths of a degree Celsius, provided the heat effect attending the change in the state of stress of the more than 30 pounds of metal forming the walls of the apparatus could be neglected. Later investigators [2] have encountered similar difficulties.

Let us consider the following somewhat idealized version of the Joule experiment. One mole of a gas, for example N_2, undergoes the change in state

$$N_2(g, V_1, T_1) = N_2(g, V_2, T_2) \tag{5.30}$$

by an adiabatic expansion from V_1, the volume of bulb A (Fig. 5.1), to V_2, the total volume of bulbs A and B. Throughout the process the boundary of the system is placed parallel to and displaced an infinitesimal distance into the walls of the two bulbs and the connecting tubing. On the assumption that the boundary of the system is rigid as well as adiabatic, we find:

$$Q = 0 \qquad W = 0 \qquad \therefore E = 0 \tag{5.31}$$

Thus the energy of a fluid remains strictly constant in a Joule expansion.

In general, the temperature of a fluid changes in a free expansion. The Joule coefficient μ_J and the isothermal variation of energy with volume λ_J of a fluid in a specified state are defined by the following relations

$$\mu_J \equiv \lim_{\Delta V \to 0} \left(\frac{\Delta T}{\Delta V}\right)_E = \left(\frac{\partial T}{\partial V}\right)_E = -\frac{(\partial E/\partial V)_T}{(\partial E/\partial T)_V} = -\frac{(\partial E/\partial V)_T}{C_V} \tag{5.32}$$

$$\lambda_J \equiv \left(\frac{\partial E}{\partial V}\right)_T = -\mu_J C_V \tag{5.33}$$

Although μ_J has not been successfully measured, it has been calculated from other thermodynamic data (Section 5.6). It usually enters thermodynamic calculations as the product $\mu_J C_V$ which we have denoted by λ_J.

The latter is an intensive state property of a system with the units of pressure, for example atm; but in some tabulations it is expressed in the units of energy volume^{-1}, for example cal liter^{-1}. It is of great importance in the thermodynamic theory of real and perfect gases.

Some values* of λ_J/p^2 for air are listed in Table 5.1. As evidenced from these values, the ratio λ_J/p^2 varies only slowly with pressure at constant temperature up to moderately high pressures. Hence μ_J/p^2 for a gas changes slowly in this pressure region at constant temperature, since C_V for a gas shows a similar behavior.

Therefore along an isotherm λ_J and μ_J approach zero, as the pressure decreases, in the same manner as a constant times the square of the pressure.

Along an isopiestic p, λ_J and μ_J pass through zero at the same temperature T_J, called the Joule inversion temperature of the given gas for the pressure p.

When a gas is at a temperature below T_J on the isopiestic p, λ_J is positive (the energy of the gas increases in an isothermal expansion) and μ_J is negative (the temperature of the gas decreases in a free expansion).

When a gas is at a temperature above T_J on the isopiestic p, λ_J is negative (the energy of the gas decreases in an isothermal expansion) and μ_J is positive (the temperature of the gas increases in a free expansion).

Values** of the Joule inversion temperature T_J at 1 atm for several gases are as follows:

Gas	He	H_2	N_2
T_J at 1 atm (K)	195	715	> 2,000

As shown above, we know from experiment that λ_J, or $(\partial E/\partial V)_T$, for a real gas approaches zero at each temperature as the pressure approaches zero. Hence we might expect that λ_J for a perfect gas would be zero at all pressures and temperatures (see Section 2.9). But we cannot prove this result by thermodynamic reasoning for a gas whose thermodynamic behavior is defined solely by the laws of Boyle and Gay-Lussac. Hence we complete the definition of a perfect gas by stating that it obeys Joule's law (for gases): The energy of a perfect gas is independent of its density at constant temperature. Thus we find:

TABLE 5.1

SOME VALUES OF λ_J/p^2 ($\times 10^4$)[a] FOR AIR

Pressure (atm)	Temperature (°C)			
	−100	0	100	200
1	61.82	27.79	10.78	7.95
20	76.80	28.00	13.62	7.70
60		27.74	13.55	7.36
100		26.05	13.42	7.18

[a] Units, atm^{-1}.

*Computed from the values of λ_J given by Roebuck [3].
**Calculated by Keyes [4].

$$\left.\begin{array}{l}\left(\dfrac{\partial E}{\partial V}\right)_T = 0, \lambda_J = 0, \mu_J = 0 \\[1em] E = E(T), C_V = \dfrac{dE}{dT} \quad \text{(perfect gas)}\end{array}\right\} \tag{5.34}$$

We now have three laws for perfect gases:

$$pV = f(T) \quad \text{(Boyle's law)} \tag{5.35a}$$

$$\frac{(pV)_1}{(pV)_2} = \frac{T_1}{T_2} \quad \text{(Gay-Lussac's law)} \tag{5.35b}$$

$$\left(\frac{\partial E}{\partial V}\right)_T = 0 \quad \text{(Joule's law)} \tag{5.35c}$$

Here V is the volume of a fixed mass of gas and T is the perfect gas absolute temperature. No one of these laws can be derived from the other two by use of any first-law equation. However, if we define a perfect gas as one that obeys the laws of Boyle and Joule for all values of p and T, we can derive Gay-Lussac's law and hence the perfect gas law, eqn. (2.41), by use of an equation resulting from the second law of thermodynamics; and we can attribute special significance to the perfect gas absolute temperature scale.

5.5. Porous-plug experiment, Joule–Thomson effect. W. Thomson became interested in Joule's expansion experiment because it was a direct test of Mayer's hypothesis (Section 4.1). This proposition states that the energy of a gas is independent of pressure at constant temperature; and, if true, it would simplify the reduction of the temperature scale of a constant-volume gas thermometer to the scale known as the Kelvin (thermodynamic) scale. Thomson recognized that the free expansion experiment is not, on account of inherent experimental difficulties, sufficiently sensitive to verify or refute Mayer's hypothesis. He proposed the porous-plug experiment as a more likely method of producing results capable of settling this matter. Joule implemented Thomson's suggestion. The result was a series of joint papers [5] on the flow of fluids through a porous plug.

In a properly executed Joule–Thomson experiment, a fluid (gas or liquid) flows down a pipe which contains a throttle consisting of a porous plug, and thereby undergoes an irreversible expansion by a steady-flow process (Section 3.11 F) under the following conditions: (1) no heat effect is produced in the surroundings, and the heat flow across the plug is negligible; (2) the work effect W_X produced in the surroundings by the increase in the velocity of the gas downstream from the plug over its velocity upstream is negligible (kinetic energy effect inappreciable); and (3) the effects of shear and turbulence are negligible at a short distance away from the plug. Although the accurate determination of the change in temperature of a fluid in a porous-plug experiment is difficult, it is not intrinsically impossible, because the fluid on each side of the plug is continually being replaced by fluid of the correct pressure and temperature. As a result, each part of the apparatus can be brought to the temperature of the fluid in that part during an induction period before the final steady-flow measurements are made.

The essential features of a modern version of Joule's apparatus [6] are shown in Fig. 5.2. In practice it is placed with its axis vertical, and is immersed in a thermostat regulated at the initial temperature of the experiment. Gas maintained at a constant pressure p_1 and temperature T_1 enters the apparatus through the tube I, and flows around the thermometer T_1 into the high-pressure cell A, which is filled with a fibrous material to reduce turbulence, eliminate radiation to the outer case, and decrease convection. The gas then flows radially through the porous plug P (made of unglazed porcelain) into the low-pressure cell B, where it first flows upward in the annular space around a tapered metal flow guide G, and then downward in the cylindrical annular space around the thermometer T_2. At the bottom of G the gas mixes with other gas which has entered B through the lower part of the plug. The low-pressure gas then passes through a valve V, which controls the pressure p_2 in B, and leaves the apparatus through the tube O. The pressure lead p_1 runs to a barostat and a manometer for maintaining the pressure in A constant and measuring it, and a lead p_2 serves a similar purpose for the cell B.

In a single series of runs with the pressure p_1 and temperature T_1 of the gas in the cell A maintained constant at, for example, 300°C and 200 atm, the pressure p_2 of the gas in the cell B was held constant at each of a series of pressures, for example 180, 160, ..., 40, 20, 10, and 5 atm. After steady-flow conditions had been established in each run, (p_1, T_1), (p_2, T_2) and the rate of flow of fluid through the apparatus were measured. The plug was then replaced by one of different porosity, and another series of runs made with (p_1, T_1) having as nearly as possible the same values and p_2 having approximately the same series of values as in the first series, but with rates of flow differing from the corresponding rates of the first series. We shall show that, if the conditions mentioned above for a properly executed Joule—Thomson experiment have been fulfilled, a single curve in a pT-

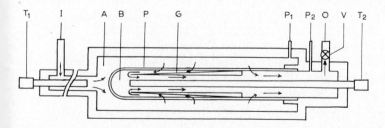

Fig. 5.2. Joule—Thomson expansion apparatus (after Roebuck [6]). T_1, T_2, thermometers; I, gas inlet; A, high-pressure cell; B, low-pressure cell; P, porous-plug of unglazed porcelain; G, flow guide; p_1, p_2, pressure leads to barostats and manometers for cells A, B respectively; O, gas outlet containing a valve V to regulate the pressure in cell B.

diagram passing through the point (p_1, T_1) will pass through all of the points representing the values of (p_2, T_2) determined in all series starting from (p_1, T_1), within the experimental error of measuring these quantities.

Let a unit mass of a fluid, for example one mole of N_2, undergo the change in state

$$N_2(g, p_1, T_1) = N_2(g, p_2, T_2) \tag{5.36}$$

by an adiabatic, steady-flow process without the production of non-expansion work W_X. Then

$$Q = 0 \qquad W = \Delta(pV) \qquad \therefore \Delta E = -\Delta(pV) \quad \Delta H = 0 \tag{5.37}$$

Thus the enthalpy of a fluid remains strictly constant in a Joule–Thomson expansion.

In general, the temperature of a fluid changes in a porous-plug expansion. The Joule–Thomson coefficient μ_{JT} and the isothermal variation of enthalpy with pressure Λ_{JT} of a fluid in a specified state are defined by the following equations:

$$\mu_{JT} \equiv \lim_{\Delta p \to 0} \left(\frac{\Delta T}{\Delta p}\right)_H = \left(\frac{\partial T}{\partial p}\right)_H = -\frac{(\partial H/\partial p)_T}{(\partial H/\partial T)_p} = -\frac{(\partial H/\partial p)_T}{C_p} \tag{5.38}$$

$$\Lambda_{JT} \equiv \left(\frac{\partial H}{\partial p}\right)_T = -\mu_{JT} C_p \tag{5.39}$$

Compare these relations with eqns. (5.32) and (5.33) where H is replaced by E and p by V.

The Joule–Thomson coefficient μ_{JT} usually enters thermodynamic relations as the product $\mu_{JT} C_p$ which we have denoted by Λ_{JT}. The latter is an extensive state property of a system with the units of volume, for example cm^3; but in some tabulations it is expressed in the units of energy pressure^{-1}, for example cal atm^{-1}. The quantity Λ_{JT} has great importance in industry, and in the thermodynamic theory of real and perfect gases.

Since the enthalpy of a fluid remains constant in a porous-plug expansion, points in a pT-diagram representing the pressures and temperatures of all states of a fluid produced in all series of Joule–Thomson expansions starting from the same initial state lie in the same curve of constant enthalpy. The slope $(\partial T/\partial p)_H$ of this curve at any point (p, T) is the Joule–Thomson coefficient μ_{JT} of the fluid in the corresponding state.

Some values [7] of the Joule–Thomson coefficient for nitrogen are listed in Table 5.2. We note that in general they do not approach zero as the pressure approaches zero at constant temperature.

The temperature at a point (p, T) where the slope of an isenthalp, i.e. a curve of constant enthalpy, in the pT-diagram (and hence μ_{JT}) passes through zero and changes sign, is called the Joule–Thomson inversion temperature T_{JT} at the pressure p. Figure 5.3 is a plot of T_{JT} against p for nitrogen [7]. The dashed line is the vapor pressure curve of liquid nitrogen.

When a fluid is in a state represented by a point in the area enclosed by the

TABLE 5.2

JOULE—THOMSON COEFFICIENTS (μ_{JT})[a] FOR NITROGEN

Pressure (atm)	Temperature (°C)				
	−100	0	100	200	300
1	0.6280	0.2570	0.1250	0.0540	0.0135
20	0.5785	0.2420	0.1140	0.0460	0.0095
60	0.4430	0.2040	0.0955	0.0365	−0.0010
100	0.2810	0.1660	0.0760	0.0260	−0.0070
200	0.0620	0.0900	0.0415	0.0075	−0.0160

[a]Units, deg atm^{-1}.

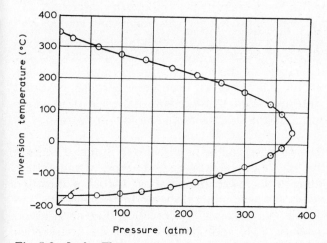

Fig. 5.3. Joule—Thomson inversion curve for nitrogen (from the smoothed values of Roebuck and Osterberg [7]). The dashed line is the liquid—vapor saturation curve of nitrogen.

inversion curve and the axis of zero pressure, μ_{JT} is positive (the temperature of the fluid decreases in a throttled expansion), and Λ_{JT} is negative (the enthalpy of the fluid increases in an isothermal expansion).

When the fluid is in a state represented by a point in the area outside of the inversion curve, μ_{JT} is negative (the temperature of the fluid increases in a throttled expansion), and Λ_{JT} is positive (the enthalpy of the fluid decreases in an isothermal expansion).

The values [4] of the upper Joule—Thomson inversion temperature T_{JT} at 1 atm for several gases are as follows:

Gas	He	H_2	N_2
T_{JT} at 1 atm (K)	45	200	600

From Boyle's law, eqn. (2.33), and Joule's law, eqn. (5.34), we find that the enthalpy H of a perfect gas is a function of T alone. Hence the following relations hold for a perfect gas:

$$\left.\begin{array}{l}\left(\dfrac{\partial H}{\partial p}\right)_T = 0,\ \Lambda_{JT} = 0,\ \mu_{JT} = 0 \\[2mm] H = H(T),\ C_p = \dfrac{dH}{dT},\ C_p - C_V = nR\ \text{(perfect gas)}\end{array}\right\} \quad (5.40)$$

The last equation in this group is obtained by evaluating the differential coefficients of eqn. (5.19) or (5.20) from eqns. (2.41) and (5.34).

We have noted that when a gas in a state corresponding to a point inside the Joule—Thomson inversion curve (Fig. 5.3) undergoes an expansion through a throttling valve its temperature decreases. This is the basis of one method of liquefying gases. The compressed gas, at a pressure of 30–200 atmospheres and precooled as far as possible by the means available, flows through a partially opened valve whereby its temperature decreases. By means of a counter-current heat interchanger this somewhat cooled low-pressure gas is used to precool the entering high-pressure gas, whose temperature falls to a still lower value in the throttling process. As the process continues, the temperature of the high-pressure gas and that of the low-pressure gas continue to fall until an isenthalp through a point representing the state of the precooled high-pressure gas on a pT-diagram cuts the vapor pressure line and liquid appears on the low-pressure side of the throttle.

From Fig. 5.3 we find that nitrogen can be liquefied by a Joule—Thomson expansion beginning with the high-pressure gas at room temperature. Hydrogen requires precooling to a lower temperature before a throttled expansion of the high-pressure gas produces a decrease in temperature, and helium requires even a lower temperature. In practice, the compressed hydrogen is precooled to about 80 K with liquid air and the compressed helium to about 15 K with liquid hydrogen.

A more efficient method of lowering the temperature of a gas is to expand it in a piston and cylinder or in a turbine where shaft work W_X is produced in the surroundings. If the expansion is adiabatic and essentially reversible, $\Delta H = -W_X$ (Table 4.4). Hence, the enthalpy of the gas actually decreases in the process, and consequently its temperature drop in a single pass through the apparatus is greater than that produced in a Joule—Thomson expansion, where the enthalpy change is zero. In some gas-liquefaction processes, part of the available pressure drop occurs in this type of expansion, and the remaining part in a Joule—Thomson expansion.

5.6. Relation of the Joule coefficient to the Joule—Thomson coefficient.

From eqns. (5.3) we can derive the expression

$$\left(\dfrac{\partial E}{\partial V}\right)_T = \left(\dfrac{\partial H}{\partial p}\right)_T \left(\dfrac{\partial p}{\partial V}\right)_T - \left(\dfrac{\partial (pV)}{\partial V}\right)_T \quad (5.41)$$

Substitution from eqns. (5.33) and (5.39) gives the following relations:

$$\lambda_J = \Lambda_{JT} \left(\frac{\partial p}{\partial V}\right)_T - \left(\frac{\partial (pV)}{\partial V}\right)_T \qquad (5.42\text{a})$$

$$\mu_J C_V = \left[\mu_{JT} C_p + \left(\frac{\partial (pV)}{\partial p}\right)_T\right] \left(\frac{\partial p}{\partial V}\right)_T \qquad (5.42\text{b})$$

The partial derivatives appearing in the above equations can be evaluated from an equation of state for the fluid under investigation.

5.7. Isothermal Joule—Thomson experiment. The quantity $(\partial H/\partial p)_T$ [see eqn. (5.39)] has been measured [8] calorimetrically by a method which is applicable to a gas in the pressure and temperature region where its Joule—Thomson coefficient is positive, that is, the temperature of the gas decreases in an (adiabatic) Joule—Thomson expansion. The apparatus is somewhat similar to that shown in Fig. 5.2. The gas under investigation flows from a high-pressure cell A where its temperature and pressure are maintained constant at (p_1, T_1), through a throttle consisting of a metal capillary, partially plugged with an electrical heater, to a low-pressure cell B, where its pressure is maintained constant at p_2. The apparatus is immersed in a thermostat regulated at the temperature T_1. Practically all of the pressure drop between A and B takes place in the narrow annular space around the case of the heater. A sufficiently large electrical current is passed through the heater to maintain the temperature of the gas in cell B equal to the temperature of the gas in cell A. After steady-flow conditions have been established, the pressures and temperatures of the gas in cells A and B, the rate of flow of the gas, and the electrical power input to the heater are measured. Thus we can determine electrically the quantity of heat Q withdrawn from its surroundings by a known mass of gas as it passes from a state (p_1, T_1) to a state (p_2, T_1) by a steady-flow process. This experiment is just as difficult to carry out successfully as is the (adiabatic) Joule—Thomson experiment.

Let a unit mass of a gas, for example one mole of N_2, undergo the change in state

$$N_2(g, p_1, T_1) = N_2(g, p_2, T_1) \qquad (5.43)$$

by an isothermal, steady-flow process in which the quantity of heat Q is withdrawn from the surroundings, and non-expansion work W_x is absent. Then the following equations hold:

$$Q = Q \qquad W = \Delta(pV) \qquad \therefore \Delta E = Q - \Delta(pV) \qquad \Delta H = Q \qquad (5.44)$$

Thus in an isothermal Joule—Thomson experiment the increase in enthalpy of the gas equals the quantity of heat withdrawn from the surroundings.

From eqns. (5.44) and the change in state (5.43) of the gas we find

$$\left(\frac{\Delta H}{\Delta p}\right)_T = \left(\frac{Q}{\Delta p}\right)_T \qquad (\Delta p = p_2 - p_1) \qquad (5.45)$$

If it is deemed necessary, we can determine Q in a series of runs with (p_1, T_1) fixed and p_2 having a series of decreasing values, such as in the Joule–Thomson (adiabatic) experiment. Then:

$$\left(\frac{\partial H}{\partial p}\right)_T = \lim_{\Delta p \to 0} \left(\frac{Q}{\Delta p}\right)_T \tag{5.46}$$

Thus the quantity $(\partial H/\partial p)_T$ at p_1 and T_1 is determined calorimetrically.

In order to show the relation of this experiment to the (adiabatic) Joule–Thomson experiment, we suppose the change in state (5.43) is brought about by the following two-step process: (1) the gas initially in the state (p_1, T_1) undergoes an (adiabatic) Joule–Thomson expansion whereby it reaches the state (p_2, T_2); (2) the gas is then brought to its initial temperature by a reversible, isopiestic process in which the quantity of heat Q is withdrawn from the surroundings and its final state is (p_2, T_1). The changes in state and the corresponding values of ΔH are as follows:

$$N_2(g, p_1, T_1) = N_2(g, p_2, T_2) \qquad \Delta H_a = 0 \tag{5.47a}$$

$$N_2(g, p_2, T_2) = N_2(g, p_2, T_1) \qquad \Delta H_b = \int_{p_2\, T_2}^{T_1} C_{p_2}\, dT \tag{5.47b}$$

$$N_2(g, p_1, T_1) = N_2(g, p_2, T_1) \qquad \Delta H_c = \Delta H_a + \Delta H_b \tag{5.47c}$$

Here C_{p_2} is the molar heat capacity of the gas at p_2.

Now the change in state (5.47c) is the same as the change (5.43). Hence, for this latter change in state, we find

$$\left.\begin{aligned} Q = \Delta H_c &= -\bar{C}_{p_2}(T_2 - T_1) \\ &= -\bar{C}_{p_2}\left(\frac{T_2 - T_1}{p_2 - p_1}\right)_H (p_2 - p_1) \\ &= -\bar{C}_{p_2}\bar{\mu}_{JT}(p_2 - p_1) \end{aligned}\right\} \tag{5.48}$$

Here $T_2 - T_1$ and $p_2 - p_1$ are the changes in temperature and pressure experienced by the gas in the (adiabatic) Joule–Thomson expansion (5.47a) for which H is constant; $\bar{C}_{p_2}$ is the mean constant-pressure heat capacity of the gas at p_2 over the temperature range T_2 to T_1; and $\bar{\mu}_{JT}$ is the mean Joule–Thomson coefficient of the gas over the range (p_1, T_1) to (p_2, T_2). If we allow p_2 to approach p_1, the last expression of eqns. (5.48) approaches eqn. (5.39).

5.8. Perfect gases. In Section 2.10 we derived the perfect gas equation

$$pV = nRT \tag{2.41}$$

from the laws of Boyle and Gay-Lussac and the empirical definitions of the perfect gas absolute temperature scale and the mole.

Substitution from the above relation into eqns. (3.98), (3.102), and (3.106) gives the following relations for a perfect gas:

$$\alpha_p = \frac{1}{V}\left(\frac{\partial V}{\partial T}\right)_p = \frac{1}{T}, \ \alpha_V = \frac{1}{p}\left(\frac{\partial p}{\partial T}\right)_V = \frac{1}{T}, \ \beta_T = -\frac{1}{V}\left(\frac{\partial V}{\partial p}\right)_T = \frac{1}{p} \tag{5.49}$$

Here T is on the perfect gas absolute temperature scale.

In this chapter we have noted that we cannot derive from any law of thermodynamics a relation for the isothermal variation of the energy $(\partial E/\partial V)_T$ of a gas which is defined only by the laws of Boyle and Gay-Lussac. Hence we completed the definition of a perfect gas by adding Joule's law (for gases) which states that this differential coefficient is zero at all temperatures. The thermodynamic behavior of a perfect gas is completely defined by the two equations:

$$pV = nRT, \quad \left(\frac{\partial E}{\partial V}\right)_T = 0 \tag{5.50}$$

For the present we shall consider that these relations are derived from the laws of Boyle, Gay-Lussac, and Joule. By employing the results of the second law, we shall derive Gay-Lussac's law from the laws of Boyle and Joule, and we shall discover a deeper significance to the perfect gas temperature scale.

Some of the thermodynamic behavior of a perfect gas is summarized in the following relations [see eqns. (5.34) and (5.40)]:

$$\mu_J = 0, \ \lambda_J = 0, \ \left(\frac{\partial E}{\partial V}\right)_T = 0, \ E = E(T), \ C_V = \frac{dE}{dT} \tag{5.51}$$

$$\mu_{JT} = 0, \ \Lambda_{JT} = 0, \ \left(\frac{\partial H}{\partial p}\right)_T = 0, \ H = H(T), \ C_p = \frac{dH}{dT} \tag{5.52}$$

$$C_p - C_V = nR \tag{5.53}$$

When a perfect gas A undergoes the isothermal change in state

$$A(g, p_1, T_1) = A(g, p_2, T_1) \qquad \text{(perfect gas)} \tag{5.54}$$

by any process whatever, we find from the above equations that

$$\Delta E = 0 \qquad \Delta H = 0 \qquad \therefore Q = W \tag{5.55}$$

Let a system A composed of a perfect gas undergo the change in state

$$A(g, p_1, V_1, T_1) = A(g, p_2, V_2, T_2) \qquad \text{(perfect gas)} \tag{5.56}$$

by any process whatever. Then eqns. (5.7) become

$$dE = C_V dT, \ dH = C_p dT \tag{5.57}$$

Integration of these equations from T_1 to T_2 gives

$$\Delta E = \int_{T_1}^{T_2} C_V dT, \ \Delta H = \int_{T_1}^{T_2} C_p dT \tag{5.58}$$

It is not necessary to specify a path for these integrations [see eqns. (5.51) and (5.52)].

Let us suppose that the final state of the change (5.56) is reached from the initial state by a reversible, adiabatic process. Then, since the gas is perfect, eqns. (5.15) and (5.16) become

$$đQ = C_V dT + nRT \frac{dV}{V} = 0 \qquad\qquad đQ = C_p dT - nRT \frac{dp}{p} = 0 \qquad (5.59)$$

Hence

$$\frac{dT}{T} = -\frac{nR}{C_V} \frac{dV}{V}, \quad \frac{dT}{T} = \frac{nR}{C_p} \frac{dp}{p}, \quad \frac{dp}{p} = -\kappa \frac{dV}{V} \qquad (5.60)$$

where

$$\kappa \equiv \frac{C_p}{C_V} = \frac{C_V + nR}{C_V} \qquad (5.61)$$

As we define a perfect gas in thermodynamics, its constant-volume and constant-pressure heat capacities can be functions of temperature. Over a temperature range in which these heat capacities may be considered to be essentially constant, we may integrate eqns. (5.60) to give:

$$TV^{nR/C_V} = \text{constant}, \quad Tp^{-nR/C_p} = \text{constant}, \quad pV^\kappa = \text{constant} \qquad (5.62)$$

Equations for the work of expansion of a perfect gas for a number of processes are given in Section 3.11.

REFERENCES

1 J. P. Joule, The Scientific Papers of James Prescott Joule, The Physical Society, London, 1884, (1845) pp. 172—189.
2 F. G. Keyes and F. W. Sears, Proc. Natl. Acad. Sci. U.S.A., 11 (1925) 38—41.
3 J. R. Roebuck, Proc. Am. Acad. Arts Sci., 64 (1930) 287—334.
4 F. G. Keyes, Temperature, Reinhold, New York, 1941, pp. 45—59.
5 J. P. Joule and W. Thomson, Joint Scientific Papers of James Prescott Joule, The Physical Society, London, 1887 (originally published 1852—1862), pp. 215—362. See also W. Thomson, Mathematical and Physical Papers, University Press, Cambridge, 1882, Vol. 1, pp. 333—455.
6 J. R. Roebuck, Proc. Am. Acad. Arts Sci., 60 (1925) 537—596; ibid., 64 (1930) 287—334. For references to later work, see J. R. Roebuck, T. A. Murrell and E. E. Miller, J. Am. Chem. Soc., 64 (1942) 400—411.
7 J. R. Roebuck and H. Osterberg, Phys. Rev., 48 (1935) 450—457.
8 F. G. Keyes and S. C. Collins, Proc. Natl. Acad. Sci. U.S.A., 18 (1932) 328—333; S. C. Collins and F. G. Keyes, Proc. Am. Acad. Arts Sci., 72 (1937) 283—299; J. Phys. Chem., 43 (1939) 5—14.

Chapter 6

THERMOCHEMISTRY

Thermochemistry began in the latter half of the eighteenth century as a branch of chemistry concerned with the heat effects attending processes involving chemical changes. By 1840 with the publication of Hess' law of constant heat summation it had become a science, ten years before the first law of thermodynamics was announced by Clausius. Thermochemistry gradually became accepted as a branch of thermodynamics. Originally the quantity of heat Q withdrawn from the surroundings by a system undergoing a change in state by a given process was given a negative sign in thermochemistry; it has always been given a positive sign in thermodynamics. Today the thermodynamic convention is universally employed.

Unless otherwise stated, we shall limit the considerations of the present chapter to closed, simple systems in which chemical reactions may occur.

6.1. Enthalpy, energy, and heats of reaction. The expression "enthalpy of reaction at a pressure p and a temperature T" denotes the change in enthalpy, ΔH, attending a change in state in which stoichiometric quantities of a set of pure reactants, each in a specified aggregation state at p and T, are converted quantitatively into a set of pure products, each in a specified aggregation state also at p and T.

The expression "heat of reaction at constant pressure", Q_p, denotes the heat withdrawn from the surroundings in an experiment designed to bring about the above-mentioned change in state, whereby the reactants are mixed, the chemical reaction is brought about, and the products are separated by means of an isopiestic process in which the only work effect produced in the surroundings is reversible, or essentially reversible, expansion work. The process need not be reversible, or even quasistatic, in all respects. For example, an irreversible chemical reaction may occur at a finite rate.

Let the change in state

$$\sum_i \nu_i B_i \text{ (state } i\text{)} = 0 \qquad (p, T = \text{constant}) \qquad (6.1)$$

(see Section 1.8) be brought about by the process described above. Then

$$\Delta H = Q_p = \sum_i \nu_i h_i \qquad (6.2)$$

In this equation h_i is the molar enthalpy of pure B_i in the aggregation state i at the pressure p and the temperature T. Molar enthalpy is discussed in Sections 6.4 and 6.6.

In a similar manner, the expression "energy of reaction at a volume V

and a temperature T" denotes the change in energy, ΔE, attending a change in state in which stoichiometric quantities of a set of pure reactants, each in a specified aggregation state in a volume V at a temperature T, are converted quantitatively into a set of pure products, each in a specified aggregation state also in a volume V at a temperature T.

The corresponding quantity "heat of reaction at constant volume", Q_V, is the heat withdrawn from the surroundings in an experiment designed to bring about the above-mentioned change in state, whereby the reactants are mixed, the chemical reaction is brought about, and the products are separated by means of an isometric process in which no work is produced in the surroundings.

If the change in state

$$\sum_i \nu_i B_i \text{ (state } i\text{)} = 0 \qquad (V, T = \text{constant}) \tag{6.3}$$

is brought about by the above process, we find

$$\Delta E = Q_V = \sum_i \nu_i e_i \tag{6.4}$$

Here e_i is the molar energy of pure B_i in the aggregation state i at the molar volume V/ν_i and the temperature T.

In the past, the heat effects attending the mixing of the reactants and the unmixing of the products have not always been taken into account in the computation of Q_p and Q_V from calorimetric measurements. Heats of mixing at constant pressure and at constant volume are usually small in gas phases at moderate pressures. They may be large in condensed phases.

It has been found by experiment that at a constant temperature the variations of enthalpies of reaction with pressure and the variations of energies of reaction with volume can usually be neglected when the pressures of the substances taking part in the change in state do not exceed several atmospheres, and when the changes in pressure or volume do not produce a change in aggregation state in any of the reactants or products. The variations of these two quantities with temperature are more important.

Let the change in state (6.1) be brought about by a two-step process in which the first step is the change (6.3). We may indicate the changes in states involved in the following manner:

$$\left. \begin{array}{l} \text{Reactants } (p_1, V_1) \xrightarrow{1} \text{Products } (p_1, V_2) \\ \phantom{\text{Reactants } (p_1, V_1)} \searrow_2 \quad \uparrow 3 \\ \phantom{\text{Reactants } (p_1, V_1)\searrow_2} \text{Products } (p_2, V_1) \end{array} \right\} \quad (T = \text{constant}) \tag{6.5}$$

Here the term Reactants (p_1, V_1) represents a set of pure reactants, each in a specified aggregation state at p_1 and T, or at V_1 and T, depending on whether we are considering reaction 1 or reaction 2. The other symbols have corresponding significance for the products of the reaction. Application of eqn. (4.15) to the change in state 1 shown in eqns. (6.5) gives

$$\Delta H_1 = \Delta E_1 + p_1 \Delta V \tag{6.6}$$

where ΔV is the sum of the volumes of all of the products minus the sum of the volumes of all of the reactants of the change 1. Since energy is a state property, we can replace ΔE_1 by the sum of ΔE_2 and ΔE_3. Hence

$$\Delta H_1 = \Delta E_2 + \Delta E_3 + p_1 \Delta V \tag{6.7}$$

If the pressures do not exceed several atmospheres we may, as a first approximation, neglect ΔE_3 for both condensed and gaseous substances; and we can neglect the term $p_1 \Delta V$ for condensed substances and evaluate it from the perfect gas law for gaseous substances. Equation (6.7) can then be written in the following form:

$$\Delta H = \Delta E + (\Delta \nu) RT \tag{6.8}$$

where $\Delta \nu$ is the number of moles of gaseous products minus the number of moles of gaseous reactants in the chemical reaction under consideration. In eqn. (6.8) ΔH is the enthalpy of reaction of the change in state 1 at p_1 and T and ΔE is the energy of reaction of the change 2 at V_1 and T. In terms of heats of reaction at T, this equation becomes

$$Q_p = Q_V + (\Delta \nu) RT \tag{6.9}$$

We have used in this derivation one of the results of the first law of thermodynamics, which says that ΔH and ΔE for any change in state depend only on the initial and final states of a system. Hence enthalpies and energies of reaction are independent of whether a given change takes place in one step or in a number of steps. This result of the first law is called Hess' law of constant heat summation.

In chemical thermodynamics we are mostly concerned with heat effects attending reactions carried out at constant pressure, which are related to an enthalpy change. Therefore we shall deal mainly with the determination and use of enthalpies of reaction in the rest of this chapter.

6.2. *Standard states.* The concept of standard states is useful in making thermodynamic calculations. Standard states are certain real or hypothetical states of substances which make possible the tabulation of a large body of data in a form that facilitates the application of the equations of thermodynamics to numerical problems.

Each substance has, at each temperature, a number of standard states. The choice of a standard state depends on the problem under consideration.

6.2A. *Standard states of pure substances.* At each temperature T the standard state of a gaseous substance is the pure gas in the state (PG, 1 atm, T). Here PG signifies that the gas is in the state of a perfect gas. This is a hypothetical, not a real, state of the substance in which it obeys the perfect gas law at and up to a pressure of 1 atm. Hence the gas has in this state the same energy, enthalpy, and heat capacities at constant pressure and at constant volume as the corresponding real gas in the state (0 atm, T) [see equations (5.51) and (5.52)].

At each temperature T the standard state of a liquid or solid substance is the corresponding pure substance in the state (l, 1 atm, T) or (s, 1 atm, T), respectively.

We may use any one of these standard states that is appropriate, independently of the actual aggregation state of a substance. For example, for liquid water at 300 K we may use the standard state of water as a gas or as a liquid; and for ice at 270 K we may use the standard state of water as a gas, a liquid, or a solid.

6.2B. Standard states of substances in solution. At each temperature T the standard state of a constituent B_i of a gas mixture under a pressure p is the pure gas B_i in the state (PG, 1 atm, T) defined in Section 6.2A.

We may use either one of two standard states for a constituent B_i of a liquid or solid solution.

(a) At each pressure p and temperature T the standard state of a constituent B_i of a liquid or a solid solution is the pure substance B_i at p and T, and usually, but not always, in the same aggregation state as the solution. For example, the standard state of the substance B_i in a solid solution at the pressure p and temperature T is usually taken to be pure solid B_i at p and T; but in some problems it is taken to be pure liquid B_i at p and T.

(b) At each pressure p and temperature T the standard state of a constituent B_i of a liquid or a solid solution is the substance B_i in the dissolved state (DS, p, T, y_i = 1). Here DS denotes that B_i is in the state of a solute that obeys the dilute solution law; and y_i stands for any method of expressing the concentration of B_i in the solution, for example, mole fraction x_i, weight molar concentration m_i, volume molar concentration c_i, volume fraction z_i, etc. This is a hypothetical state of the solute B_i in which it obeys the dilute solution law at all concentrations at and up to $y_i = 1$, where it has (among other things) the same partial molar volume, energy, enthalpy, and heat capacities at constant pressure and at constant volume as it has at $y_i = 0$ in the given solution.

In the symmetrical system of standard states for condensed solutions, the standard state 2(a) is used for each of the constituents. It is particularly appropriate for the constituents of a solution composed of substances miscible in all proportions.

In the unsymmetrical system the standard state 2(a) is used for one constituent, usually present in excess and called the solvent; and the state 2(b) is used for each of the other constituents, called the solutes. It is usually employed for a constituent that has limited solubility.

6.3. Calorimetric determination of enthalpies of reaction. Most of the enthalpies of reaction involving chemical processes are computed from thermal measurements made in calorimeters operating near room temperature. These heat effects can then be computed for any other temperature by methods discussed in Section 6.6.

Several types of equipment are used: (1) the bomb calorimeter

(Section 3.19); (2) the open calorimeter (Section 3.20); (3) the Bunsen ice calorimeter (Section 3.22); and (4) the flow calorimeter (Section 3.24) which has been modified to permit the measurement of heat effects attending isopiestic, steady-flow chemical processes involving gaseous reactants [1].

Let us consider briefly the method of calculating standard enthalpies of reaction from measurements made in closed and open calorimeters.

In Section 3.19 we considered a bomb calorimetric method for measuring the change in energy ΔE_{T_1} attending an isothermal, constant-volume process producing the change in state described by eqn. (3.70c):

Reactants (T_1) = Products (T_1) ΔE_{T_1} (V = constant)

In the actual bomb process the Reactants and Products are often complicated mixtures of a number of substances distributed in two or more phases under a pressure of 20 atm or more. We wish to calculate from ΔE_T the value of the standard enthalpy of reaction ΔH_T^0 attending a standard change in state in which stoichiometric quantities of a set of reactants, each in its standard state at T, are quantitatively converted into a set of products, each in its standard state at T. For example, from the heat effect ΔE_{T_1} attending the combustion of benzoic acid in a bomb calorimeter, we wish to calculate ΔH_T^0 for the following change in state

$$C_6H_5COOH(s) + 7.5O_2(g) = 7CO_2(g) + 3H_2O(l) \quad \Delta H_T^0 \tag{6.10}$$

Here the symbol ΔH_T^0 indicates that each reactant and each product is in its standard state at T. In general, the temperature T of the above equation is close to, but not identical with, the temperature T_1 of the calorimetric experiment.

In order to obtain ΔH_T^0 we first compute the change in energy ΔE_T^0 attending the standard change in state (6.10) from ΔE_{T_1} for the corresponding change (3.70c). This is a painstaking, arduous task involving the computation of the increments of energy attending: (1) the preparation of the system we have called the "Reactants" of eqn. (3.70c) in the state (V, T_1) from the various substances constituting it, each initially in its standard state at T; and (2) the separation of the system called the "Products" of this change in the state (V, T_1) into its constituent substances, each in its standard state at T.

This computation may require as many as 20 separate calculations, and they may include the consideration of the effects produced by the presence in the bomb of small amounts of substances which do not appear in the change in state (6.10), but which were added to initiate the calorimetric process and to catalyze the desired chemical reaction [2].

We can then compute ΔH_T^0 for the change in state (6.10) from the value of ΔE_T^0 for the same change by using eqn. (6.8). The error introduced into the result by the use of this approximate equation is less than the error of the calorimetric measurement of ΔE_{T_1}.

The reduction of the observations made in an open or in a flow calorimeter is a much simpler calculation than that described above, since the change in state produced in the calorimetric process approximates to the standard change much more closely than that occurring in a bomb calorimeter.

6.4. Standard enthalpy of formation. The standard enthalpy of reaction attending the synthesis of one mole of a substance in a specified standard state at a temperature T from the elementary substances composing it, each in the standard state corresponding to its stable aggregation state at 1 atm and T, is called the standard molar enthalpy of formation, or, more simply, the standard molar enthalpy, of the given substance at T.

We shall use the notations ΔH_T^0 and h_T^0 interchangeably, as well as the expressions standard molar enthalpy of formation and standard molar enthalpy. Some authors use the notation ΔH_f^0 for this quantity. In the thermochemical equation

$$\sum_i \nu_i B_i \text{ (state } i\text{)} = 0 \quad \Delta H_T^0 = a \text{ kcal} \tag{6.11}$$

ν_i is the number of moles of the substance B_i taking part in the change in state and it is negative for a reactant and positive for a product. The superscript zero to ΔH indicates that each substance B_i is in its standard state at T. If we substitute the standard molar enthalpy h_i^0 for each chemical symbol in eqn. (6.11) (see Section 1.8) we find

$$\Delta H_T^0 = \sum_i \nu_i h_i^0 \tag{6.12}$$

This relation can also be written in the following form:

$$\Delta H_T^0 = H^0(P) - H^0(R) \tag{6.13}$$

Here $H^0(P)$ is the sum of the standard enthalpies of all of the products of the change in state (6.11) and $H^0(R)$ is the corresponding quantity for all of the reactants.

The standard molar enthalpy of formation of an element in its stable aggregation state at 1 atm pressure at each temperature is evidently zero. But the corresponding quantity for the element in any other aggregation state is not zero. We know from calorimetric measurements that

$$Br_2(l) = Br_2(g) \quad \Delta H_{298.15}^0 = 7.34 \text{ kcal} \tag{6.14}$$

Bromine is a liquid at 1 atm and 298.15 K. Hence

$$h_{298.15}^0(Br_2, l) = 0 \quad h_{298.15}^0(Br_2, g) = 7.34 \text{ kcal} \tag{6.15}$$

When two different aggregation states of an element are in equilibrium at $(1 \text{ atm}, T)$, the standard molar enthalpy of either form may be assigned the value zero. The corresponding quantity for the other form is determined from the heat of phase transformation.

The standard molar enthalpy of gaseous HCl at 298.15 K has been determined calorimetrically by direct synthesis from its elements in a flow calorimeter. The result can be exhibited in the thermochemical equation:

$\tfrac{1}{2} H_2(g) + \tfrac{1}{2} Cl_2(g) = HCl(g) \quad \Delta H^0_{298.15} = -22.063$ kcal (6.16)

Thus

$$\Delta H^0_T = h^0_T(HCl, g) - 0.5\, h^0_T(H_2, g) - 0.5\, h^0_T(Cl_2, g) \tag{6.17a}$$

so that

$$\Delta H^0_T = h^0_T(HCl, g) = -22.063 \text{ kcal} \quad (T = 298.15 \text{ K}) \tag{6.17b}$$

since both $h^0_T(H_2, g)$ and $h^0_T(Cl_2, g)$ are zero.

When a compound cannot be synthesized in one or more steps, each suitable for calorimetric study, we can determine its standard molar enthalpy from the measured enthalpy change attending any change in state involving the given substance and other substances of known standard molar enthalpies. From the equation

$$C_6H_5COOH(s) + 7.5 O_2(g) = 7 CO_2(g) + 3 H_2O(l)$$

$$\Delta H^0_{298.15} = -771.57 \text{ kcal} \tag{6.18}$$

and the values

$$h^0_{298.15}(CO_2, g) = -94.052 \quad h^0_{298.15}(H_2O, l) = -68.317 \text{ kcal} \tag{6.19}$$

we find

$$-771.57 = 7(-94.052) + 3(-68.317) - h^0_{298.15}(C_6H_5COOH, s) \tag{6.20a}$$

so that

$$h^0_{298.15}(C_6H_5COOH, s) = -91.74 \text{ kcal} \tag{6.20b}$$

The conversion of a standard molar enthalpy of formation into the corresponding energy of formation is so simple a calculation [see eqn. (6.8)] that only the former quantity is listed in tables.

6.5. Calculation of enthalpies of reaction from tables. Extensive tabulations of standard molar enthalpies of formation (and other thermodynamic properties) of substances have been published, arranged according to the following schemes:

1. Values of standard molar enthalpies h^0_T are listed for evenly spaced temperatures ranging from 0 K upward to as high as several thousand degrees Kelvin [3].

2. Values of $h^0_{298.15}$, together with values of the changes in standard molar enthalpy $h^0_T - h^0_{298.15}$ from 298.15 K to T K are listed [4] for evenly spaced temperatures above 298.15 K. Addition of these two values for a substance gives h^0_T.

3. Values of $h^0_{298.15}$, or in some cases $h^0_{291.15}$, are listed [5].

All of the tables mentioned above list the standard molar enthalpies of formation of a substance in two or more aggregation states at temperatures where two or more such states are in equilibrium at 1 atm pressure. The first two sets of tables also list standard constant-pressure heat capacities at evenly spaced temperatures.

We can calculate from eqn. (6.12) the value of ΔH_T^0 for any standard change in state at any temperature T involving substances whose standard molar enthalpies h_T^0 are given in any one of these tables.

Standard energies of reaction can be calculated from the corresponding standard enthalpies of reaction by use of eqn. (6.8).

Enthalpies and energies of reaction for conditions other than standard can be computed from the corresponding standard values by application of the methods presented in the next section.

6.6. Effects of pressure and temperature on enthalpies of reaction. Two methods are available for computing the variation of an enthalpy of reaction with changes in pressure and temperature.

6.6A. When no reactant or product of a change in state undergoes a change in aggregation state in the pressure and temperature interval under consideration. Suppose we wish to express the enthalpy of reaction ΔH_{pT} of the following change in state as a function of pressure and temperature:

$$CO(g) + \tfrac{1}{2}O_2(g) = CO_2(g) \qquad (p, T = \text{constant}) \qquad (6.21)$$

We are given $\Delta H_{p_1, T_1}$ for the corresponding change under the fixed conditions (p_1, T_1), together with $(\partial h/\partial T)_p$ and $(\partial h/\partial p)_T$ as functions of p and T for each of the reactants and products.

Now for the change in state (6.21) we have

$$d\Delta H = \left(\frac{\partial \Delta H}{\partial T}\right)_p dT + \left(\frac{\partial \Delta H}{\partial p}\right)_T dp \qquad (6.22)$$

Integration of this expression from the conditions (p_1, T_1) to (p_1, T) along the isopiestic p_1, and then from the conditions (p_1, T) to (p, T) along the isothermal T gives

$$\Delta H_{pT} = \Delta H_{p_1, T_1} + \int_{p_1 T_1}^{T} \left(\frac{\partial \Delta H}{\partial T}\right)_{p_1} dT + \int_{Tp_1}^{p} \left(\frac{\partial \Delta H}{\partial p}\right)_T dp \qquad (6.23)$$

The second term of the right-hand side of this equation gives the temperature variation of ΔH at the pressure p_1 and the third term gives the pressure variation of ΔH at the temperature T. We shall treat these effects separately.

If we write the change in state (6.21) in the generalized form of eqn. (6.1) and apply eqns. (6.12) and (6.13) to the present problem, we have

$$\Delta H = \Sigma_i \nu_i h_i = H(\text{P}) - H(\text{R}) \qquad (6.24)$$

Here h_i is the molar enthalpy of the substance B_i in the change in state (6.1). Differentiation gives

$$d\Delta H = \sum_i \nu_i dh_i = dH(P) - dH(R) = \Delta dH \tag{6.25}$$

Thus the operators d and Δ commute, where Δ refers to an enthalpy of reaction ΔH at a constant pressure and temperature and d refers to a change of ΔH with pressure and temperature.

Division of eqns. (6.25) by dT at constant p and then by dp at constant T gives the following relations:

$$\left(\frac{\partial \Delta H}{\partial T}\right)_p = \sum_i \nu_i \left(\frac{\partial h_i}{\partial T}\right)_p = \left[\frac{\partial H(P)}{\partial T}\right]_p - \left[\frac{\partial H(R)}{\partial T}\right]_p = \Delta \left(\frac{\partial H}{\partial T}\right)_p \tag{6.26a}$$

so that

$$\left(\frac{\partial \Delta H}{\partial T}\right)_p = \sum_i \nu_i c_{pi} = C_p(P) - C_p(R) = \Delta C_p \tag{6.26b}$$

and

$$\left(\frac{\partial \Delta H}{\partial p}\right)_T = \sum_i \nu_i \left(\frac{\partial h_i}{\partial p}\right)_T = \left[\frac{\partial H(P)}{\partial p}\right]_T - \left[\frac{\partial H(R)}{\partial p}\right]_T = \Delta \left(\frac{\partial H}{\partial p}\right)_T \tag{6.27}$$

We note that $(\partial h_i/\partial p)_T$ for a gaseous substance is the quantity Λ_{JT} defined in Section 5.5. Equations (6.26) are expressions of Kirchhoff's formula.

Substitution from eqns. (6.26) and (6.27) separately into eqn. (6.23) gives

$$\Delta H_{p_1 T} = \Delta H_{p_1 T_1} + \int_{p_1 T_1}^{T} \Delta C_p dT \tag{6.28}$$

$$\Delta H_{pT} = \Delta H_{p_1 T} + \int_{T\, p_1}^{p} \Delta \left(\frac{\partial H}{\partial p}\right)_T dp \tag{6.29}$$

Note that the elimination of the terms $\Delta H_{p_1 T}$ in the above two relations gives eqn. (6.23).

The variation of a standard enthalpy of reaction ΔH_T^0 with temperature is given by replacing ΔC_p by ΔC_p^0 in eqn. (6.28). Equation (6.29) permits us to calculate an enthalpy of reaction at any pressure p from the corresponding standard value at the same temperature. In this computation we must place $p_1 = 1$ atm for condensed substances and $p_1 = 0$ atm for gases.

In general, the variation of an enthalpy of reaction with pressure is much smaller than its variation with temperature. Moreover the values of $(\partial h/\partial p)_T$ have not been determined for many substances. Hence we seldom employ eqn. (6.29).

6.6B. *When one or more of the reactants and products of a change in state undergo a change in aggregation state in the temperature and pressure range under consideration.* Suppose we are given the enthalpy of reaction $\Delta H_{p_1 T_1}$ for the change in state in which HBr is formed from H_2 and Br_2

at the pressure p_1 and temperature T_1 where Br_2 is a liquid. We wish to calculate ΔH_{pT} for the corresponding change in state at p and T where Br_2 is a gas. We follow the procedure employed in eqns. (6.28) and (6.29). First we calculate the effect of a change in temperature from T_1 to T with p_1 constant, which gives $\Delta H_{p_1,T}$. Then we compute the effect of a change in pressure from p_1 to p with T constant, which gives ΔH_{pT}.

The first computation may be arranged in the following manner.

$$\begin{array}{l} \tfrac{1}{2}H_2(g) + \tfrac{1}{2}Br_2(g) = HBr(g) \qquad \Delta H_{p_1,T} \\ \Delta H(R) \uparrow \quad \overset{\Delta H}{\nearrow} \quad \uparrow \Delta H(P) \\ \tfrac{1}{2}H_2(g) + \tfrac{1}{2}Br_2(l) = HBr(g) \qquad \Delta H_{p_1,T_1} \end{array} \right\} \qquad (6.30)$$

Now $\Delta H(R)$ and $\Delta H(P)$ are the changes in enthalpy attending the increase in temperature of the reactants and products, respectively, from T_1 to T under the pressure p_1; and ΔH is the change in enthalpy attending the change in state of the system starting with each pure reactant at (p_1, T_1) and ending with each pure product at (p_1, T). ΔH is independent of the process. Hence

$$\Delta H = \Delta H_{p_1,T_1} + \Delta H(P) = \Delta H(R) + \Delta H_{p_1,T} \tag{6.31a}$$

so that

$$\Delta H_{p_1,T} = \Delta H_{p_1,T_1} + \Delta H(P) - \Delta H(R) \tag{6.31b}$$

Here

$$\Delta H(P) = \int_{T_1}^{T} c_p(HBr, g)\, dT \qquad (\text{at } p_1) \tag{6.32a}$$

and

$$\Delta H(R) = \tfrac{1}{2}\int_{T_1}^{T} c_p(H_2, g)\, dT + \tfrac{1}{2}\left[\int_{T_1}^{T_b} c_p(Br_2, l)\, dT + \Delta H_{vap}(Br_2, l)\right.$$

$$\left. + \int_{T_b}^{T} c_p(Br_2, g)\, dT\right] \qquad (\text{at } p_1) \tag{6.32b}$$

The quantity ΔH_{vap} is the molar heat of vaporization of liquid Br_2 at its boiling point T_b under the pressure p_1, and c_p is a molar constant-pressure heat capacity.

As noted above, ordinary variations in pressure produce a much smaller change in the enthalpy of a closed system than do ordinary changes in temperature. Moreover, very few values of $(\partial H/\partial p)_T$ have been measured.

Let us now apply eqn. (6.29) to a system undergoing a change in aggregation state. The pressure varies from p_1 to p and the temperature T is constant. The method employed to derive eqns. (6.31) gives the relation

$$\Delta H_{pT} = \Delta H_{p_1 T} + \Delta H(P) - \Delta H(R) \tag{6.33}$$

The methods we have employed in this section for calculating the effects of variations of pressure and temperature on an enthalpy of reaction are also applicable to the determination of the effects of variations of volume and temperature on an energy of reaction. We must of course replace H by E and p by V in eqns. (6.21) — (6.33).

Of more interest in chemical thermodynamics are the following computations: (1) the calculation of ΔE for a given change in state from ΔH for the same change by use of eqn. (4.15); and (2) the calculation of an energy of reaction from the corresponding enthalpy of reaction at the same temperature, in the pressure region where the approximate relation (6.8) is sufficiently accurate.

6.7. *Maximum flame and explosion temperatures.* The scheme employed in eqns. (6.30) can be modified for the purpose of computing the theoretical maximum temperature reached in a combustion or other rapid chemical reaction. The calculation is based on the following assumptions: (1) the process is adiabatic; (2) the reaction goes to completion and no decomposition of the products occurs; (3) no side reactions occur; (4) the effects of pressure on the thermal properties of substances are negligible; and (5) the effects of mixing of the constituents composing the reactants and the products may be neglected. The last two effects are zero for perfect gases. Assumption (5) is necessary since calculated enthalpies and energies of reaction refer to changes in state in which each reactant and each product is pure, whereas in a flow or a bomb process this may not be true.

Let one mole of methane be burned in 10 moles of air (assumed to have the composition 1 O_2:4 N_2) by an adiabatic, isopiestic, steady-flow process in which the only work produced in the surroundings is essentially reversible expansion work. In principle, all of the enthalpy of combustion of the methane at the temperature T_1 is available for increasing the temperature of the products (including the nitrogen) from T_1 to the theoretical maximum temperature T_2 at the applied pressure p_1. We can arrange the computation in the following manner:

$$\begin{array}{c} CO_2(g) + 2H_2O(g) + 8N_2(g) \quad (p_1, T_2) \\ \Delta H \nearrow \quad \uparrow \Delta H(P) \\ CH_4(g) + 2O_2(g) + 8N_2(g) = CO_2(g) + 2H_2O(g) + 8N_2(g) \quad (p_1, T_1) \end{array} \tag{6.34}$$

Since the process is adiabatic, ΔH is zero. Hence

$$\Delta H = \Delta H_{p_1 T_1} + \Delta H(P) = 0 \tag{6.35}$$

Therefore

$$\Delta H_{p_1 T_1} = -\Delta H(P) = -\int_{T_1}^{T_2} C_p(P)\, dT \qquad \text{(at } p_1\text{)} \tag{6.36}$$

Usually p_1 is taken to be 1 atm. Thus if we are given h_T^0 for each reactant and product, and $C_p^0(\text{P})$ for the products, we can compute T_2, the theoretical maximum temperature of combustion.

We can correct the maximum temperature T_2 given by eqn. (6.36) for the dissociation of the products if we know the dissociation constants and the corresponding heats of dissociation of the products as functions of temperature in the high-temperature region [6]. The measured maximum temperature is lower than the calculated value, even after this correction has been applied, largely because the experimental process is not adiabatic.

If an excess of air were used to insure complete combustion of the methane, the excess would appear on the right-hand side of eqn. (6.34) and thus tend to reduce the maximum temperature. On the other hand, a much higher temperature could be reached by burning the methane in pure oxygen.

We can employ the same general procedure to compute the theoretical maximum temperature of an explosion in a bomb. Here the volume of the system is essentially constant and, in principle, the change in state is brought about by an adiabatic, isometric process. We must replace H by E and p by V in eqns. (6.34)—(6.36). The error introduced into the calculation by neglecting the effects of the mixing of the reactants and the products is larger than in the flow process because of the higher pressures involved.

6.8. Enthalpies of reaction in solution. We shall discuss enthalpies of reaction for changes involving substances in the dissolved state in the next chapter.

REFERENCES

1 See Experimental Thermochemistry, Interscience, New York: Vol. I, 1956 edited by F. D. Rossini, and Vol. II, 1962, edited by H. A. Skinner.
2 E. W. Washburn, J. Res. Natl. Bur. Stand., 10 (1933) 525—558; W. N. Hubbard, D. W. Scott and G. Waddington, ref. 1, Vol. I, Chapter 5; W. N. Hubbard, ref. 1, Vol. II, Chapter 6.
3 (a) F. D. Rossini, K. S. Pitzer, R. L. Arnett, R. M. Braun and G. C. Pimentel, Selected Values of Properties of Hydrocarbons and Related Compounds, Carnegie Institute of Technology, Pittsburgh, PA, 1954; (b) J. P. Coughlin, U.S. Bur. Mines, Bull., 542 (1954); (c) D. R. Stull et al., JANAF Thermochemical Tables, Dow Chemical Company, Midland, MI, 1964; (d) H. Borchers, H. Hausen, K. H. Hellwege, K. L. Schäfer and E. Schmidt (Series Eds.), Landolt-Börnstein Zahlenwerte und Funktionen, 6th. edn., Vol. IV, Part 4, edited by H. Hausen, Springer, Berlin, 1967, pp. 161—944.
4 K. K. Kelley, U.S. Bur. Mines, Bull., 584 (1960).
5 (a) E. W. Washburn (Ed.), International Critical Tables, McGraw-Hill, New York, 1929—1930: Vol. V, pp. 79—213; Vol. VII, pp. 224—312; (b) F. D. Rossini, The Thermochemistry of the Chemical Substances, Reinhold, New York, 1936; (c) W. M. Latimer, The Oxidation States of the Elements and Their Potentials in Aqueous Solution, 2nd. edn., Prentice-Hall, New York, 1952; (d) K. K. Kelley and E. G. King, U.S. Bur. Mines, Bull., 592 (1961); (e) F. D. Rossini, D. D. Wagman, W. H. Evans, S. Levine and I. Jaffe, Selected Values of Thermodynamic Properties, Natl. Bur. Stand. (U.S.), Circ., 500 (1961); D. D. Wagman, W. H. Evans, V. B. Parker, I. Halow, S. M. Bailey and R. H. Schumm, Natl. Bur. Stand. (U.S.), Tech. Note, 270-3 (1968); (f) R. A. Robie and D. R. Waldbaum, U.S. Geol. Surv., Bull., 1259 (1968).
6 G. W. Jones, B. Lewis, J. B. Friauf and G. St. J. Perrott, J. Am. Chem. Soc., 53 (1931) 869—883.

Chapter 7

PARTIAL MOLAR PROPERTIES

Associated with each extensive property of a simple, homogeneous system of c components is a set of c partial differential coefficients called partial molar properties (or quantities). They are intensive properties of the system and they express the variation of the corresponding extensive property with the composition of the solution at constant pressure and temperature.

Partial molar quantities are introduced into thermodynamics because of their immediate application to chemical and phase equilibrium in systems having one or more solution phases.

The relations developed in the present chapter hold for gaseous, liquid, or solid solutions unless otherwise stated.

7.1. Definition of a partial molar property. A general extensive property F of a simple, homogeneous system of c components $B_1, B_2, \ldots, B_c$ may be considered to be a function of the pressure p, the temperature T, and the numbers of moles $n_1, n_2, \ldots, n_c$ of the components. At constant p and T, F is a homogeneous function of the first degree in the mole numbers of the components (Section 1.6):

$$F(p, T, \lambda n_1, \lambda n_2, \ldots, \lambda n_c) = \lambda F(p, T, n_1, n_2, \ldots, n_c) \tag{7.1}$$

The total differential of F is

$$dF = \left(\frac{\partial F}{\partial p}\right)_{Tn} dp + \left(\frac{\partial F}{\partial T}\right)_{pn} dT + \sum_{i=1}^{c} \bar{f}_i dn_i \tag{7.2}$$

where

$$\bar{f}_i \equiv \left(\frac{\partial F}{\partial n_i}\right)_{pTn} \qquad (i = 1, 2, \ldots, c) \tag{7.3}$$

The subscript n to a partial derivative denotes that all of the mole numbers are to be held constant except the one occurring in the denominator. The quantity $\bar{f}_i$ is the partial molar property of the component B_i associated with the extensive property F. It is an intensive property and at constant p and T it is a homogeneous function of the zeroth degree in the mole numbers:

$$\bar{f}_i(p, T, \lambda n_1, \lambda n_2, \ldots, \lambda n_c) = \bar{f}_i(p, T, n_1, n_2, \ldots, n_c) \tag{7.4}$$

Thus $\bar{f}_i$ may be considered to be a function of $p, T, x_1, x_2, \ldots, x_{c-1}$ where x_i is the mole fraction of the component B_i in the solution.

The definition of eqn. (7.3) has operational significance: $\bar{f}_i$ is the change in the property F of a solution attending the addition of one mole of component B_i to a very large (infinite) quantity of the solution at a fixed pressure and temperature.

It is sometimes convenient to employ the mean molar property f of a solution, which is defined by the relation

$$f \equiv F/\Sigma_i n_i \tag{7.5}$$

It is an intensive property. For a one-component system it is identical with the corresponding partial molar property.

If n_i in eqns. (7.3) and (7.5) is measured in grams, $\bar{f}_i$ is the partial specific property of the component B_i related to F, and f is the specific property of the solution derived from F.

7.2. Relations among partial molar properties. Application of Euler's theorem on homogeneous functions to the property F gives the relation

$$F = \sum_i^c n_i \bar{f}_i \tag{7.6}$$

Division by $\Sigma_i n_i$ gives

$$f = \sum_i^c x_i \bar{f}_i \tag{7.7}$$

The total differential of F from eqn. (7.6) is

$$dF = \sum_i^c n_i d\bar{f}_i + \sum_i^c \bar{f}_i dn_i \tag{7.8}$$

Comparison of eqns. (7.2) and (7.8) yields

$$\sum_i^c n_i d\bar{f}_i = \left(\frac{\partial F}{\partial p}\right)_{Tn} dp + \left(\frac{\partial F}{\partial T}\right)_{pn} dT \tag{7.9}$$

Division by $\Sigma_i n_i$ gives

$$\sum_i^c x_i d\bar{f}_i = \left(\frac{\partial f}{\partial p}\right)_{Tn} dp + \left(\frac{\partial f}{\partial T}\right)_{pn} dT \tag{7.10}$$

When pressure and temperature are held constant, we find:

$$\sum_i^c n_i d\bar{f}_i = 0 \qquad (p, T \text{ constant}) \tag{7.11}$$

$$\sum_i^c x_i d\bar{f}_i = 0 \qquad (p, T \text{ constant}) \tag{7.12}$$

Division of eqn. (7.12) by dx_k at constant p, T, and $x_1, x_2, \ldots, x_{k-1}, x_{k+1}, \ldots x_{c-1}$ yields

$$\sum_i^c x_i \left(\frac{\partial \bar{f}_i}{\partial x_k}\right)_{pTx} = 0 \qquad (k = 1, 2, \ldots, c) \qquad (7.13)$$

The subscript x to the partial derivative denotes that all of the $c-1$ independent mole fractions are to be held constant except the one occurring in the denominator.

7.3. Effect of pressure and temperature on partial molar properties.
The variations of a partial molar property $\bar{f}_i$ with pressure and temperature are given by the following relations:

$$\left(\frac{\partial \bar{f}_i}{\partial p}\right)_{Tn} = \left[\frac{\partial}{\partial p}\left(\frac{\partial F}{\partial n_i}\right)_{pTn}\right]_{Tn} = \left[\frac{\partial}{\partial n_i}\left(\frac{\partial F}{\partial p}\right)_{Tn}\right]_{pTn} \qquad (7.14)$$

$$\left(\frac{\partial \bar{f}_i}{\partial T}\right)_{pn} = \left[\frac{\partial}{\partial T}\left(\frac{\partial F}{\partial n_i}\right)_{pTn}\right]_{pn} = \left[\frac{\partial}{\partial n_i}\left(\frac{\partial F}{\partial T}\right)_{pn}\right]_{pTn} \qquad (7.15)$$

In the remainder of this chapter we shall limit our discussion to binary solutions.

7.4. Relations for binary solutions.
The following relations derived from those of Section 7.2 hold for a system of two components B_1 and B_2:

$$F = n_1 \bar{f}_1 + n_2 \bar{f}_2 \qquad (7.16)$$

$$f = x_1 \bar{f}_1 + x_2 \bar{f}_2 \qquad (7.17)$$

$$n_1 d\bar{f}_1 + n_2 d\bar{f}_2 = 0 \qquad (p, T \text{ constant}) \qquad (7.18)$$

$$n_1 \left(\frac{\partial \bar{f}_1}{\partial n_2}\right)_{pTn_1} + n_2 \left(\frac{\partial \bar{f}_2}{\partial n_2}\right)_{pTn_1} = 0 \qquad (7.19)$$

$$x_1 d\bar{f}_1 + x_2 d\bar{f}_2 = 0 \qquad (p, T \text{ constant}) \qquad (7.20)$$

$$x_1 \left(\frac{\partial \bar{f}_1}{\partial x_2}\right)_{pT} + x_2 \left(\frac{\partial \bar{f}_2}{\partial x_2}\right)_{pT} = 0 \qquad (7.21)$$

Here F is considered to be a function of p, T, n_1, n_2; while $f, \bar{f}_1$ and $\bar{f}_2$ may be considered to be functions of these same variables, or of p, T, x_2.

Integration of eqn. (7.20) from a composition x_2' to a composition x_2'' gives

$$\bar{f}_1'' - \bar{f}_1' = -\int_{x_2'}^{x_2''} \frac{x_2}{1-x_2} d\bar{f}_2 \qquad (p, T \text{ constant}) \qquad (7.22)$$

Thus if the partial molar property $\bar{f}_2$ of component B_2 is known as a function of x_2 over a range of compositions, and the value of $\bar{f}_1$ is known for a single composition in this range, we can determine $\bar{f}_1$ as a function of x_2, either analytically or graphically.

When $\bar{f}_1$ and $\bar{f}_2$ are each plotted against the mole fraction x_2 at constant pressure and temperature, the slope of one curve at each composition is related to the slope of the other curve by eqn. (7.21). In particular we note the following: (1) the slopes of the two curves at each value of x_2 have opposite algebraic signs; (2) at $x_2 = 0.5$ the slopes of the two curves are equal in absolute magnitude but of opposite sign; (3) at a value of x_2 where one curve has a relative maximum the other curve has a relative minimum; (4) when $x_2 = 0$ either $(\partial \bar{f}_1/\partial x_2)_{pT}$ is zero (curve horizontal), or $(\partial \bar{f}_2/\partial x_2)_{pT}$ is positively or negatively infinite (curve vertical); (5) when $x_2 = 1$ either $(\partial \bar{f}_1/\partial x_2)_{pT}$ is positively or negatively infinite (curve vertical), or $(\partial \bar{f}_2/\partial x_2)_{pT}$ is zero (curve horizontal).

Although eqns. (7.14)–(7.22) may be applied in a formal manner to a property which contains an arbitrary constant, such as enthalpy, they must be modified for use in numerical calculations, since we can measure only *differences* in enthalpy.

Consider a change in state in which $n_1 + n_2$ moles of a binary solution $n_1B_1 \cdot n_2B_2$ of enthalpy H are produced from n_1 moles of pure B_1 of molar enthalpy h_1^0 and n_2 moles of pure B_2 of molar enthalpy h_2^0, both reactants and products having the same pressure and temperature. The corresponding change in enthalpy ΔH_F is called the enthalpy (or heat) of formation of the solution. The change in state and equation for ΔH_F are:

$$n_1B_1(a_1) + n_2B_2(a_2) = (n_1B_1 \cdot n_2B_2)(a_3) \quad (p, T) \tag{7.23}$$

$$\left.\begin{aligned}\Delta H_F &= H(\text{solution}) - (n_1 h_1^0 + n_2 h_2^0) \\ &= (n_1 \bar{h}_1 + n_2 \bar{h}_2) - (n_1 h_1^0 + n_2 h_2^0) \\ &= n_1(\bar{h}_1 - h_1^0) + n_2(\bar{h}_2 - h_2^0)\end{aligned}\right\} \tag{7.24}$$

Here a_1, a_2, and a_3 denote the aggregation states of B_1, B_2, and the solution, respectively. The quantity ΔH_F is the quantity of heat Q withdrawn from the surroundings when the change in state (7.23) is brought about by an isopiestic process in which only reversible expansion work is produced.

Comparison of eqn. (7.24) with eqn. (7.16) indicates the modifications to be made in eqns. (7.14)–(7.22) for quantities that contain an arbitrary constant.

These changes are:
1. For F write ΔH
2. For f write Δh, where $\Delta h = \Delta H/\Sigma_i n_i$
3. For $\bar{f}_i$ write $\bar{h}_i - h_i^0 \quad (i = 1, 2)$

The quantity $\bar{h}_i - h_i^0$ is the difference between the partial molar enthalpy of a component B_i in a given solution and the molar enthalpy of the pure substance at the pressure and temperature of the solution. This is the standard state of each component in eqns. (7.23) and (7.24). A second standard state is discussed in Section 7.10 and 7.14 below (see also Section 6.2B).

We shall find that there are four types of extensive properties F in thermodynamics:

1. Those which are defined absolutely for a system in a given state, and for which the corresponding partial molar properties are always finite, e.g., the volume V.

2. Those which are defined except for an additive arbitrary constant for a system in a given state, and for which the corresponding partial molar properties are always finite, e.g., the energy E and enthalpy H.

3. Those which are defined absolutely for a system in a given state, and for which the corresponding partial molar properties are not always finite, e.g., the entropy S which we shall define in Chapter 8.

4. Those which are defined except for an additive arbitrary constant for a system in a given state, and for which the corresponding partial molar properties are not always finite, e.g., the Helmholtz work function A and the Gibbs free energy G. These properties are usually combinations of types 2 and 3, for example $A \equiv E - TS$ and $G \equiv H - TS$.

7.5. Determination of partial molar properties. A number of methods are available for the experimental determination of partial molar properties. Among these [1] are the following, in which we use volume as an example of an extensive property which is defined absolutely for a system in a given state, and enthalpy as one which is defined except for an additive arbitrary constant.

7.5A. From the definition. Suppose we have determined the volume V of a series of solutions as a function of n_2 with p, T, n_1 held constant. Differentiation of this function with respect to n_2 and use of eqn. (7.16) give

$$\bar{v}_2 = \left(\frac{\partial V}{\partial n_2}\right)_{pTn_1} ; \bar{v}_1 = \frac{V - n_2 \bar{v}_2}{n_1} \tag{7.25}$$

The derivative in these relations can be determined analytically if the function $V(n_2)$ is given analytically, or it can be found graphically if it is given in tabular form (see Fig. 7.1).

Application of these methods may be facilitated by employing an auxiliary function known as the apparent molar volume ϕ_{v_2} of component B_2,

$$\phi_{v_2} \equiv \frac{V - n_1 v_1^0}{n_2} = \frac{n_1}{n_2}(\bar{v}_1 - v_1^0) + \bar{v}_2 \tag{7.26}$$

where v_1^0 is the molar volume of the pure component B_1. Differentiation of this equation with respect to n_2 at constant p, T, n_1, use of eqn. (7.19), and substitution of the result into eqn. (7.26) give:

$$\bar{v}_2 = \phi_{v_2} + n_2 \left(\frac{\partial \phi_{v_2}}{\partial n_2}\right)_{pTn_1} ; \bar{v}_1 = v_1^0 - \frac{n_2^2}{n_1}\left(\frac{\partial \phi_{v_2}}{\partial n_2}\right)_{pTn_1} \tag{7.27}$$

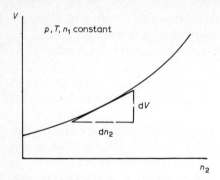

Fig. 7.1. Determination of the partial molar volume $\bar{v}_2$ graphically from the definition.

When investigating a property which contains an arbitrary constant, such as the enthalpy H, we measure values of ΔH_F [eqn. (7.24)] in a series of experiments with n_2 varied but p, T, n_1 held constant. Differentiation of eqn. (7.24) with respect to n_2 at constant p, T, n_1, use of eqn. (7.19), and substitution of the result into eqn. (7.24) give

$$\bar{h}_2 - h_2^0 = \left(\frac{\partial \Delta H_F}{\partial n_2}\right)_{pTn_1} ; \bar{h}_1 - h_1^0 = \frac{1}{n_1} [\Delta H_F - n_2 (\bar{h}_2 - h_2^0)] \qquad (7.28)$$

The derivative in these relations may be evaluated by analytical or graphical methods.

In order to define the apparent molar enthalpy ϕ_{h_2}, we write eqn. (7.26) in terms of enthalpies and subtract h_2^0 from both sides,

$$\phi_{h_2} - h_2^0 = \frac{n_1}{n_2} (\bar{h}_1 - h_1^0) + (\bar{h}_2 - h_2^0) = \frac{\Delta H_F}{n_2} \qquad (7.29)$$

[see eqn. (7.24)]. Differentiation of this equation with respect to n_2 at constant p, T, n_1, use of eqn. (7.19), and substitution of the result into eqn. (7.29) give

$$\left.\begin{array}{l} \bar{h}_2 - h_2^0 = (\phi_{h_2} - h_2^0) + n_2 \left[\dfrac{\partial(\phi_{h_2} - h_2^0)}{\partial n_2}\right]_{pTn_1} \\[2ex] \bar{h}_1 - h_1^0 = -\dfrac{n_2^2}{n_1} \left[\dfrac{\partial(\phi_{h_2} - h_2^0)}{\partial n_2}\right]_{pTn_1} \end{array}\right\} \qquad (7.30)$$

The derivative may be evaluated either analytically or graphically.

7.5B. Method of intercepts. Suppose we have determined the molar volume v of a series of solutions for different compositions at constant pressure p and temperature T. From eqn. (7.17) we have

$$v = x_1 \bar{v}_1 + x_2 \bar{v}_2 \qquad (7.31)$$

Differentiation of this equation with respect to x_2 at constant p and T, and use of eqn. (7.21) yield

$$\left(\frac{\partial v}{\partial x_2}\right)_{pT} = \bar{v}_2 - \bar{v}_1 \tag{7.32}$$

From the last two equations we find

$$\bar{v}_1 = v - x_2 \left(\frac{\partial v}{\partial x_2}\right)_{pT}; \bar{v}_2 = v + (1 - x_2) \left(\frac{\partial v}{\partial x_2}\right)_{pT} \tag{7.33}$$

In Fig. 7.2, the molar volume v of a solution is plotted against x_2 at constant p and T. The tangent to this curve at any point $x_2 = a$ cuts the axis $x_2 = 0$ at $\bar{v}_1$, and it cuts the axis $x_2 = 1$ at $\bar{v}_2$, both for the solution with $x_2 = a$.

When dealing with a property which contains an arbitrary constant such as the enthalpy, we proceed in the manner described in Section 7.5A, but compute the change in enthalpy Δh attending the formation of one mole of solution at constant p and T. In eqns. (7.23) and (7.24) n_1, n_2 are replaced by x_1, x_2, and ΔH_F by Δh. Differentiation of the equation for Δh with respect to x_2 at constant p and T, and use of eqn. (7.21) give:

$$\bar{h}_1 - h_1^0 = \Delta h - x_2 \left(\frac{\partial \Delta h}{\partial x_2}\right)_{pT}; \bar{h}_2 - h_2^0 = \Delta h + (1 - x_2) \left(\frac{\partial \Delta h}{\partial x_2}\right)_{pT} \tag{7.34}$$

In order to evaluate the partial derivative on the right-hand side of these equations, we employ a graphical method similar to that illustrated in Fig. 7.2, but with v replaced by Δh.

We shall now discuss enthalpy changes attending certain changes in state involving solutions. The nomenclature and methods used here can be applied to the treatment of any extensive property of a solution, for example, volume, energy, entropy, free energy, etc.

7.6. Integral enthalpy of solution. The change in enthalpy attending the solution of one mole of solute (component B_2) in n_1 moles of a solvent (component B_1), both reactants and products having the same pressure and temperature, is called the integral enthalpy (or heat) of solution ΔH_S of the

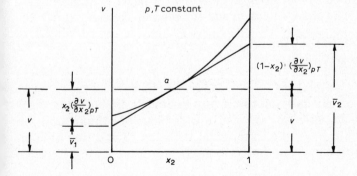

Fig. 7.2. Determination of partial molar volumes by the method of intercepts.

solute in the given solution. In a particular experiment let water be the solvent and NaCl be the solute. The change in state and the equation for ΔH_S are:

$$NaCl(s) + n_1 H_2O(l) = NaCl \cdot n_1 H_2O(l) \quad (p, T) \tag{7.35}$$

$$\Delta H_S = H(\text{solution}) - (n_1 h_1^0 + h_2^0) = n_1(\bar{h}_1 - h_1^0) + (\bar{h}_2 - h_2^0) \tag{7.36}$$

We note that $\Delta H_S(T, p, x_1)$ is equal to $\Delta H_F/n_2 (T, p, x_1)$; see eqn. (7.24).

The composition of the solution in eqn. (7.35) may be expressed in terms of the weight formal concentration of the solute.

7.7. Integral enthalpy of dilution. The change in enthalpy attending the addition of $(n_1'' - n_1')$ moles of solvent to a solution of 1 mole of solute in n_1' moles of solvent to give a solution of 1 mole of solute in n_1'' moles of solvent, both reactants and products having the same pressure and temperature, is called the integral enthalpy (or heat) of dilution ΔH_D of the solution from the initial to the final composition. The change in state and equation for ΔH_D are:

$$(n_1'' - n_1') H_2O(l) + NaCl \cdot n_1' H_2O(l) = NaCl \cdot n_1'' H_2O(l) \quad (p, T) \tag{7.37}$$

$$\Delta H_D = (n_1'' \bar{h}_1'' + \bar{h}_2'') - [(n_1'' - n_1') h_1^0 + (n_1' \bar{h}_1' + \bar{h}_2')] = \Delta H_S'' - \Delta H_S' \tag{7.38}$$

Equation (7.38) can be written in the form

$$\Delta H_S'' = \Delta H_S' + \Delta H_D \quad (p, T \text{ constant}) \tag{7.39}$$

Thus, if we are given the integral enthalpy of solution for a pair of substances for any one composition and the integral heats of dilution from this solution to a series of solutions of other compositions, we can find ΔH_S as a function of composition over the range investigated.

7.8. Differential enthalpy of dilution. The change in enthalpy attending the addition of one mole of the solvent to an infinite quantity of a solution of a given composition, both reactants and products having the same pressure and temperature, is called the differential enthalpy (or heat) of dilution ΔH_d of the given solution. The change in state and the relation for ΔH_d are:

$$H_2O(l) = H_2O (\text{in } NaCl \cdot n_1 H_2O, l) \quad (p, T) \tag{7.40}$$

$$\Delta H_d = \bar{h}_1 - h_1^0 = \left(\frac{\partial \Delta H_S}{\partial n_1}\right)_{pT} \tag{7.41}$$

When the terms solvent and solute are not pertinent because the components of a solution are miscible over such a wide range of composition, we shall call the quantity $\bar{h}_1 - h_1^0$ the differential enthalpy of addition of component B_1 to the solution and use the symbol ΔH_d for it.

7.9. Differential enthalpy of solution. The change in enthalpy attending the addition of one mole of the solute to an infinite quantity of a solution of a given composition, both reactants and products having the same pressure and temperature, is called the differential enthalpy (or heat) of solution ΔH_s of the solute in the given solution. The change in state and equation for ΔH_s are:

$$\text{NaCl(s)} = \text{NaCl (in NaCl} \cdot n_1 \text{H}_2\text{O, l)} \qquad (p, T) \qquad (7.42)$$

$$\Delta H_s = \bar{h}_2 - h_2^0 = \Delta H_S - n_1 \Delta H_d \qquad (7.43)$$

As n_1 increases without limit, we find from experiment that

$$\lim_{n_1 \to \infty} n_1 \Delta H_d = 0 \qquad (7.44)$$

whence, from eqn. (7.43), we find

$$\lim_{n_1 \to \infty} \Delta H_s = \lim_{n_1 \to \infty} \Delta H_S \equiv \Delta H_S^* \qquad (7.45)$$

Here ΔH_S^* is the integral enthalpy of solution of the solute in an infinite quantity of the pure solvent. Although ΔH_s and ΔH_S have approximately the same value in dilute solution, they may differ widely in concentrated solutions.

When the distinction between solvent and solute is not pertinent, we shall call the quantity $(\bar{h}_2 - h_2^0)$ the differential enthalpy of addition of component B_2 to the solution and use the symbol ΔH_s for it.

7.10. *Differential enthalpy of transfer of solvent and solute*. The change in enthalpy attending the transfer of one mole of the solvent (component B_1) from an infinite quantity of a solution of one composition to an infinite quantity of a solution of another composition (but the same components), both reactants and products having the same pressure and temperature, is called the differential enthalpy (or heat) of transfer ΔH_{tr_1} of the solvent from one solution to the other. The change in state and the equation for ΔH_{tr_1} are:

$$\text{H}_2\text{O (in NaCl} \cdot n_1' \text{H}_2\text{O, l)} = \text{H}_2\text{O (in NaCl} \cdot n_1'' \text{H}_2\text{O, l)} \qquad (p, T) \qquad (7.46)$$

$$\Delta H_{tr_1} = \bar{h}_1'' - \bar{h}_1' = \Delta H_d'' - \Delta H_d' \qquad (7.47)$$

The corresponding equations for the change in state and differential enthalpy (or heat) of transfer ΔH_{tr_2} of the solute (component 2) are:

$$\text{NaCl (in NaCl} \cdot n_1' \text{H}_2\text{O, l)} = \text{NaCl (in NaCl} \cdot n_1'' \text{H}_2\text{O, l)} \qquad (p, T) \qquad (7.48)$$

$$\Delta H_{tr_2} = (\bar{h}_2'' - \bar{h}_2') = \Delta H_s'' - \Delta H_s' \qquad (7.49)$$

If $n_1' = \infty$ in eqn. (7.46), ΔH_{tr_1} becomes $\bar{h}_1'' - h_1^0$. This is the differential enthalpy of addition of component 1 to the solution NaCl$\cdot n_1''$H$_2$O, written ΔH_d.

Correspondingly, if $n_1' = 0$ in eqn. (7.48), ΔH_{tr_2} becomes $\bar{h}_2'' - h_2^0$. This is the differential enthalpy of addition of component 2 to the solution NaCl$\cdot n_1''$H$_2$O, written ΔH_s.

So far, the only standard state we have used for a component B_i of a binary condensed solution is the pure substance of molar enthalpy h_i^0. In Section 7.14 we define a second standard state in which the enthalpy of the component B_i is equal to its partial molar enthalpy $\bar{h}_i^0$ in a solution infinitely dilute in this substance.

If $n_1' = \infty$ in eqn. (7.48), ΔH_{tr_2} becomes $\bar{h}_2'' - \bar{h}_2^0$. This is the difference between the partial molar enthalpy of component B_2 in the solution $\text{NaCl} \cdot n_1'' \text{H}_2\text{O}$ and its partial molar enthalpy at infinite dilution.

Finally, let us suppose that the component NaCl in eqn. (7.46) is replaced by a substance in which H_2O is soluble, for example, ethanol. Then if $n_1' = 0$ in this revised eqn. (7.46), ΔH_{tr_1} becomes $\bar{h}_1'' - \bar{h}_1^0$, which is the difference between the partial molar enthalpy of component B_1 in the solution $C_6H_5OH \cdot n_1'' H_2O$ and its partial molar enthalpy at infinite dilution.

7.11. Effect of pressure and temperature on partial molar enthalpies. On making the modifications listed in Section 7.4 in eqns. (7.14) and (7.15), we find the following relations for the effect of pressure and temperature on the difference $\bar{h}_i - h_i^0$ between the partial molar enthalpy of a component B_i in a given solution and the molar enthalpy of the pure substance:

$$\left[\frac{\partial(\bar{h}_i - h_i^0)}{\partial p}\right]_{Tn} = \left[\frac{\partial}{\partial n_i}\left(\frac{\partial \Delta H}{\partial p}\right)_{Tn}\right]_{pTn} = \left(\frac{\partial \Delta H_{tr_i}}{\partial p}\right)_{Tn} \quad (i = 1, 2) \qquad (7.50)$$

$$\left[\frac{\partial(\bar{h}_i - h_i^0)}{\partial T}\right]_{pn} = \left(\frac{\partial \Delta H_{tr_i}}{\partial T}\right)_{pn} = \bar{c}_{p_i} - c_{p_i}^0 \quad (i = 1, 2) \qquad (7.51)$$

We may replace $(\bar{h}_i - h_i^0)$ and $(\bar{c}_{p_i} - c_{p_i}^0)$ in these equations by $(\bar{h}_i - \bar{h}_i^0)$ and $(\bar{c}_{p_i} - \bar{c}_{p_i}^0)$, respectively; see Section 7.10, where the significance of ΔH_{tr_i} is also treated.

7.12. Calculation of the partial molar enthalpy of one component from that of the other. Since the enthalpy function contains an arbitrary constant, we must write eqn. (7.22) in the following form,

$$(\bar{h}_1'' - h_1^0) - (\bar{h}_1' - h_1^0) = -\int_{x_2'}^{x_2''} \frac{x_2}{1-x_2} d(\bar{h}_2 - h_2^0) \quad (p, T \text{ constant}) \qquad (7.52)$$

Here $\bar{h}_i - h_i^0$ is the difference between the partial molar enthalpy of a component B_i in a given solution and the molar enthalpy of the pure substance. The single and double prime signs denote solutions of different concentrations.

We may replace $\bar{h}_i - h_i^0$ in this equation by $\bar{h}_i - \bar{h}_i^0$, the difference between the partial molar enthalpy of B_i in the given solution and its partial molar enthalpy at infinite dilution (see Section 7.10).

7.13. The quantities $\bar{h}_1$, $\bar{h}_2$, and ϕ_{h_2}. From eqns. (7.29), (7.36), (7.41), and (7.43) we find that the partial molar enthalpy, $\bar{h}_1$, of the solvent of a condensed binary solution, the corresponding enthalpy, $\bar{h}_2$, of the solute, and the apparent molar enthalpy, ϕ_{h_2}, of the solute are given by the relations

$$\bar{h}_1 = \Delta H_d + h_1^0 = \left(\frac{\partial \Delta H_S}{\partial n_1}\right)_{pT} + h_1^0 \tag{7.53}$$

$$\bar{h}_2 = \Delta H_s + h_2^0 = \Delta H_S - n_1 \left(\frac{\partial \Delta H_S}{\partial n_1}\right)_{pT} + h_2^0 \tag{7.54}$$

$$\phi_{h_2} = H - n_1 h_1^0 = \Delta H_S + h_2^0 \tag{7.55}$$

Here H is the enthalpy of a solution composed of one mole of solute and n_1 moles of solvent, and h_1^0 and h_2^0 are the enthalpies of the pure solvent and solute at the pressure and temperature of the solution.

Now $\bar{h}_1 - h_1^0$, $\bar{h}_2 - h_2^0$, and $\phi_{h_2} - h_2^0$ represent, respectively, the quantities of heat ΔH_d, ΔH_s, and ΔH_S withdrawn from the surroundings when the changes in state (7.40), (7.42), and (7.35), respectively, are brought about by isopiestic processes in each of which only reversible expansion work is produced. All can be determined by analytical or graphical methods if we know ΔH_S as a function of x_2.

On the other hand, the values of $\bar{h}_1$, $\bar{h}_2$, and ϕ_{h_2} cannot be derived solely from experimental measurements. Each depends on an arbitrary constant, h_1^0 or h_2^0, to which we may assign any value consistent with the problem under consideration.

The variations of $\bar{h}_1 - h_1^0$ and $\bar{h}_2 - h_2^0$ with pressure and temperature are given by eqns. (7.50) and (7.51), respectively; and the pressure and temperature coefficients of $\phi_{h_2} - h_2^0$ are the same as the corresponding quantities for ΔH_S. As mentioned in Section 6.6, the effects of pressure on these quantities may be neglected, except at high pressure.

7.14. Standard states and standard molar enthalpies. The standard states chosen for the components of a solution and the numerical values assigned the standard molar enthalpies depend on the problem under consideration.

The following two standard states have been used for a component B_i of a condensed binary solution at a pressure p and a temperature T (Section 6.2B).

1. The standard state of a component B_i is the pure substance B_i in the state (a_i, p, T) where it has the enthalpy h_i^0. Here a_i denotes the aggregation state of B_i.

This standard state may be used for each component of a solution, or it may be used for only one of the components.

2. The standard state of a component B_i is the hypothetical dissolved state (DS, p, T, $x_i = 1$) where its enthalpy, written $\bar{h}_i^0$, is equal to its partial molar enthalpy in a solution infinitely dilute in B_i. We indicate the latter state for an aqueous solution of B_i by writing B_i (aq).

When the components of a solution are miscible in all proportions, this standard state may be used for each substance, or it may be used for only one of the substances. When one component has limited solubility, this standard state may be used for the solute, but it cannot be used for the solvent.

In the symmetrical system of standard states for condensed binary solutions either the first or the second of the above standard states may be used for both components. In the unsymmetrical system, the first standard state is used for one component and the second is used for the other.

In order to find the relation between h_2^0 and $\bar{h}_2^0$ for the component B_2, we determine the change in enthalpy attending the solution of 1 mole of it in an infinite quantity of pure B_1, both the reactants and products having the same pressure and temperature. Let the change in state be

NaCl (s) = NaCl (aq) (7.56)

Then

$$\bar{h}_2^0 = h_2^0 + \Delta H_s^* \tag{7.57}$$

[see eqn. (7.45)]. If the component B_1 is soluble in component B_2, an analogous expression, but with the subscript 2 replaced by 1, holds for $\bar{h}_1^0$.

The assignment of numerical values to the standard molar enthalpies h_i^0 and $\bar{h}_i^0$ of the component B_i may be made on the basis of the following considerations.

1. When B_i does not take part in a chemical reaction during a change in state in which the reactants and products have the same pressure and temperature, its standard molar enthalpy either cancels out of the equation for the corresponding change in enthalpy, or it enters only as the difference $\bar{h}_i - h_i^0$ or $\bar{h}_i - \bar{h}_i^0$. These quantities are the differential enthalpies of transfer of 1 mole of B_i from its standard state — pure B_i, or B_i in the hypothetical dissolved state — to the given solution. They are the quantities ΔH_d and ΔH_s, or ΔH_{tr_1} and ΔH_{tr_2}; see Sections 7.8, 7.9, and 7.10. Their values can be determined from calorimetric measurements.

If the standard state of each component is taken to be the pure substance, eqns. (7.53) to (7.55) may be written in the following forms:

$$\bar{h}_1 - h_1^0 = \Delta H_d \tag{7.58a}$$

$$\bar{h}_2 - h_2^0 = \Delta H_s \tag{7.58b}$$

$$\phi_{h_2} - h_2^0 = \Delta H_s \tag{7.58c}$$

If the standard state of each component is the hypothetical dilute solution state the above equations become:

$$\bar{h}_1 - \bar{h}_1^0 = \Delta H_d + h_1^0 - \bar{h}_1^0 \tag{7.59a}$$

$$\bar{h}_2 - \bar{h}_2^0 = \Delta H_s - \Delta H_s^* \tag{7.59b}$$

$$\phi_{h_2} - \bar{h}_2^0 = \Delta H_s - \Delta H_s^* \tag{7.59c}$$

Lewis and Randall [1] write
$$\bar{L}_1 = \bar{h}_1 - \bar{h}_1^0, \quad \bar{L}_2 = \bar{h}_2 - \bar{h}_2^0 \qquad (7.60)$$
and call $\bar{L}_i$ the relative molar enthalpy of B_i. This saves writing the more cumbersome expression $\bar{h}_i - \bar{h}_i^0$ at the expense of the introduction of a new symbol.

We may use eqns. (7.58) or (7.59) for both components of a solution; or we may employ eqns. (7.58) for one component and (7.59) for the other.

2. When a substance B_i takes part in a chemical reaction during a change in state, its standard enthalpy h_i^0 or $\bar{h}_i^0$ at pressure p and temperature T must be referred to its standard enthalpy of formation $h_i^0(B_i, a_i)$ at T, where a_i is its aggregation state. If we neglect the effect of pressure on enthalpy, we find that

$$h_i^0 = h_T^0(B_i, a_i) \qquad (i = 1, 2) \qquad (7.61)$$

The corresponding equation for $\bar{h}_i^0$ has $\Delta H_{S_i}^*$ subtracted from the right-hand side of the above relation.

If we use eqn. (7.61) for both components of a solution made of a component B_1 in the aggregation state a_1 and a component B_2 in the aggregation state a_2 at a temperature T, eqns. (7.58) become:

$$\bar{h}_1 = h_T^0(B_1, a_1) + \Delta H_d \qquad (7.62a)$$
$$\bar{h}_2 = \bar{h}_T^0(B_2, a_2) + \Delta H_s \qquad (7.62b)$$
$$\phi_{h_2} = h_T^0(B_2, a_2) + \Delta H_s \qquad (7.62c)$$

Here we have neglected the effect of pressure on enthalpies of formation.

7.15. Enthalpies of reaction in solution. We shall consider a chemical change in state involving finite amounts of solution, and one involving essentially an infinite quantity of solution. Equations (7.16), (7.36), and (7.55) are three expressions for the enthalpy H of a solution composed of n_1 formula weights of solvent B_1 and one formula weight of solute B_2:

$$H = n_1 \bar{h}_1 + \bar{h}_2 \qquad (7.63)$$
$$H = n_1 h_1^0 + h_2^0 + \Delta H_S \qquad (7.64)$$
$$H = \phi_{h_2} + n_1 h_1^0 \qquad (7.65)$$

All three of these equations reduce to the following expression if we employ eqns. (7.62) and (7.43):

$$H = n_1 h_T^0(B_1, a_1) + h_T^0(B_2, a_2) + \Delta H_S \qquad (7.66)$$

In the following change in state, finite quantities of solution are involved

$$\text{NaOH} \cdot 50\,H_2O\,(l') + \text{HCl} \cdot 50\,H_2O\,(l'') = \text{NaCl} \cdot 101\,H_2O\,(l''') \qquad (p, T) \qquad (7.67)$$

Substitution from eqn. (7.66) gives

$$\Delta H = \Delta H_s''' - (\Delta H_s' + \Delta H_s'') + \Delta h_T^0 \qquad (7.68\text{a})$$

where

$$\Delta h_T^0 = h_T^0(\text{H}_2\text{O},\text{l}) + h_T^0(\text{NaCl},\text{s}) - [h_T^0(\text{NaOH},\text{s}) + h_T^0(\text{HCl},\text{g})] \qquad (7.68\text{b})$$

Substitution from eqn. (7.65) gives

$$\Delta H = \phi_{h_2}''' - (\phi_{h_2}' + \phi_{h_2}'') + h_T^0(\text{H}_2\text{O},\text{l}) \qquad (7.69)$$

In the above relations $\Delta H_s'$, $\Delta H_s''$, and $\Delta H_s'''$ are, respectively, the integral enthalpies of solution of NaOH(s), HCl(g), and NaCl(s) in the solutions identified by the prime signs, and ϕ_{h_2}', ϕ_{h_2}'', and ϕ_{h_2}''' are the corresponding apparent molar enthalpies, while h_T^0 is a standard enthalpy of formation at T.

In many applications of thermodynamics a substance B_i appearing in a change in state is considered to be produced in, or withdrawn from, an infinite amount of solution. In this case the partial molar enthalpy $\bar{h}_i$ of the substance, given by eqns. (7.62), appears in the corresponding equation for the enthalpy change ΔH. For example, during the discharge of the lead storage cell, the following change in state takes place when two faradays of electricity pass through a cell which has a large (infinite) amount of electrolyte,

$$\text{Pb(s)} + \text{PbO}_2(\text{s}) + 2\text{H}_2\text{SO}_4(x_2) = 2\text{PbSO}_4(\text{s}) + 2\text{H}_2\text{O}(x_1) \quad (p,T) \qquad (7.70)$$

$$\Delta H = 2h(\text{PbSO}_4,\text{s}) + 2\bar{h}_1 - [h(\text{PbO}_2,\text{s}) + 2\bar{h}_2] \qquad (7.71)$$

No term for the enthalpy of solid lead occurs in this relation because the standard enthalpy of formation of an element in its stable aggregation state is, by definition, zero at every temperature, and the effect of pressure on this enthalpy may be neglected. The quantities $\bar{h}_1$ and $\bar{h}_2$ are the partial molar enthalpies of H_2O and H_2SO_4 in the solution. Substitution from eqns. (7.62) into eqn. (7.70) gives

$$\Delta H = 2\Delta H_d - 2\Delta H_s + \Delta h_T^0 \qquad (7.72\text{a})$$

where

$$\Delta h_T^0 = 2h_T^0(\text{PbSO}_4,\text{s}) + 2h_T^0(\text{H}_2\text{O},\text{l}) - [h_T^0(\text{PbO}_2,\text{s}) + 2h_T^0(\text{H}_2\text{SO}_4,\text{l})] \qquad (7.72\text{b})$$

Here ΔH_d and ΔH_s are, respectively, differential heats of dilution and solution of the sulfuric acid solution.

REFERENCES

1 See, for example, G. N. Lewis and M. Randall, Thermodynamics, McGraw-Hill, New York, 1923, pp. 33—46; and also the revised version of this book, K. S. Pitzer and L. Brewer, 1961, pp. 205—213.

Chapter 8

THE SECOND LAW OF THERMODYNAMICS

In Chapter 4 we found that the first law of thermodynamics formulated by Clausius (1850) and Thomson (1851), and based largely on the work of Joule, led to the definition of a state property of a system called energy. In the present chapter we shall show how Carnot's principle (1824) was used by Thomson (1849) as a basis of a scale of temperature which is independent of the properties of any substance. This principle became the second law of thermodynamics in the hands of Clausius (1850) and Thomson (1851), and led to the definition of a new state property of a system which Clausius called entropy.

8.1. Historical introduction. In the early years of the nineteenth century scientists and engineers were seeking the answers to several questions regarding the efficiency of the steam engine: Is there an upper limit to the amount of work that can be obtained from a fixed quantity of heat? If such a limit exists, on what variables does it depend? Is there a more efficient working substance than water, for example, air? Carnot [1] proposed to "examine the principle of the production of motion from heat", and he gave the name "heat engine" to any mechanism which accomplished this result.

Carnot introduced the concepts of the cycle and the reversible process; and he invented the reversible cycle known by his name (see Section 8.8). From a consideration of this cycle he concluded that "The motive power of heat is independent of the agents employed to develop it; its quantity is determined solely by the temperatures of the bodies between which, in the final result, the transfer of caloric occurs". This is, in effect, the first statement of the second law of thermodynamics; until 1850 it was known as Carnot's principle. Carnot attempted to prove this result from the principles of the conservation of caloric (heat), and the impossibility of perpetual motion of the first kind. Although Carnot's proof is not valid, his principle is one of the most basic tenets of science.

Clapeyron [2] derived from Carnot's principle correct analytical differential expressions for the latent heat of expansion of a homogeneous system and for the latent heat of vaporization of a liquid. The latter is known as the Clapeyron equation.

Thomson [3] (1848), realizing that Carnot's principle enables us to define scales of temperature which are independent of the properties of any substance, proposed a scale (see Section 8.12), known as Kelvin's first scale.

As late as 1849, Thomson could not reconcile Joule's principle with Carnot's theory, since he believed that the latter was based on the principle of the conservation of heat. He decided to ignore Joule's work, and proceeded to correlate Regnault's [4] newly published measurements on the saturation properties of water by means of Clapeyron's equation, which he derived anew from Carnot's principle.

Clausius [5] (1850) gave the solution to the puzzle of reconciling the work of Joule with that of Carnot (Section 4.1). To illustrate his method, Clausius applied first Joule's principle and then Carnot's principle to each of two heat engines operating on a Carnot cycle between two fixed temperatures, one engine employing a permanent gas and the other a condensible vapor as the working fluid. He showed that the resulting equations did not contradict each other. He realized that Joule's principle and Carnot's principle are two separate and independent laws. He gave two statements of the latter, now known as the second law of thermodynamics:

"Heat cannot, of itself, pass from a colder body to a hotter body."

"A passage of heat from a colder body to a hotter body cannot take place without compensation."

By "compensation" Clausius meant that there must take place a simultaneous passage of heat from a hotter to a colder body, or else some change which cannot be reversed except by such an opposite passage of heat.

Clausius introduced a new state property of a system, which he called entropy; and he proved that the entropy of an isolated system is a nondecreasing function of time. Thus he wrote (1865):

"Die Energie der Welt ist konstant."

"Die Entropie der Welt strebt einem Maximum zu."

Thomson (1851) finally reconciled the work of Joule with that of Carnot, and clearly stated the first law (Section 4.1) and the second law as propositions not subject to proof. He credited the second law to Carnot and Clausius; and he gave two statements of this law:

"It is impossible, by means of inanimate material agency, to derive mechanical effect from any portion of matter by cooling it below the temperature of the coldest of the surrounding objects."

"It is impossible for a self-acting machine, unaided by any external agency, to convey heat from one body to another at a higher temperature."

Planck [6] (1897) stated the second law as follows:

"It is impossible to construct an engine which will work in a complete cycle and produce no other effect except the raising of a weight and the cooling of a heat reservoir."

Carathéodory [7], employing the axiomatic method, founded his treatment of the first and second laws of thermodynamics on the concepts of work and an adiabatic wall, without introducing the concept of heat. His discussion of the second law is based on Carathéodory's theorem: "If a Pfaffian equation

$$dx_0 + X_1 dx_1 + X_2 dx_2 + \cdots + X_n dX_n = 0$$

is given, where the X_i's are finite, continuous, differentiable functions of the x_i's, and if in every neighborhood of an arbitrary point P in the space of the x_i's there are points that cannot be reached along curves which satisfy this equation, then the Pfaffian expression must necessarily have a multiplier that makes it an exact differential."

Carathéodory's statement of the second law is:

"In every neighborhood of an arbitrary specified initial state, there are states which cannot be reached as closely as we please by adiabatic processes."

It has been shown [8] that Carathéodory's principle can be derived from a physical statement of the second law, such as that of Kelvin or Clausius.

We shall base our discussion of the second law of thermodynamics on a physical statement of the law. This is then put into mathematical form. The Carnot cycle is described, and a number of corollaries of the second law are derived. A thermodynamic temperature scale which is independent of the properties of any substance is presented, and a new state variable, entropy, is defined and its properties investigated. Finally, the equation combining the two laws of thermodynamics is derived.

8.2. Heat engine. Any closed system operating in a cycle may be called a heat engine.

8.3. Heat reservoir. A heat reservoir is a system in a state of thermodynamic equilibrium, in which the intensive properties of each phase remain essentially constant when any required quantity of heat passes through its boundary to or from an external system, such as a heat engine. The external system is said to be connected to the heat reservoir. No work is transmitted by way of this connection to or from the given external system.

A heat reservoir may, for example, consist of a large mass of liquid water in a state of thermodynamic equilibrium, or liquid water and water vapor in equilibrium at a fixed pressure, or ice, liquid water, and water vapor at equilibrium in a fixed volume.

8.4. Second law of thermodynamics. The first law of thermodynamics states that, when one form of energy is converted into another, there is exact equivalence between the amount of energy that appears and the amount that disappears. No other condition is placed on the interconvertibility of the various forms of energy. We have learned from experience that, while the various forms of work can be completely converted into one another and into heat, the conversion of heat into work is subject to certain restrictions. These are embodied in the second law which may be stated in the following form:

It is impossible for a system connected to a single heat reservoir to produce net positive work in its surroundings in a cyclical process.

The second law is the expression of the results of many attempts to construct systems which violate it. Its correctness is confirmed by the agreement of the deductions made from it with well-established principles and facts. Among these is the principle of the impossibility of perpetual motion of the second kind which may be stated as follows: It is impossible for a system operating in a cycle and kept in surroundings at one temperature to produce net positive work in its surroundings perpetually.

A heat engine which violates the second law need not violate the first law. While operating in a cycle, it could withdraw a positive quantity of heat from its environment (the sea, the earth, or the atmosphere) and produce an exactly equivalent amount of work in its surroundings, as required by the first law. Although this device would not strictly be a perpetual motion machine, it would for all practical purposes be just as advantageous as one which violates the first law, because a free and almost limitless source of heat is always available. Moreover, virtually all of the work produced would in the end be dissipated in the environment by friction, fluid viscosity, impact, and other such effects. All seriously proposed perpetual motion machines are attempts to break the second law.

8.5. The second-law equation. Let a system which is connected to a single heat reservoir undergo a cyclical process. According to the second law, the net amount of work produced in the surroundings in this cyclical process is either zero or negative,

$$\oint dW \leqslant 0 \qquad \text{(a single heat reservoir)} \tag{8.1}$$

On substitution from this expression into the first-law equation (4.4) we find that the quantity of heat withdrawn from the surroundings (the heat reservoir) in this cyclical process is also either zero or negative,

$$\oint dQ \leqslant 0 \qquad \text{(a single heat reservoir)} \tag{8.2}$$

Equation (8.1) is the second-law equation; eqn. (8.2) applies in virtue of this relation and the first law.

8.6. The second-law equation for reversible cycles. If a system which is connected to a single heat reservoir undergoes a reversible cyclical process, the sign of equality holds in eqn. (8.1). Conversely, if the sign of equality holds in this equation for a cycle executed by a system which is connected to a single heat reservoir, the process is reversible.

We may prove these theorems in the following manner. Let a system connected to a single heat reservoir undergo a reversible cycle, and then execute the reverse of this cycle. According to eqns. (8.1) and (8.2), the work produced in the surroundings and the heat withdrawn therefrom are each zero or negative for the direct, and also for the reverse, cycle. When the cycle is reversed, each element of work dW and each element of heat dQ of the direct process change algebraic sign without change in absolute magnitude (Section 1.11). These last two statements are consistent only if the sign of equality is used in eqn. (8.1).

Conversely, if the sign of equality holds in eqn. (8.1) for a given cycle executed by a system connected to a single heat reservoir, it will also apply in a cycle in which each element of work đW and each element of heat đQ of the given cycle have been changed in algebraic sign, but not in absolute magnitude. This implies that the process is reversible.

Therefore, for every reversible cycle executed by a system connected to a single heat reservoir, the second law requires that

$\oint đW_R = 0$ (reversible cycle) (8.3)

$\oint đQ_R = 0$ (reversible cycle) (8.4)

Here the subscript R denotes that the process is reversible.

8.7. The second-law equation for irreversible cycles. If a system which is connected to a single heat reservoir undergoes an irreversible cyclical process, the sign of inequality holds in eqn. (8.1). Conversely, if the sign of inequality holds in this equation for a cycle executed by a system which is connected to a single heat reservoir, the process is irreversible.

We may prove these theorems as follows. Equations (8.1) and (8.2) hold for any cycle executed by a system connected to a single heat reservoir. In Section 8.6 we found that when a cycle is reversible the signs of equality hold in the above equations. From the discussion of Section 1.11, it is evident that a cycle is either reversible or irreversible. Hence, when the above system undergoes an irreversible cycle, the signs of inequality must be used in eqns. (8.1) and (8.2).

Conversely, since the sign of equality in eqn. (8.1) implies that the cycle undergone by a system connected to a single heat reservoir is reversible (Section 8.6), the sign of inequality implies that a process executed by this system is irreversible. Hence, for every irreversible cycle executed by a system connected to a single heat reservoir, the second law requires that

$\oint đW < 0$ (irreversible cycle) (8.5)

$\oint đQ < 0$ (irreversible cycle) (8.6)

Let W and Q be the work and heat effects attending a cyclical process undergone by a system connected to a single heat reservoir. We have found that: (1) if W and Q are both zero, the process is reversible, and there has been no conversion of heat into work, or of work into heat; (2) if W and Q are both negative, the process is irreversible and work has been converted into an equivalent quantity of heat; (3) the cycle for which W and Q are both positive, which corresponds to the conversion of heat into an equivalent amount of work, is impossible by the second law.

Thus the second law makes a clear distinction between a reversible and an irreversible process, and between work and heat. The first law imposes the condition of equivalence, eqn. (4.4), on the interconversion of heat and work, but it makes no distinction between them; it is completely silent in regard to reversibility.

8.8. The Carnot cycle. In the last two sections, the second law of thermodynamics was applied to systems which are connected to a single heat reservoir. In order to remove this limitation, we shall employ the Carnot cycle [1], in which two heat reservoirs at different temperatures are used.

In the Carnot cycle, a system composed of a fixed mass of a fluid, initially in a volume V_1 and at a temperature t_1 (measured on any temperature scale), undergoes four *reversible* processes in the following order:

1. A reversible, adiabatic expansion from a state 1 to a state 2, during which its temperature decreases from t_1 to t_2:

$$\text{Fluid }(V_1, t_1) = \text{Fluid }(V_2, t_2) \qquad Q = 0 \qquad (8.7)$$

2. A reversible, isothermal compression at t_2 from the state 2 to a state 3, during which a quantity of heat Q_2 is withdrawn by the system from a heat reservoir at t_2:

$$\text{Fluid }(V_2, t_2) = \text{Fluid }(V_3, t_2) \qquad Q = Q_2 \qquad (8.8)$$

3. A reversible, adiabatic compression from the state 3 to a state 4, whereby its temperature increases to t_1:

$$\text{Fluid }(V_3, t_2) = \text{Fluid }(V_4, t_1) \qquad Q = 0 \qquad (8.9)$$

4. A reversible, isothermal expansion at t_1 from the state 4 to the initial state 1, during which the quantity of heat Q_1 is withdrawn from a heat reservoir at t_1:

$$\text{Fluid }(V_4, t_1) = \text{Fluid }(V_1, t_1) \qquad Q = Q_1 \qquad (8.10)$$

A Carnot engine is a *reversible* heat engine that operates on a Carnot cycle. Any system which undergoes a Carnot cycle may be called a Carnot engine.

Figure 8.1 is the graph of the Carnot cycle in the pV-plane for a closed system composed of a gas. The net work produced in the surroundings during

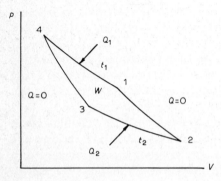

Fig. 8.1. The Carnot cycle on a pV-diagram.

the cycle is represented by the area enclosed by the curve. Figure 8.2 is a schematic representation of the cycle. Here R is a Carnot engine which, during one cycle, produces the net work W_R in the surroundings and withdraws the quantities of heat Q_1 and Q_2 from heat reservoirs at t_1 and t_2, respectively.

We recall (Sections 3.2 and 3.13) that work and heat are algebraic quantities. Thus, if a given Carnot engine should, per cycle, produce 10 J of work in the surroundings, withdraw 25 J of heat from the reservoir at t_1, and reject 15 J of heat to the reservoir at t_2, we would write: $W_R = 10$ J, $Q_1 = 25$ J, $Q_2 = -15$ J. If this Carnot engine is reversed so that it is acting as a refrigerator, Figs. 8.1 and 8.2 need not be altered, but now $W_R = -10$ J, $Q_1 = -25$ J, and $Q_2 = 15$ J.

The Carnot cycle is not limited to processes involving the work of expansion and compression of a fluid. It may, for example, be composed of processes of charging and discharging a capacitor or a reversible galvanic cell, in which case volume is replaced by charge; or it may be composed of processes of magnetization and demagnetization of a paramagnetic substance, in which case volume is replaced by magnetization.

We have considered the Carnot cycle to be composed of two reversible, isothermal processes and two reversible, adiabatic processes, such as those shown in Fig. 8.1. The paths of two reversible, isothermal processes cannot pass through the same state of a system since a given state cannot have two different uniform temperatures*. The corresponding intersection of two reversible, adiabatic paths is forbidden by the second law. For, if such paths can intersect, we could construct a heat engine which operates on the cycle $1 \rightarrow 2 \rightarrow 3 \rightarrow 1$ shown in Fig. 8.3, and, although connected to a single heat reservoir (at t), it produces the positive work W in its surroundings in each cycle. This is a second-law violation.

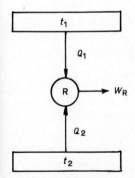

Fig. 8.2. Schematic diagram of a Carnot engine.

---

*Isothermal curves drawn on the pVT-surface of a substance do not intersect, but the projection of these curves on the pV-plane may cross for a liquid, such as water which has a temperature of maximum density.

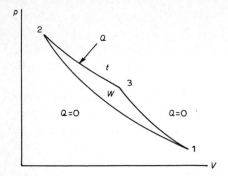

Fig. 8.3. Proof that two reversible adiabatics cannot intersect.

8.9. Efficiency of the Carnot cycle. The efficiency ϵ of a Carnot engine which produces the work W in its surroundings and withdraws the quantity of heat Q_1 from a heat reservoir at t_1 and Q_2 from one at t_2 ($t_1 > t_2$) is defined by the relation

$$\epsilon \equiv \frac{W}{Q_1} = \frac{Q_1 + Q_2}{Q_1} = 1 + \frac{Q_2}{Q_1} \tag{8.11}$$

When a Carnot engine produces net negative work in its surroundings during a cycle, we may use another measure of its effectiveness, called its coefficient of performance, c, defined by the relation

$$c = \frac{Q_2}{-W} = -\frac{Q_2}{Q_1 + Q_2} = \frac{1-\epsilon}{\epsilon} \tag{8.12}$$

Here the Carnot engine is acting as a refrigerating machine.

8.10. Some corollaries of the second law. In the following discussions, we shall make use of Carnot heat engines (Section 8.8) which are always reversible, and other heat engines which may or may not be reversible. All engines operate between two heat reservoirs. The results below follow from the first and second laws.

1. In a Carnot cycle, Q_1 and Q_2 cannot both be positive or both be negative, nor can Q_2 be zero.

Let us suppose that Q_1 and Q_2 are both positive for the Carnot engine R of Fig. 8.4. Then, by the first law, W is positive. Let us connect the heat reservoir at t_1 to that at t_2 ($t_1 > t_2$) by a thermal conductor through which the quantity of heat Q_2 flows from the reservoir at t_1 to the one at t_2 during each cycle of the engine. Thus the reservoir at t_2 undergoes no change in state during the cycle. The Carnot engine R and the heat reservoir at t_2 constitute a composite system (enclosed in broken lines in Fig. 8.4) which is connected to a single heat reservoir (at t_1) and produces the net positive work W in the surroundings in each cycle. This violates the second law. Hence Q_1 and Q_2 cannot both be positive in a Carnot cycle.

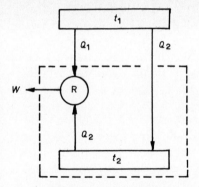

Fig. 8.4. Proof that Q_1 and Q_2 cannot both be positive.

Let us now suppose that Q_1 and Q_2 in Fig. 8.4 are both negative, and therefore W is negative. We reverse the Carnot engine. Then Q_1, Q_2, and W are all positive quantities. We have shown this to be a second-law violation. Hence Q_1 and Q_2 cannot both be negative for a Carnot cycle.

Finally let Q_2 in Fig. 8.4 be zero. Under these conditions, we may remove the thermal conductor leading from the heat reservoir at t_1 to that at t_2. Then the heat reservoir at t_2 undergoes no change in state during the cycle, and the composite system consisting of the Carnot engine R and the heat reservoir at t_2 violates the second law, since it is connected to a single heat reservoir and produces the net positive work W in the surroundings in each cycle. Hence Q_2 cannot be zero in a Carnot cycle.

Since Q_1 and Q_2 have different algebraic signs and Q_2 cannot be zero for a Carnot cycle, we must find that the ratio Q_2/Q_1 is negative for every such cycle:

$$Q_2/Q_1 < 0 \qquad \text{(Carnot cycle)} \qquad (8.13)$$

Hence, by eqn. (8.11), the efficiency ϵ of a Carnot engine is less than unity,

$$\epsilon < 1 \qquad (8.14)$$

When W and Q_1 are both positive and hence Q_2 is negative for a Carnot cycle, net positive work is produced in the surroundings of the Carnot engine and the quantity of heat $-Q_2$ is transferred from the heat reservoir at t_1 to that at t_2 $(t_1 > t_2)$.

When W and Q_1 are both negative and hence Q_2 is positive for a Carnot cycle, net negative work is produced in the surroundings and the quantity of heat Q_2 is transferred from the heat reservoir at t_2 to that at t_1. The Carnot engine is acting as a refrigerator.

We can prove by methods similar to those used above that W and Q_2 cannot have the same algebraic sign for a Carnot cycle.

2. No heat engine has a greater efficiency than a reversible heat engine when both operate between the same two heat reservoirs.

Let any heat engine A and a Carnot engine R (Fig. 8.5a) operate between a heat reservoir at t_1 and one at t_2 ($t_1 > t_2$) in such a direction that W_A and Q_{A_1} and also W_R and Q_{R_1} are all positive quantities. Then Q_{A_2} and Q_{R_2} are negative. Let R be adjusted so that for one cycle of each heat engine

$$Q_{A_1} = Q_{R_1} \tag{8.15}$$

Now reverse the Carnot engine R. Each of the quantities W_R, Q_{R_1} and Q_{R_2} change sign without change in absolute magnitude (Fig. 8.5b). Hence the heat reservoir at t_1 does not undergo any change in state when A and R have each completed one cycle. The composite system consisting of A, R, and the heat reservoir at t_1 is connected to a single heat reservoir (at t_2) and undergoes a cyclical process. By the second-law equation (8.1) the net work $W_A - W_R$ of this cycle (Fig. 8.5b) must be zero or negative. Hence, for the heat engines A and R (Fig. 8.5a), we have

$$W_A - W_R \leqslant 0 \tag{8.16}$$

Therefore when A and R have each completed one cycle we have

$$W_A \leqslant W_R \tag{8.17}$$

Division by eqn. (8.15) and substitution from eqn. (8.11) give

$$\epsilon_A \leqslant \epsilon_R \tag{8.18}$$

In this proof we have supposed that the Carnot engine was adjusted so that the ratio Q_{A_1}/Q_{R_1} is unity. The same result would be found if this ratio were equal to the ratio n_R/n_A of any two integers. Then we can apply the method used above to one cycle of the composite system consisting of n_A cycles of A and n_R cycles of R.

3. All reversible heat engines operating between the same two heat

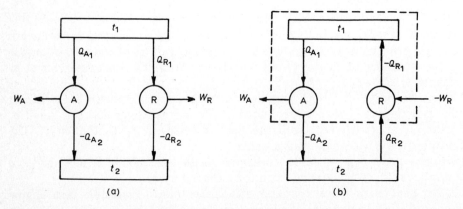

Fig. 8.5. Proof that the efficiency of any heat engine A cannot exceed that of a Carnot engine R when both operate between the same two heat reservoirs.

reservoirs have the same efficiency; but the efficiency of an irreversible heat engine is less than that of a reversible heat engine when both operate between the same two heat reservoirs.

The signs of equality and inequality in eqn. (8.18) are the same as those in the second-law equation (8.1). In Section 8.6 we found that the sign of equality holds if the composite heat engine undergoes a reversible cycle. Therefore if A is a reversible heat engine R' the following relation holds for the efficiencies $\epsilon_{R'}$ and ϵ_R of R' and R, respectively,

$$\epsilon_{R'} = \epsilon_R \tag{8.19}$$

In Section 8.7 we showed that the sign of inequality holds in eqn. (8.1) and hence in eqn. (8.18), if the composite heat engine undergoes an irreversible cycle. Hence if A is an irreversible heat engine, I, we find

$$\epsilon_I < \epsilon_R \tag{8.20}$$

4. The efficiency of a reversible heat engine operating between a heat reservoir at t_1 and one at t_2 is a function only of these two temperatures.

In the derivation of eqn. (8.19), the only condition made was that each of the two Carnot engines operated between the same two heat reservoirs, one at t_1 and the other at t_2. Hence the efficiency, ϵ, of a Carnot engine depends only on these two temperatures:

$$\epsilon = W/Q_1 = F(t_1, t_2), \text{ or } Q_1/Q_2 = f(t_1, t_2) \tag{8.21}$$

where $f = 1/(\epsilon - 1)$.

8.11. Thermodynamic temperature scales. In Fig. 8.6 [9] the temperatures of the three heat reservoirs are measured on any proper temperature scale (Section 2.6), and $t_1 > t_2 > t_0$. The Carnot engines R', R'', and R''' operate between the various heat reservoirs as shown, and they produce the work and heat effects indicated. Let the Carnot engines operate in such a direction that Q_1', W'', and W''' are each positive.

Let us adjust R'' and R''' so that, when each of the three Carnot engines has completed one cycle, we have

$$Q_2' + Q_2'' = 0 \text{ and } Q_1''' = Q_1' \tag{8.22}$$

Then, during each cycle of R', R'', and R''', the heat reservoir at t_2 does not undergo any change in state. We can consider that R', R'', and the heat reservoir at t_2 constitute a composite Carnot engine operating between heat reservoirs at t_1 and t_0, as does R'''. From eqns. (8.19) and (8.22), and the first law we find

$$W' + W'' = W''' \text{ and } Q_0'' = Q_0''' \tag{8.23}$$

By eqn. (8.21):

$$\frac{Q_1'}{Q_2'} = f(t_1, t_2), \quad \frac{Q_2''}{Q_0''} = f(t_2, t_0), \quad \frac{Q_1'''}{Q_0'''} = f(t_1, t_0) \tag{8.24}$$

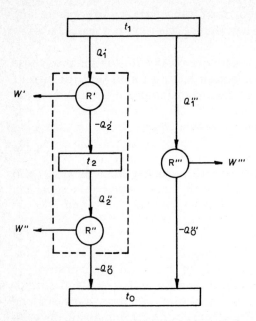

Fig. 8.6. Derivation of absolute thermodynamic temperature scales.

From eqns. (8.22) and (8.23) we find

$$\frac{Q_1'}{Q_2'} = -\frac{Q_1'''/Q_0'''}{Q_2''/Q_0''} \tag{8.25}$$

Substitution from eqn. (8.24) gives

$$f(t_1, t_2) = -\frac{f(t_1, t_0)}{f(t_2, t_0)} = -\frac{\phi(t_1)}{\phi(t_2)} \tag{8.26}$$

The last equality in the above expressions must hold true since $f(t_1, t_2)$ does not contain t_0 and therefore t_0 must cancel in the middle member.

The form of the function $\phi(t)$ is not determined by eqn. (8.26), but the ratios $\phi(t_1)/\phi(t_2)$ for two fixed temperatures are, since Q_1'/Q_2' are heat quantities which may be measured. Each form chosen for $\phi(t)$ defines a ϕ-scale of temperature which is independent of the properties of any substance, and is called a thermodynamic temperature scale. We shall consider two of these in the next section.

Let a Carnot heat engine withdraw the quantity of heat Q_1 from a heat reservoir at t_1 and Q_2 from one at t_2, and produce the work W in the surroundings. Then we find

$$\frac{Q_1}{Q_2} = -\frac{\phi(t_1)}{\phi(t_2)}; \frac{Q_1}{\phi(t_1)} + \frac{Q_2}{\phi(t_2)} = 0 \tag{8.27}$$

$$W = Q_1 \frac{\phi(t_1)-\phi(t_2)}{\phi(t_1)} = Q_2 \frac{\phi(t_2)-\phi(t_1)}{\phi(t_2)} \tag{8.28}$$

$$\epsilon = \frac{\phi(t_1) - \phi(t_2)}{\phi(t_1)}; \; c = \frac{\phi(t_2)}{\phi(t_1) - \phi(t_2)} \qquad (8.29)$$

Here ϵ is the efficiency of a Carnot engine operating between heat reservoirs at t_1 and t_2, and c is the coefficient of performance of a Carnot refrigerator operating between the same two temperatures.

8.12. The Carnot function and Kelvin's temperature scales. According to Carnot's principle [1] the efficiency $đW_R/Q_t$ of a Carnot engine operating between heat reservoirs at t and $t-dt$ has the form $\mu_C(t)dt$. The temperature function μ_C is called the Carnot function. Employing eqn. (8.29), we find

$$\mu_C dt = \frac{đW_R}{Q_t} = \frac{\phi(t) - \phi(t-dt)}{\phi(t)} = d \ln \phi(t) \qquad (8.30)$$

whence

$$\mu_C = \frac{d \ln \phi(t)}{dt} \qquad (8.31)$$

Kelvin [3] (1848) first proposed a scale of temperature (θ) in which "all degrees have the same value", that is, a Carnot engine operating between heat reservoirs at θ and $\theta - 1$ would have the same efficiency whatever the value of θ. He defined this scale by placing μ_C equal to unity:

$$\mu_C = \frac{d \ln \phi(\theta)}{d\theta} = 1; \; \phi(\theta) = ae^\theta \qquad (8.32)$$

Here a is an integration constant. Substitution into eqn. (8.29) with $\theta_1 = \theta$ and $\theta_2 = \theta - 1$ gives

$$\epsilon = 1 - e^{-1} \qquad (8.33)$$

which does not depend on θ.

Kelvin's second scale of temperature (T) is an absolute thermodynamic scale defined by placing μ_C equal to the reciprocal of T:

$$\mu_C = \frac{d \ln \phi(T)}{dT} = \frac{1}{T}; \; \phi(T) = bT \qquad (8.34)$$

The integration constant b determines the size of the degree (or thermodynamic unit of temperature, the kelvin) on this scale. When this is fixed by assigning the value 273.16 kelvins to the temperature of the triple point of water (Section 2.7) the scale is called the Kelvin temperature scale (T K).

We shall show (Section 9.1) that the perfect gas absolute temperature scale (Section 2.9) coincides with the absolute thermodynamic scale defined by eqn. (8.34).

In terms of the Kelvin temperature scale, eqns. (8.27)–(8.29) become

$$\frac{Q_1}{Q_2} = -\frac{T_1}{T_2}, \frac{Q_1}{T_1} + \frac{Q_2}{T_2} = 0 \qquad (T_1 > T_2) \qquad (8.35)$$

$$W = Q_1 \left(\frac{T_1 - T_2}{T_1}\right) = Q_2 \left(\frac{T_2 - T_1}{T_2}\right) \qquad (8.36)$$

$$\epsilon = \frac{T_1 - T_2}{T_1}; c = \frac{T_2}{T_1 - T_2} \qquad (8.37)$$

The second law requires that Q_1 and Q_2 in eqn. (8.35) have different algebraic signs (Corrolary 1, Section 8.10). Hence T_1 and T_2 must have the same signs. The temperature of the triple point of water is assigned a positive value on the Kelvin scale. Hence a thermodynamic system cannot have a negative Kelvin temperature.

If in eqn. (8.36) T_2 is zero and T_1 and Q_2 are positive finite quantities for a Carnot engine operating as a refrigerator, $W = -\infty$. Hence, in the absence of a perfect heat insulator, a finite system cannot exist for a finite time at 0 K (see ref. 10).

8.13. The Clausius inequality. Whenever a system undergoes a cyclical process, the integral of đQ/T for the cycle is equal to or less than zero,

$$\oint \frac{dQ}{T} \leq 0 \qquad \text{(any cycle)} \qquad (8.38)$$

To prove this corollary of the second law [11], we suppose that the system A (Fig. 8.7) undergoes any cyclical process whatever. In any infinitesimal part of the cycle, the system produces the work đW_A in the surroundings, and withdraws therefrom the heat đQ_A which is delivered by a Carnot engine R operating between a part of the system having the temperature T and a heat reservoir at T_0. While the system is undergoing this infinitesimal part of the cycle, the Carnot engine (which need not be infinitesimal itself) undergoes an infinitesimal Carnot cycle in which it produces the work đW_R in the surroundings and withdraws the heat đQ_R from the system. Now đQ_A and đQ_R are equal in magnitude but have opposite signs, since a quantity of heat regarded as positive by the system is regarded as negative by the Carnot engine, and vice versa. Thus whichever the direction of heat flow, we find

$$dQ_A + dQ_R = 0 \qquad (8.39)$$

Application of the second-law equation (8.36) to the Carnot engine gives

$$dW_R = \frac{T - T_0}{T} dQ_R = \frac{T_0 - T}{T} dQ_A \qquad (8.40)$$

Application of the first law to system A yields

$$\oint dW_A = \oint dQ_A \qquad (8.41)$$

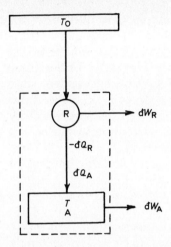

Fig. 8.7. Proof of the Clausius inequality.

When A has performed one cycle, the Carnot engine has undergone very many small cycles.

Now the composite system A + R is connected to a single heat reservoir. Hence, application of the second-law equation (8.1) to this system gives

$$\oint dW_A + \oint dW_R \leqslant 0 \tag{8.42}$$

Here the first integral is for one cycle of the system A, and the second is for the corresponding many cycles of R. Substitution from eqns. (8.40) and (8.41) into eqn. (8.42), and division by the constant temperature T_0, gives

$$\oint \frac{dQ_A}{T} \leqslant 0 \qquad \text{(any cycle of A)} \tag{8.43}$$

This is the Clausius inequality. We note that T is the temperature of the place in the system A where the heat dQ_A is delivered by the Carnot engine.

The signs of equality and inequality in eqn. (8.43) are the same as those in eqn. (8.1). We showed in Sections 8.6 and 8.7 that the sign of equality holds in eqn. (8.1) for a reversible cycle, and that the inequality holds for an irreversible cycle. Now the system A in Fig. 8.7 executed any cycle. Hence we have the following results:

1. Whenever a system undergoes a reversible cycle, the integral of dQ_R/T for the cycle is zero

$$\oint \frac{dQ_R}{T} = 0 \qquad \text{(any reversible cycle)} \tag{8.44}$$

Mathematically, eqn. (8.44) means that $1/T$ is an integrating factor for dQ_R.

We may interpret dQ_R in eqn. (8.44) as the quantity of heat received by the system at a temperature T during an infinitesimal part of a reversible cycle. Hence, eqn. (8.44) is valid even when the part of the process occurring in the surroundings is irreversible, provided the part of the process taking place inside the boundary of the system is reversible [12]. Under these conditions, we can express $đQ_R/T$ as a function of a set of independent state variables of the system, and evaluate the integral of eqn. (8.44) without resort to measurements made in the surroundings where the irreversibility occurs. We note that the work produced in the surroundings during the process does not appear explicitly in eqns. (8.43) or (8.44).

2. Whenever a system undergoes an irreversible cycle, the integral of $đQ/T$ for the cycle is negative

$$\oint \frac{đQ}{T} < 0 \qquad \text{(any irreversible cycle)} \tag{8.45}$$

8.14. Definition of the entropy of a closed system. Let a closed system A undergo an infinitesimal change in state. The change in the entropy dS attending this change in state is defined by the relation

$$dS = \frac{đQ_R}{T} \qquad \text{(definition of } S\text{)} \tag{8.46}$$

where $đQ_R$ is the heat withdrawn from the surroundings by the system in any reversible process connecting its initial and final states. Then, by eqn. (8.44), the following relation must hold for every cycle executed by the system A

$$\oint dS = 0 \qquad \text{(property of } S\text{)} \tag{8.47}$$

Therefore entropy is a state property of a system.

On integrating eqn. (8.46) for the finite change in state

$$A \text{ (state 1)} = A \text{ (state 2)} \tag{8.48}$$

we find

$$\int_1^2 dS = \int_1^2 \frac{đQ_R}{T} \tag{8.49}$$

$$\int_1^2 dS = S_2 - S_1 = \Delta S \tag{8.50}$$

where $đQ_R/T$ is integrated along the path of any reversible process connecting states 1 and 2 of the system.

We may evaluate the right-hand side of eqn. (8.49): (1) from measurements of the temperature of the system and the heat withdrawn from the surroundings for a reversible process; or (2) by expressing $đQ_R/T$ as a function of a set of independent state variables and integrating this function along the path of a reversible process leading from the initial to the final state of the system. The latter method is the procedure usually used.

As noted immediately under eqn. (8.44), this relation, and hence eqns. (8.47), (8.49), and (8.50) are valid when the part of the process occurring inside the boundary of the system is reversible, even if the part of the process taking place exterior to this boundary is irreversible. We would then use the second method of determining $đQ_R$ given in the last paragraph.

Since entropy is a state property, ΔS depends only on the change in state of a system, not on the process. We cannot compute ΔS from measurements of the heat withdrawn from the surroundings during an irreversible process. But we can compute the value of ΔS by employing one of the methods presented above, in which the process, or at least that part of the process occurring inside the boundary of the system, is reversible.

The work produced in the surroundings during a reversible process does not appear explicitly in eqn. (8.49) for the change in entropy of the system.

We note from eqn. (8.49) that the entropy of a system remains constant in a reversible, adiabatic process.

Equations (8.46) and (8.49) define the differential of entropy and its line integral. Equations (8.47) and (8.50) assert the existence of the entropy function. The entropy S of a closed system is a single-valued function of a set of independent state variables of a system, and dS is the total differential of this function.

The second law defines only the difference between the entropies of two states of a system, not the entropy of each state. The third law, which is treated in a later chapter, fixes the molar entropy of a pure substance when in a certain condition at 0 K. This is advantageous from the standpoint of numerical computations; but it is not necessary for the development of thermodynamic theory. We may, if it seems expedient, complete the definition of the entropy of a system by assigning to its entropy when in a specified standard state any arbitrary value consistent with the problem under consideration, such as we do for the energy and enthalpy (Sections 4.4, 4.8, and 6.2).

If there should be a state of a system which cannot be reached from a standard state by a reversible process for which the integral of $đQ_R/T$ can be evaluated, we cannot calculate its entropy by thermodynamic methods.

From eqn. (8.49) and the discussion of the measure of heat in Section 3.13, we find that entropy is an extensive property, as is energy (Section 4.4). It has the units of energy temperature^{-1}, for example cal deg^{-1} or J deg^{-1}, where deg is the kelvin.

When the effects of long-range forces and shape (capillarity) may be neglected, the entropy of a heterogeneous system is equal to the sum of the entropies of its individual phases.

8.15. Relation of entropy changes to heat effects attending processes. Suppose that the change in state (8.48) is brought about by any process P, and then the system is restored to its initial state by a reversible process R. From the Clausius inequality, eqn. (8.43), we find the following relation for this cycle,

$$\int_1^2 \frac{dQ}{T} + \int_2^1 \frac{dQ_R}{T} \leqslant 0 \tag{8.51}$$

Substitution from eqns. (8.49) and (8.50) and rearrangement give the following equation for the change in entropy ΔS of the system,

$$\Delta S \geqslant \int_{P\,1}^{2} \frac{dQ}{T} \quad \text{(any process P)} \tag{8.52}$$

We have shown in Section 8.13 that the sign of equality holds in eqn. (8.43) and hence in eqn. (8.52) for a reversible process, and that the sign of inequality holds for an irreversible process. Hence if ΔS is the change in entropy attending a finite change in state and dS refers to an infinitesimal change, we find

$$\Delta S > \int_1^2 \frac{dQ}{T} \text{ and } dS > \frac{dQ}{T} \quad \text{(an irreversible process)} \tag{8.53}$$

$$\Delta S = \int_1^2 \frac{dQ_R}{T} \text{ and } dS = \frac{dQ_R}{T} \quad \text{(a reversible process)} \tag{8.54}$$

$$\Delta S < \int_1^2 \frac{dQ}{T} \text{ and } dS < \frac{dQ}{T} \quad \text{(no process whatever)} \tag{8.55}$$

8.16. Principle of the increase in entropy. Let an isolated system undergo any change in state by any process. Then each element of heat dQ in eqn. (8.52) is zero, and this expression becomes

$$\Delta S \geqslant 0 \quad \text{(any process in an isolated system)} \tag{8.56}$$

Since the signs of equality and inequality in this relation are the same as those in eqn. (8.52), we find

$\Delta S > 0$ (an irreversible process in an isolated system) (8.57)
$\Delta S = 0$ (a reversible process in an isolated system) (8.58)
$\Delta S < 0$ (no process whatever in an isolated system) (8.59)

Thus the entropy of an isolated system cannot decrease. It remains constant when the process is reversible and increases in an irreversible process.

8.17. Increase in entropy in an irreversible process. Let a system A undergo a specified change in state: (1) by a reversible process; and then (2) by an irreversible process. The change in entropy ΔS_A of the system is the same in (1) as in (2), since the change in state is the same.

Let the system and that portion of its environment M (called the medium) which is appreciably disturbed by changes occurring in A be enclosed in a boundary that isolates the composite system A + M from the rest of the universe. From eqns. (8.57) and (8.58) we find

$$\Delta S_A + \Delta S_M = 0 \quad \text{(a reversible process in A)} \tag{8.60}$$

$$\Delta S_A + \Delta S_M > 0 \quad \text{(an irreversible process in A)} \tag{8.61}$$

Since ΔS_A has the same value in these two equations, ΔS_M does not. Hence the medium has not undergone the same change in state in process (1) as in process (2).

Equations (8.60) and (8.61) show that entropy is conserved in a reversible process; it is created in an irreversible process. We recall (Section 4.5) that energy is conserved in all processes.

8.18. Review of the properties of entropy. The following properties of entropy are consequences of eqns. (8.46) and (8.47), and the definition of heat (Sections 3.12, 3.13, and 3.26). They may be compared with the properties of energy (Section 4.7).

1. The quantity dS is an exact differential.

2. The increment of entropy ΔS attending a specified change in state depends only on the initial and final states of the system. No matter how this change is actually accomplished, we can compute ΔS as the integral of $đQ_R/T$ for any reversible process bringing about the required change. For this calculation, only the part of the process occurring inside the boundary of the system need be reversible (Section 8.14).

The change in entropy ΔS attending a change in state is greater than the integral of $đQ/T$ for an irreversible process producing this change. There is no process for which ΔS is less than the integral of $đQ/T$.

The entropy of a system remains constant in a reversible adiabatic process.

The work produced in the surroundings does not enter explicitly into the equation for the change in entropy ΔS of a system.

3. The second law defines only changes in entropy between two states of a system, not the entropy in each state. Hence, according to this law, entropy is an extensive state property of a system, determined except for an additive constant of integration. Its value in a given state may be fixed by selecting some one state of the system and specifying its entropy, or we may invoke the third law which fixes this integration constant. Thus S is a single-valued function of a set of independent thermodynamic variables, dS is the total

differential of this function, and ΔS is the difference between the values of this function for two finitely different states of the system.

4. The entropy of an isolated system cannot decrease. It is constant when the system undergoes a reversible process, and it increases if the system undergoes an irreversible process.

5. When a system undergoes a change in state by any process, the sum of the changes in entropy of the system and its environment is never negative. This sum is equal to zero if the process is reversible, and entropy is conserved in such a process. It is positive if the process is irreversible.

All processes in nature are irreversible. In all natural processes, the sum of the entropy of the system and its environment increases: entropy is continually being created. This is in contrast to energy which is conserved in all processes.

6. When the effects of long-range forces and shape may be neglected, the entropy of a heterogeneous system is the sum of the entropies of its individual phases.

7. We cannot calculate by thermodynamic methods the entropy of a system in a state which cannot be reached from a standard state by a reversible process.

8.19. First- and second-law equation for a closed system. Substitution from the second-law equation for closed systems

$$đQ \leqslant TdS \tag{8.62}$$

into the first-law equation for closed systems

$$dE = đQ - đW \tag{8.63}$$

gives the first- and second-law equation for closed systems

$$dE \leqslant TdS - đW \tag{8.64}$$

The sign of equality holds for a reversible process and the sign of inequality for an irreversible process.

Equation (8.64) takes the following form for a reversible process

$$dE = TdS - pdV - đW_{XR} \tag{8.65}$$

Whence

$$dH = TdS + Vdp - đW_{XR} \tag{8.66}$$

Here W_{XR} is the reversible non-expansion work of the process.

REFERENCES

[1] S. Carnot, Refléxions sur la Puissance Motrice du Feu et sur les Machines Propres à Développer cette Puissance, Bachelier, Paris, 1824; see The Second Law of Thermodynamics, translated and edited by W. F. Magie, Harper and Brothers, New York, 1899, pp. 3—61.

2 E. Clapeyron, Ann. Phys. Chem., 59 (1843) 446—451, 566—592; translated from J. Ec. Polytech., 14 (1834) 70.
3 W. Thomson, Mathematical and Physical Papers, University Press, Cambridge, 1882, Vol. 1, (1848) pp. 100—106; (1849) 113—155; (1851—1854, 1878) 174—332, (with J. P. Joule), (1852—1862) 333—455.
4 H. V. Regnault, Rel. Expériences, 1 (1847) 635—748; Ann. Phys. Chem., 78 (1848) 196—216, 523—563.
5 R. Clausius, Ann. Phys. Chem., 79 (1850) 368—397, 500—524; ibid., 93 (1854) 481—506; ibid., 125 (1865) 353—400; see The Mechanical Theory of Heat, translated by W. R. Browne, Macmillan and Co., London, 1879; see also W. F. Magie, ref. 1, pp. 65—108.
6 M. Planck, Treatise on Thermodynamics, translated by A. Ogg, Longmans, Green and Co., London, 1921, p. 86.
7 C. Carathéodory, Math. Ann., 67 (1909) 355—386; see also H. A. Buchdahl, Am. J. Phys., 17 (1949) 41—46, 212—218.
8 J. G. Kirkwood and I. Oppenheim, Chemical Thermodynamics, McGraw-Hill, New York, 1961, Chapter 4.
9 G. H. Bryan, Thermodynamics, B. G. Teubner, Leipzig, 1907, p. 18.
10 J. H. Keenan, Thermodynamics, Wiley, New York, 1941, p. 77.
11 H. B. Phillips, 1919, personal communication. See also reference 10, p. 79.
12 G. N. Hatsopoulos and J. H. Keenan, Principles of General Thermodynamics, Wiley, New York, 1965, p. 154.

Chapter 9

APPLICATIONS OF THE SECOND LAW

Unless otherwise stated we shall be concerned in this chapter with closed, simple systems at rest with respect to their environments, or moving so slowly that their thermodynamic properties are not appreciably affected. Under these conditions eqns. (8.65) and (8.66) become

$$dE = TdS - pdV \tag{9.1}$$

$$dH = TdS + Vdp \tag{9.2}$$

We note that these equations can be integrated only for reversible processes. This is in contrast to the corresponding first-law equations of Table 4.2 which may be integrated for any process.

9.1. Perfect gas temperature scale. In Section 5.8 we defined a perfect gas by the two equations

$$pV = nRT, \quad \left(\frac{\partial E}{\partial V}\right)_T = 0 \tag{5.50}$$

which we derived from the laws of Boyle, Gay-Lussac, and Joule. Here T is on the perfect gas absolute temperature scale.

Let us now define a perfect gas as a fluid that obeys the laws of Boyle [eqn. (2.33)] and Joule [eqn. (5.34)]. For a fixed mass n of gas these laws are

$$pV = f(T) \quad (n = \text{constant}) \tag{9.3}$$

$$E = E(T) \quad (n = \text{constant}) \tag{9.4}$$

No particular temperature scale is specified by these laws, since they are based on isothermal experiments. Hence we may measure T on any scale we please. Let us choose to measure T on the Kelvin scale [eqn. (8.34)].

Equation (9.1) may be written in the following form

$$dS = \frac{dE}{T} + \frac{p}{T}dV \tag{9.5}$$

Substitution from eqn. (9.4) gives

$$dS = \frac{E'(T)}{T}dT + \frac{p}{T}dV \quad (n = \text{constant}) \tag{9.6}$$

where $E'(T)$ is the derivative of $E(T)$. Application of Euler's criterion of integrability (cross differentiation) to eqn. (9.6) gives

$$\left[\frac{\partial [E'(T)/T]}{\partial V}\right]_T = \left[\frac{\partial (p/T)}{\partial T}\right]_V = 0 \qquad (n = \text{constant}) \qquad (9.7)$$

The last equality holds because $E'(T)/T$ is a function of T alone. Integration gives

$$\frac{p}{T} = \phi(V) \qquad (n = \text{constant}) \qquad (9.8)$$

In this equation T is the absolute thermodynamic temperature, since it is the same T which occurs in the first- and second-law equation (9.5).

On equating the values of p from eqns. (9.3) and (9.8), we have

$$\frac{f(T)}{T} = V\phi(V) = k \qquad (n = \text{constant}) \qquad (9.9)$$

Here k is a constant, since the first equality holds for all values of V and T, and $f(T)/T$ is a function of T alone, while $V\phi(V)$ contains only V. Thus, from eqns. (9.3) and (9.9) we find

$$pV = f(T) = kT \qquad (n = \text{constant}) \qquad (9.10)$$

Use of the empirical definition of the mole (Section 2.10) gives

$$pV = nRT \qquad (T \text{ is on the Kelvin scale}) \qquad (9.11)$$

Thus we have proved that the absolute perfect gas temperature scale defined by eqn. (2.36) is identical with the Kelvin thermodynamic scale defined by eqn. (8.34).

We note that in each of the eqns. (5.50) and (9.11) two separate postulates are used to define a perfect gas.

The laws of Boyle and Gay-Lussac are direct consequences of the two postulates (9.11). We shall show in Section 9.14 that Joule's law also follows from these postulates.

9.2. *Variations of entropy with temperature.* From eqns. (9.1) and (9.2), and the two eqns. (5.5) we find the following relations

$$\left(\frac{\partial S}{\partial T}\right)_V = \frac{1}{T}\left(\frac{\partial E}{\partial T}\right)_V = \frac{C_V}{T} \qquad (9.12)$$

$$\left(\frac{\partial S}{\partial T}\right)_p = \frac{1}{T}\left(\frac{\partial H}{\partial T}\right)_p = \frac{C_p}{T} \qquad (9.13)$$

Actually, these equations are merely specializations of the following relation which comes from the definitions (3.45) and (8.46)

$$\left(\frac{dS}{dT}\right)_C = \frac{1}{T}\left(\frac{dQ}{dT}\right)_C = \frac{C_C}{T} \qquad (9.14)$$

Here the subscript C denotes the path of any reversible process.

9.3. Variations of entropy with volume and pressure. Three expressions for the total differential of entropy of a closed, simple, homogeneous system are

$$dS = \left(\frac{\partial S}{\partial T}\right)_V dT + \left(\frac{\partial S}{\partial V}\right)_T dV \tag{9.15}$$

$$dS = \left(\frac{\partial S}{\partial T}\right)_p dT + \left(\frac{\partial S}{\partial p}\right)_T dp \tag{9.16}$$

$$dS = \left(\frac{\partial S}{\partial p}\right)_V dp + \left(\frac{\partial S}{\partial V}\right)_p dV \tag{9.17}$$

On substitution from eqns. (9.15) and (9.16) into (9.1) and (9.2), we find

$$dE = T\left(\frac{\partial S}{\partial T}\right)_V dT + \left[T\left(\frac{\partial S}{\partial V}\right)_T - p\right] dV \tag{9.18}$$

$$dH = T\left(\frac{\partial S}{\partial T}\right)_p dT + \left[T\left(\frac{\partial S}{\partial p}\right)_T + V\right] dp \tag{9.19}$$

The four Maxwell relations may be obtained by cross differentiating eqns. (9.1), (9.2), (9.18), and (9.19), and noting that the higher derivatives of entropy are independent of the order of differentiation

$$\left(\frac{\partial p}{\partial S}\right)_V = -\left(\frac{\partial T}{\partial V}\right)_S, \text{ or } \left(\frac{\partial S}{\partial p}\right)_V = -\left(\frac{\partial V}{\partial T}\right)_S \tag{9.20}$$

$$\left(\frac{\partial V}{\partial S}\right)_p = \left(\frac{\partial T}{\partial p}\right)_S, \text{ or } \left(\frac{\partial S}{\partial V}\right)_p = \left(\frac{\partial p}{\partial T}\right)_S \tag{9.21}$$

$$\left(\frac{\partial S}{\partial V}\right)_T = \left(\frac{\partial p}{\partial T}\right)_V \tag{9.22}$$

$$\left(\frac{\partial S}{\partial p}\right)_T = -\left(\frac{\partial V}{\partial T}\right)_p \tag{9.23}$$

9.4. Relations for reversible, adiabatic processes. If we place $dS = 0$ in eqns. (9.15)–(9.17) and use (9.12), (9.13), (9.22), and (9.23), we find the following relations for isentropic processes in a closed, simple, homogeneous system

$$\left(\frac{\partial V}{\partial T}\right)_S = -\frac{(\partial S/\partial T)_V}{(\partial S/\partial V)_T} = -\frac{C_V}{T}\left(\frac{\partial T}{\partial p}\right)_V \tag{9.24}$$

$$\left(\frac{\partial p}{\partial T}\right)_S = -\frac{(\partial S/\partial T)_p}{(\partial S/\partial p)_T} = \frac{C_p}{T}\left(\frac{\partial T}{\partial V}\right)_p \tag{9.25}$$

$$\left(\frac{\partial V}{\partial p}\right)_S = -\frac{(\partial S/\partial T)_V(\partial T/\partial p)_V}{(\partial S/\partial T)_p(\partial T/\partial V)_p} = \frac{C_V}{C_p}\left(\frac{\partial V}{\partial p}\right)_T \tag{9.26}$$

9.5. Relations for the calorimetric coefficients. The six calorimetric coefficients were defined in Section 3.28, and some were used in the equations of Chapter 5. Two of these are the heat capacities C_V and C_p; two are the latent heats l_T and Λ_T; and two are the quantities Γ_V and η_p. On substituting TdS for the reversible heat dQ in eqns. (3.87)–(3.92) and using eqns. (9.20)–(9.25), we find eqns. (9.12), (9.13), and the following four expressions for closed, simple, homogeneous systems

$$l_T = T\left(\frac{\partial S}{\partial V}\right)_T = T\left(\frac{\partial p}{\partial T}\right)_V \tag{9.27}$$

$$\Lambda_T = T\left(\frac{\partial S}{\partial p}\right)_T = -T\left(\frac{\partial V}{\partial T}\right)_p \tag{9.28}$$

$$\Gamma_V = T\left(\frac{\partial S}{\partial p}\right)_V = C_V\left(\frac{\partial T}{\partial p}\right)_V \tag{9.29}$$

$$\eta_p = T\left(\frac{\partial S}{\partial V}\right)_p = C_p\left(\frac{\partial T}{\partial V}\right)_p \tag{9.30}$$

In eqns. (9.42) and (9.43) we shall show that the difference between C_p and C_V can be expressed in terms of quantities that may be evaluated from an equation of state of the phase under consideration. Hence eqns. (9.12), (9.13), and (9.27)–(9.30) express all of the differential coefficients of eqns. (9.15)–(9.17) in terms of quantities that may be evaluated if we are given: (1) C_p as a function of T along one isopiestic (for example, the standard value C_p^0); and (2) the equation of state of the phase.

9.6. Changes in entropy attending finite changes in state. On substituting from eqns. (9.12), (9.13), and (9.27)–(9.30) into eqns. (9.15)–(9.17), we find the following relations for a closed, simple, homogeneous system

$$dS = \frac{C_V}{T}dT + \left(\frac{\partial p}{\partial T}\right)_V dV \tag{9.31}$$

$$dS = \frac{C_p}{T}dT - \left(\frac{\partial V}{\partial T}\right)_p dp \tag{9.32}$$

$$dS = \frac{C_V}{T}\left(\frac{\partial T}{\partial p}\right)_V dp + \frac{C_p}{T}\left(\frac{\partial T}{\partial V}\right)_p dV \tag{9.33}$$

Let the change in state

$$n\ NH_3(g, p_1, V_1, T_1) = n\ NH_3\ (g, p_2, V_2, T_2) \tag{9.34}$$

where only two of the three variables (p, V, T) are independent, be brought about by any process whatever. If we employ the two reversible two-step processes outlined in Section 5.1, we find from eqns. (9.31) and (9.32) the following relations for the change in entropy ΔS attending the change in state (9.34)

$$\Delta S = \int_{V_1\,T_1}^{T_2} \frac{C_V}{T} dT + \int_{T_2\,V_1}^{V_2} \left(\frac{\partial p}{\partial T}\right)_V dV \tag{9.35}$$

$$\Delta S = \int_{p_1\,T_1}^{T_2} \frac{C_p}{T} dT - \int_{T_2\,p_1}^{p_2} \left(\frac{\partial V}{\partial T}\right)_p dp \tag{9.36}$$

In these expressions, the first term gives the change in entropy with temperature at constant volume (or pressure); and the second term gives the isothermal variation of entropy with volume (or pressure).

A relation similar to the above in terms of the independent variables (p, V) can be derived by integration of eqn. (9.33).

9.7. *Isothermal variations of C_V and C_p.* The isothermal variations of C_V with volume and of C_p with pressure of a closed, simple, homogeneous system can be derived by differentiation of eqns. (9.12) and (9.13), and substitution from (9.22) and (9.23)

$$\left(\frac{\partial C_V}{\partial V}\right)_T = T\left[\frac{\partial}{\partial V}\left(\frac{\partial S}{\partial T}\right)_V\right]_T \qquad \left(\frac{\partial C_p}{\partial p}\right)_T = T\left[\frac{\partial}{\partial p}\left(\frac{\partial S}{\partial T}\right)_p\right]_T$$

$$= T\left[\frac{\partial}{\partial T}\left(\frac{\partial S}{\partial V}\right)_T\right]_V \qquad = T\left[\frac{\partial}{\partial T}\left(\frac{\partial S}{\partial p}\right)_T\right]_p$$

$$= T\left(\frac{\partial^2 p}{\partial T^2}\right)_V \qquad = -T\left(\frac{\partial^2 V}{\partial T^2}\right)_p \tag{9.37}$$

9.8. *Relations among heat capacities.* In general, the most direct method for deriving thermodynamic relations for thermal quantities is to use an entropy relation, particularly those of Sections 9.2—9.5. The six eqns. (9.40)—(9.45) given below can all be obtained by dividing eqns. (9.15) or (9.16) by dT subject to a given condition, and substituting from eqns. (9.22) or (9.23).

The same six relations may also be derived from the equations of Section 5.2, in which the results were expressed in relations which contained the differential coefficients $(\partial E/\partial V)_T$ or $(\partial H/\partial p)_T$. By use of eqns. (9.1), (9.2), (9.22), and (9.23) we can now express these differential coefficients in terms of quantities that can be evaluated from an equation of state for the phase under consideration

$$\left(\frac{\partial E}{\partial V}\right)_T = T\left(\frac{\partial p}{\partial T}\right)_V - p = -\left[\frac{\partial(p/T)}{\partial(1/T)}\right]_V = T^2\left[\frac{\partial(p/T)}{\partial T}\right]_V \tag{9.38}$$

$$\left(\frac{\partial H}{\partial p}\right)_T = V - T\left(\frac{\partial V}{\partial T}\right)_p = \left[\frac{\partial (V/T)}{\partial (1/T)}\right]_p = -T^2\left[\frac{\partial (V/T)}{\partial T}\right]_p \qquad (9.39)$$

These relations, sometimes called thermodynamic equations of state, may be substituted into eqns. (5.17)–(5.22) to give the expressions listed below.

The equations expressing the heat capacity C_C of a closed, simple, homogeneous system along a given path C in terms of C_V and C_p are

$$C_C = C_V + T\left(\frac{\partial p}{\partial T}\right)_V \left(\frac{dV}{dT}\right)_C \qquad (9.40)$$

$$C_C = C_p - T\left(\frac{\partial V}{\partial T}\right)_p \left(\frac{dp}{dT}\right)_C \qquad (9.41)$$

The differences between C_p and C_V are

$$C_p - C_V = -T\frac{(\partial p/\partial T)_V^2}{(\partial p/\partial V)_T} \qquad (9.42)$$

$$C_p - C_V = -T\frac{(\partial V/\partial T)_p^2}{(\partial V/\partial p)_T} = \frac{TV\alpha_p^2}{\beta_T} \qquad (9.43)$$

where α_p and β_T are given by eqns. (3.98) and (3.106).

The saturation heat capacity C_{SAT} of each phase of a univariant system is related to the corresponding heat capacities C_V and C_p by the expressions

$$C_{SAT} = C_V + T\left(\frac{\partial p}{\partial T}\right)_V \left(\frac{dV}{dT}\right)_{SAT} \qquad (9.44)$$

$$C_{SAT} = C_p - T\left(\frac{\partial V}{\partial T}\right)_p \left(\frac{dp}{dT}\right)_{SAT} \qquad (9.45)$$

9.9. Measurement of heat capacity. Constant-pressure and saturation heat capacities of substances have been measured in calorimeters described in Sections 3.21–3.24. Two methods for gases which do not involve the direct measurement of a heat quantity are based on eqns. (9.24) and (9.25) which may be written in the forms

$$C_p = T\left(\frac{\partial V}{\partial T}\right)_p \left(\frac{\partial p}{\partial T}\right)_S \qquad (9.46)$$

$$C_V = -T\left(\frac{\partial p}{\partial T}\right)_V \left(\frac{\partial V}{\partial T}\right)_S \qquad (9.47)$$

Both require a knowledge of the equation of state of the gas.

In the adiabatic expansion method of Lummer and Pringsheim, the value of $(\partial T/\partial p)_S$ of eqn. (9.46) is measured (see ref. 1).

In the second method, the quantity $(\partial V/\partial T)_S$ of eqn. (9.47) is evaluated from measurement of the velocity of sound in the gas (see ref. 2).

9.10. Variation of heat capacity with temperature. The normal course of the variation of the standard molar constant-pressure heat capacity c_p^0 of a homogeneous substance versus temperature is shown in Fig. 3.8. Curves illustrating several types of anomalies which occur for some substances are given in Figs. 3.9—3.11.

The value of c_p^0 for a homogeneous substance has been represented over the temperature range 300—1500 K, and higher, by the truncated series

$$c_p^0 = a + bT + cT^2 + dT^3 \tag{9.48}$$

and also by the equation

$$c_p^0 = a' + b'T - c'T^{-2} \tag{9.49}$$

where a, b, c, d and a', b', c', are constants [3]. Power series similar to eqn. (9.48) have been used in restricted temperature ranges below 273 K.

Other relations which have been found useful for representing the temperature variation of heat capacities in the range 0—300 K are mentioned in Chapter 11.

9.11. Entropy of phase transformation. Let us now consider a closed, simple, univariant system (see Sections 3.16 and 5.3). An example of such a system is one consisting of two phases of a pure substance in equilibrium. Let the change in state

$$H_2O(l) = H_2O(g) \qquad (p, T) \tag{9.50}$$

be brought about by any process. Integration of eqn. (9.2) for a reversible, isothermal, isopiestic process gives

$$\Delta S = \Delta H/T \qquad (p, T = \text{constant}) \tag{9.51}$$

Here ΔS and ΔH are, respectively, the molar entropy and molar heat (enthalpy) of vaporization of the liquid. Corresponding equations hold for transition, sublimation, and fusion, as well as for the reverse of each of these processes.

9.12. Joule and Joule—Thomson coefficients. The free expansion and porous-plug experiments were discussed in Sections 5.4 and 5.5. On substituting the value of $(\partial E/\partial V)_T$ from eqn. (9.38) into (5.33) and the value of $(\partial H/\partial p)_T$ from eqn. (9.39) into (5.39), we find

$$\mu_J C_V = p - T\left(\frac{\partial p}{\partial T}\right)_V = \left[\frac{\partial (p/T)}{\partial (1/T)}\right]_V \tag{9.52}$$

$$\mu_{JT} C_p = T\left(\frac{\partial V}{\partial T}\right)_p - V = -\left[\frac{\partial (V/T)}{\partial (1/T)}\right]_p \tag{9.53}$$

These relations express the Joule coefficient μ_J and Joule—Thomson coefficient μ_{JT} in terms of a heat capacity and quantities that can be evaluated from an equation of state.

9.13. Correction of the indications of gas thermometers to the Kelvin scale.

In Section 2.8 we found relations for calculating absolute temperatures on the constant-volume scale (T_V) and the constant-pressure scale (T_p) of a gas thermometer filled with a real gas. We shall write them in the following forms:

$$T_V = 273.16 \frac{p_T}{p_{tr}} \qquad (\text{tr} \equiv 273.16) \tag{2.23}$$

$$T_p = 273.16 \frac{\bar{V}_T}{\bar{V}_{tr}} \tag{2.29}$$

One method of reducing these temperatures to the perfect gas absolute scale, which coincides with the absolute thermodynamic scale known as the Kelvin scale (Section 9.1), is to use the extrapolation procedure of eqns. (2.24) and (2.32).

A second method [4] is to employ Joule and Joule–Thomson coefficients (Section 9.12) together with constant-volume and constant-pressure heat capacities. Integrating eqn. (9.52) at constant volume and eqn. (9.53) at constant pressure from 273.16 K to T K, we find

$$\frac{p_T}{T} - \frac{p_{tr}}{273.16} = \int_{273.16}^{T} \mu_J C_V \, d\left(\frac{1}{T}\right) \qquad (V = \text{constant}) \tag{9.54}$$

$$\frac{\bar{V}_T}{T} - \frac{\bar{V}_{tr}}{273.16} = - \int_{273.16}^{T} \mu_{JT} C_p \, d\left(\frac{1}{T}\right) \qquad (p = \text{constant}) \tag{9.55}$$

On solving the first equation for p_T/p_{tr} and the second for $\bar{V}_T/\bar{V}_{tr}$ and substituting into eqns. (2.23) and (2.29), we obtain the relations

$$T = T_V - \frac{273.16 \, T}{p_{tr}} \int_{273.16}^{T} \mu_J C_V \, d\left(\frac{1}{T}\right) \tag{9.56}$$

$$T = T_p + \frac{T p_{tr}}{R} \int_{273.16}^{T} \mu_{JT} C_p \, d\left(\frac{1}{T}\right) \tag{9.57}$$

In the first equation p_{tr} and $\mu_J C_V$ may be expressed in atm. In the second we may express $p_{tr} = 273.16 \, R/\bar{V}_{tr}$ in atm, $\mu_{JT} C_p$ in liter mole^{-1}, and R in liter atm mole^{-1} deg^{-1}. For the low pressures used in gas thermometric measurements with the permanent gases, we may use the gas temperature scales T_V and T_p in evaluating the integrals of the correction term — at least for the first approximation.

A third method [5] of reducing the indications of gas thermometers to the Kelvin scale is based on the fact that the equations of state (2.43) and (2.44) approach the perfect gas law (9.11) as the pressure of a gas approaches zero, and hence the temperature T in these equations is on the Kelvin temperature scale (Section 9.1). Note that by eqn. (2.45) B' equals B for a given gas. In a constant-volume gas thermometer V in eqn. (2.44) is constant and equal to $\bar{V}_{tr}$ throughout a series of measurements having the same triple-point pressure; and in a constant-pressure gas thermometer p in eqn. (2.43) is constant and equal to p_{tr} throughout a series with the same triple-point molar volume. By division we can find an expression for p_T/p_{tr} at constant volume from eqn. (2.44), and an expression for $\bar{V}_T/\bar{V}_{tr}$ at constant pressure from (2.43). Substitution of these values into eqns. (2.23) and (2.29), respectively, gives

$$T = T_V - (B_T - B_{tr}) \frac{T}{273.16\,\text{R}} p_{tr} \tag{9.58}$$

$$T = T_p - \left(\frac{B_T}{T} - \frac{B_{tr}}{273.16}\right) \frac{T}{\text{R}} p_{tr} \tag{9.59}$$

If we express R in liter atm mole^{-1} deg^{-1}, then B is in liter mole^{-1} and p in atm.

9.14. Perfect gases. We shall now define a perfect gas by eqn. (9.11) and the statement that T in the perfect gas law is on the absolute thermodynamic temperature scale. All of the equations given in Section 5.8 may be derived from these postulates. In particular, we find the following results on substituting from eqn. (9.11) into (9.38), (9.39), (9.52) and (9.53)

$$\left(\frac{\partial E}{\partial V}\right)_T = 0 \quad E = E(T) \quad \mu_J = 0 \tag{9.60a}$$

$$\left(\frac{\partial H}{\partial p}\right)_T = 0 \quad H = H(T) \quad \mu_{JT} = 0 \tag{9.60b}$$

From eqns. (9.11)–(9.13), and (9.37) we find

$$C_V = \left(\frac{\partial E}{\partial T}\right)_V \quad \left(\frac{\partial C_V}{\partial V}\right)_T = 0 \quad C_V = C_V(T) \tag{9.61a}$$

$$C_p = \left(\frac{\partial H}{\partial T}\right)_p \quad \left(\frac{\partial C_p}{\partial p}\right)_T = 0 \quad C_p = C_p(T) \tag{9.61b}$$

From eqns. (9.11), (4.13), and (9.43) we have

$$H = E + n\text{R}T \quad C_p = C_V + n\text{R} \tag{9.62}$$

In addition, we learn from eqn. (9.11) and expressions derived in Section 9.13 that no corrections are required to reduce the indications of constant-volume and constant-pressure gas thermometers filled with a perfect gas to the thermodynamic absolute temperature scale.

From eqns. (9.11)–(9.13) and (9.27)–(9.30) we find the following expressions for n moles of a perfect gas

$$\left(\frac{\partial S}{\partial T}\right)_V = \frac{nc_V}{T}, \quad \left(\frac{\partial S}{\partial V}\right)_T = \frac{nR}{V} \tag{9.63}$$

$$\left(\frac{\partial S}{\partial T}\right)_p = \frac{nc_p}{T}, \quad \left(\frac{\partial S}{\partial p}\right)_T = -\frac{nR}{p} \tag{9.64}$$

$$\left(\frac{\partial S}{\partial p}\right)_V = \frac{nc_V}{p}, \quad \left(\frac{\partial S}{\partial V}\right)_p = \frac{nc_p}{V} \tag{9.65}$$

Here c_p and c_V are molar heat capacities.

Equations (9.31)–(9.33) take the following forms for n moles of a perfect gas

$$dS = \frac{nc_V}{T} dT + nR \, d \ln V \tag{9.66}$$

$$dS = \frac{nc_p}{T} dT - nR \, d \ln p \tag{9.67}$$

$$dS = nc_V \, d \ln p + nc_p \, d \ln V \tag{9.68}$$

Let a system composed of n moles of a perfect gas B undergo the change in state

$$nB \, (g, T_1) = nB \, (g, T_2) \qquad (V, \text{ or } p) \tag{9.69}$$

by any process. From eqns. (9.66) and (9.67) we find the following relations for the change in entropy ΔS attending this change in state

$$\Delta S = \int_{T_1}^{T_2} \frac{nc_V}{T} dT \qquad (V = \text{constant}) \tag{9.70}$$

$$\Delta S = \int_{T_1}^{T_2} \frac{nc_p}{T} dT \qquad (p = \text{constant}) \tag{9.71}$$

From eqns. (9.61) we learn that c_V and c_p are functions only of temperature. Hence the values of ΔS computed from eqns. (9.70) and (9.71) depend only on T_1 and T_2. Thus from eqns. (9.48) and (9.71) we find

$$\Delta S = na \ln \frac{T_2}{T_1} + nb \, (T_2 - T_1) + \frac{nc}{2} (T_2^2 - T_1^2) + \frac{nd}{3} (T_2^3 - T_1^3)$$

$$(p = \text{constant}) \tag{9.72}$$

From eqns. (9.62), (9.70), and (9.71) we find that ΔS for the constant-pressure change in state (9.69) is greater than ΔS for the corresponding constant-volume change by the term $n\mathrm{R}\ln(T_2/T_1)$.

Now let n moles of a perfect gas B undergo the isothermal change in state

$$n\mathrm{B}\,(\mathrm{g},\,V_1,\,\text{or}\,p_1) = n\mathrm{B}\,(\mathrm{g},\,V_2,\,\text{or}\,p_2) \qquad (T) \qquad (9.73)$$

by any process. From eqns. (9.66) and (9.67) we find

$$\Delta S = n\mathrm{R}\ln\frac{V_2}{V_1} \qquad (T = \text{constant}) \qquad (9.74)$$

$$\Delta S = n\mathrm{R}\ln\frac{p_1}{p_2} \qquad (T = \text{constant}) \qquad (9.75)$$

If the change in state (9.73) is brought about by a reversible process, the change in entropy of the environment of the system (called the medium) is the negative of the entropy change ΔS of the system. Hence entropy has been conserved in the process, in accordance with eqn. (8.60).

When a perfect gas undergoes a Joule or Joule–Thomson expansion (see Sections 5.4 and 5.5) its temperature does not change [see eqns. (9.60)]. The entropy change ΔS of the system in each expansion is positive as shown by eqns. (9.74) and (9.75); but the medium does not undergo any change in entropy since each process is adiabatic. Thus, in each of these irreversible processes entropy has been created, which is in accordance with eqn. (8.61).

9.15. Reversible, adiabatic processes in perfect gases. Let n moles of a perfect gas B undergo the following change in state by a reversible, adiabatic process

$$n\mathrm{B}\,(p_1,\,V_1,\,T_1) = n\mathrm{B}\,(p_2,\,V_2,\,T_2) \qquad (S) \qquad (9.76)$$

Only two of the variables (p, V, T) are independent. Since the entropy of the gas is constant, we place $dS = 0$ in eqns. (9.66)–(9.68). On integrating these relations over a temperature range in which c_p may be considered to be essentially constant, we find the following expressions for the above change in state:

$$\ln\frac{V_2}{V_1} + \frac{c_V}{\mathrm{R}}\ln\frac{T_2}{T_1} = 0 \qquad (S = \text{constant}) \qquad (9.77)$$

$$\ln\frac{p_2}{p_1} - \frac{c_p}{\mathrm{R}}\ln\frac{T_2}{T_1} = 0 \qquad (S = \text{constant}) \qquad (9.78)$$

$$\ln\frac{p_2}{p_1} + \kappa\ln\frac{V_2}{V_1} = 0 \qquad (S = \text{constant},\,\kappa \equiv c_p/c_V) \qquad (9.79)$$

These relations may be written as

$$TV^{\mathrm{R}/c_V} = \text{constant};\ Tp^{-\mathrm{R}/c_p} = \text{constant};\ pV^{\kappa} = \text{constant} \qquad (9.80)$$

which correspond to eqns. (5.62).

9.16. Reversible, polytropic processes in perfect gases. Let n moles of a perfect gas B undergo the following change in state by a reversible, polytropic process [see eqn. (3.32)]

$$n\text{B}\,(g, p_1, V_1, T_1) = n\text{B}\,(g, p_2, V_2, T_2) \qquad (pV^k = \text{constant}) \qquad (9.81)$$

Substitution from eqn. (9.11) into the equation for a polytropic process (pV^k = constant) gives

$$TV^{k-1} = \text{constant}; \; Tp^{-(k-1)/k} = \text{constant} \qquad (9.82)$$

From the first relation and eqns. (9.70) and (9.74), we find

$$\Delta S = \frac{n(c_p - kc_V)}{k-1} \ln \frac{T_1}{T_2} \qquad (9.83)$$

Since the process is reversible, the entropy of the medium has changed by $-\Delta S$.

9.17. Mixtures of perfect gases. In order to express the thermodynamic properties of a mixture of perfect gases in terms of the corresponding properties of the pure constituent gases, we must make an assumption regarding the nature of such a mixture which involves the concept of equilibrium between a pure perfect gas and a mixture of it with other perfect gases. We shall use a proposition concerning the equilibrium pressure of a gas in a mixture, which is defined as the pressure of the pure gas when in equilibrium with the mixture through a rigid, thermally conducting wall permeable to that gas alone.

We shall base our treatment of the thermodynamic properties of mixtures of perfect gases on the following postulate: At all pressures and temperatures, the equilibrium pressure p_{ei} of a gas B_i in a mixture of perfect gases is given by the relation

$$p_{ei} = px_i \qquad (9.84)$$

where p is the pressure of the mixture and x_i is the mole fraction of B_i in the mixture [6].

In the following discussion, the subscript k refers to a pure perfect gas B_k and the subscript i denotes the same gas when it is in a mixture of perfect gases.

Consider a system composed of a gaseous mixture of c perfect gases $B_1, B_2, \ldots, B_c$ in the state $(V, T, n_1, n_2, \ldots, n_c)$ where its pressure is p; and c systems each containing one of the pure constituent gases B_k in the state (p_{ek}, V, T) where p_{ek} is the equilibrium pressure of B_k in the mixture. If we suppose that p_{ek} is given by eqn. (9.84), the following relations apply

$$p_{ek} = px_i \qquad (i \equiv k = 1, 2, \ldots, c) \qquad (9.85\text{a})$$

$$T(\text{each pure gas}) = T(\text{mixture}) \qquad (9.85\text{b})$$

$$V(\text{each pure gas}) = V(\text{mixture}) \qquad (9.85\text{c})$$

Gibbs [7] has shown that, when the pressure of a gas mixture is equal to the sum of the equilibrium pressures of the constituent gases, certain additive relations hold. Applying the same methods, we shall prove [8] in a later chapter that, when eqns. (9.85) hold for a mixture of perfect gases, the following relations are true:

1. The density n_i/V of the gas B_i in the mixture when in the state $(V, T, n_1, n_2, \ldots, n_c)$ is equal to the density n_{ek}/V of the corresponding pure gas B_k when in the state (px_i, V, T)

$$n_i/V = n_{ek}/V \qquad (i \equiv k = 1, 2, \ldots, c) \tag{9.86}$$

2. The energy E and the entropy S of the gas mixture in the state $(V, T, n_1, n_2, \ldots, n_c)$ are equal, respectively, to the sum of the energies and the sum of the entropies of the pure constituent gases B_k each in the state (px_i, V, T)

$$E = \sum_k E_{ek} = \sum_k E_k \tag{9.87}$$

$$S = \sum_k S_{ek} \tag{9.88}$$

Here, each summation extends over all of the c gases in the mixture. By virtue of eqn. (9.60) we have replaced the energy E_{ek} in eqn. (9.87) by E_k which is the energy of n_{ek} moles of pure B_k at T and any pressure. Evidently a relation similar to (9.87) holds for enthalpy.

We note that when eqn. (9.84) applies to each gas in a mixture of gases "every gas is as a vacuum to every other gas" [7]. But from this statement alone we cannot derive eqn. (9.84) by thermodynamic reasoning.

From eqns. (9.85) and (9.11) we find

$$p = \sum_k p_{ek} = \sum_k \frac{n_{ek}}{V} RT \tag{9.89}$$

On substitution from eqn. (9.86) and using (9.85), we find

$$pV = \sum_i n_i RT \tag{9.90}$$

Here n_i is the number of moles of the perfect gas B_i in a mixture of c such gases, which has the pressure p, volume V, and Kelvin temperature T. Equation (9.90) is the equation of state of a perfect gas mixture.

9.18. *Entropy of mixing perfect gases.* Let a long, rectangular, thermally insulated vessel of fixed volume V be divided into c sections of volume $v_1, v_2, \ldots, v_c$ by $c-1$ removable, gas-tight partitions of negligible thickness. Initially each section is evacuated. A sufficient number of moles n_k of the perfect gas B_k is introduced into the section of volume v_k ($k = 1, 2, \ldots, c$) to bring the gas pressure to p at a temperature T, p and T having the same value in every section. When the partitions are removed, the gases mix to form a mixture in which the mole fraction of B_i is x_i.

By the conditions of the experiment, $\Delta V = 0$, $Q = 0$, $W = 0$ for the system of c gases. Thus $\Delta E = 0$ for the process. From this result and eqns. (9.89) and (9.90) we learn that the final temperature and pressure of the gas mixture are the same as the initial temperature and pressure of each pure gas. Hence the change in state produced by this mixing process, in which ΔV, Q, and W are all zero, is

$$n_1 B_1(g) + n_2 B_2(g) + \cdots + n_c B_c(g) = [n_1 B_1 \cdot n_2 B_2 \cdots n_c B_c] \text{ (g, mixed)} (p, T)$$
(9.91)

In order to calculate the value of ΔS for the change in state (9.91) we bring about this change by the following two-step process.

1. Let each pure gas B_k undergo the following change in state by a reversible, isothermal expansion

$$n_i B_k(g, p) = n_i B_k(g, p x_i) \qquad (T) \qquad (i \equiv k = 1, 2, \ldots, c) \qquad (9.92)$$

Here x_i is the mole fraction and $p x_i$ is the equilibrium pressure of B_k in the final gas mixture. By eqn. (9.89), each gas B_k at the end of this expansion process occupies a volume equal to the final volume V of the gas mixture in eqn. (9.91). Applying eqn. (9.75) to each pure gas and summing over the c gases, we obtain the following equation for the change in entropy ΔS_1 of the system in this process

$$\Delta S_1 = \sum_i \left(n_i R \ln \frac{p}{p x_i} \right) = - \sum_i (n_i R \ln x_i) \qquad (T = \text{constant}) \qquad (9.93)$$

2. By virtue of eqn. (9.88) we can produce the change in state

$$n_1 B_1 (g, p x_1) + n_2 B_2 (g, p x_2) + \ldots + n_c B_c (g, p x_c) = (\Sigma n_i) [x_1 B_1 \cdot x_2 B_2 \cdots x_c B_c]$$
$$(g, p, \text{mixed}) (T) \qquad (9.94)$$

by a reversible, adiabatic process in which the gases pass through semipermeable membranes into a mixing chamber. Throughout this process we can maintain constant the pressure $p x_i$ and temperature T of each pure gas B_k, and the pressure p, temperature T, and composition $x_1, x_2, \ldots, x_c$ of the mixture by controlling the rates of motion of frictionless pistons in the separate cylinders containing the various gases and the mixture. The entropy change ΔS_2 in this reversible, adiabatic mixing process is zero

$$\Delta S_2 = 0 \qquad (9.95)$$

The change in entropy attending the complete change in state (9.91) is the sum of ΔS_1 and ΔS_2. Hence ΔS for the formation of *one mole of the mixture* of eqn. (9.91) at constant p and T is

$$\Delta S(1 \text{ mole of mixture}) = -\sum_i x_i R \ln x_i \qquad (p, T = \text{constant}) \qquad (9.96)$$

where the summation extends over all c gases in the mixture. Note that ΔS refers to one mole of mixture, and that it is independent of both p and T, provided that each pure gas in its initial state and the mixture in its final state have the same pressure and the same temperature. This entropy change is referred to as the entropy of mixing perfect gases at constant pressure and temperature, and it is always positive.

When two perfect gases B_1 and B_2 undergo the change in state

$$\tfrac{1}{2}B_1(g) + \tfrac{1}{2}B_2(g) = [\tfrac{1}{2}B_1 \cdot \tfrac{1}{2}B_2] \quad (g, \text{mixed}) \quad (p, T) \tag{9.97}$$

the change in entropy ΔS of the system is

$$\Delta S = R \ln 2 \tag{9.98}$$

9.19. The Gibbs Paradox. In the last Section we found that the entropy of mixing perfect gases at constant pressure and temperature [see eqn. (9.96)] is independent of the pressure, temperature, and kinds of gases mixed, except that the gases must be dissimilar. On the other hand, the entropy of mixing two or more portions of the same perfect gas at constant pressure and temperature is zero. The reason for the difference in these two results is that we can devise a reversible process, based on a diversity of properties, for separating the mixture of dissimilar perfect gases into its pure constituents. This process is necessarily attended by a net decrease in entropy of the system of gases. Hence when the gases are mixed by the reverse process the entropy of the system must increase [see eqn. (9.96)]. But we do not admit that a mixture of two or more masses of the same gas, whether or not perfect, can be separated into its original parts by any process, real or ideal, since there is no difference in properties of the several portions as initially constituted — they are identical, except perhaps for their masses. Hence we cannot conceive of a reversible process attended by a net decrease in entropy for separating them. We must conclude that the mixing of two or more portions of the same gas at constant pressure and temperature is not attended by a change in the entropy of the gas.

We note that the entropy of mixing two different perfect gases at constant pressure and temperature does not depend on the *degree* of dissimilarity between the gases. Gibbs [7] points out that as far as the general laws of such gases go "there does not appear to be any limit to the resemblance which there might be between such kinds of gas". He imagined the case that the two perfect gases B_1 and B_2 of eqn. (9.97) are "absolutely identical" in all of their physical properties, but that B_1 reacts chemically with a third substance C, while B_2 is inert towards C. When B_1 and B_2 are mixed in accordance with eqn. (9.97) the change in entropy is $R \ln 2$. When two portions of one of these gases are mixed in a similar manner the entropy change is zero. Yet "the process of mixture, dynamically considered, might be absolutely identical in its minutest details (even with respect to the precise path of each atom)" in the two experiments. According to modern theories of the structure of matter, the continuous approach to identity of two gases does not occur: two gases are either identical in the thermodynamic sense, or they are finitely dissimilar. Thus the Gibbs Paradox does not arise in nature.

9.20. Standard molar entropy of a pure substance. The standard molar entropy s^0 of a pure solid, liquid, or gas at a temperature T is its molar entropy in the state (s, 1 atm, T), (l, 1 atm, T), or (PG, 1 atm, T), respectively (see Section 6.2). The first two are real states, but the third is a hypothetical state of a substance in which it obeys the perfect gas law (9.11) at (1 atm, T). The molar entropy $s(p, T)$ of many pure substances in the real state (p, T), where p is about 1 atm, has been calculated from calorimetric data. Equation (9.32) shows that $s(p, T)$ varies slowly with p for a solid or liquid, and hence the measured value is essentially equal to s^0 at T. At low pressures $s(p, T)$ varies approximately as R ln p for a gas and the measured value must be corrected to 1 atm and for the gas imperfection to give s^0 at T.

In the change in state

$$NH_3(g, p) = NH_3(g, p^\star) \qquad (T) \qquad (9.99)$$

both states of the gas are real, and $p^\star$ is a small pressure. Addition of the term R ln $p/p^\star$ to each side of eqn. (9.36) gives the following relation for the change in entropy attending the above change in state

$$[s(p^\star) + R \ln p^\star] - [s(p) + R \ln p] = \int_p^{p^\star} \left[\frac{R}{p} - \left(\frac{\partial v}{\partial T}\right)_p\right] dp \quad (T = \text{constant})$$

(9.100)

Substitution from the pressure virial equation of state (2.44) and integration gives

$$s(p^\star) + R \ln p^\star + \frac{dB'}{dT} p^\star + \frac{dC'}{dT} \frac{p^{\star 2}}{2} + \ldots = s(p) + R \ln p + \frac{dB'}{dT} p + \frac{dC'}{dT} \frac{p^2}{2}$$
$$+ \ldots \text{ (at } T) \qquad (9.101)$$

Now let $p^\star$ approach zero at constant temperature; and on the right-hand side of the equation let $p = 1$ atm and $B' = C' = \ldots = 0$, so that the gas behaves as a perfect gas in the state (1 atm, T). Then $s(p, T)$ is simply $s(\text{PG}, 1\text{ atm}, T)$, and eqn. (9.101) becomes

$$\lim_{p^\star \to 0} [s(p^\star) + R \ln p^\star] = s^0 \qquad (\text{at } T) \qquad (9.102)$$

From eqns. (9.101) and (9.102) we find

$$s^0 = s(p) + R \ln p + \frac{dB'}{dT} p + \frac{dC'}{dT} \frac{p^2}{2} + \ldots (\text{at } T) \qquad (9.103)$$

This is the exact equation for expressing the standard molar entropy of a gas at T in terms of its molar entropy $s(p, T)$ in the real state (p, T) and the pressure virial coefficients B', C',...of eqn. (2.44). For pressures up to 1 atm we may omit terms of higher order than the first power in p without sensible error.

For the present purpose it is usual to express the parameter B' of eqn. (9.103) as a function of temperature which contains the constants a and b of the Berthelot equation of state, since they can be evaluated [9] for any gas from its critical pressure p_c and temperature T_c. Berthelot's equation written in the pressure virial form (2.44), and the corresponding relation for B' are

$$v = \frac{RT}{p} + \frac{9}{128}\frac{RT_c}{p_c}\left(1 - 6\frac{T_c^2}{T^2}\right) + 0(p) \tag{9.104a}$$

$$B' = \frac{9}{128}\frac{RT_c}{p_c}\left(1 - 6\frac{T_c^2}{T^2}\right) \tag{9.104b}$$

Substitution of the above expression for B' into eqn. (9.103) and omission of terms of higher order than the first in p gives

$$s^0 = s(p) + R \ln p + \frac{27}{32}\frac{RT_c^3}{p_c T^3} p \qquad \text{(at } T\text{)} \tag{9.105}$$

This is an approximate relation for evaluating s^0 of a gas at T when $s(p, T)$ and the critical constants are known.

Let one mole of a pure substance undergo a change in temperature from $(p, 0\,\text{K})$ to (p, T) by a constant pressure process at about 1 atm. Application of eqns. (9.36) and (9.51) to the requisite measured constant-pressure heat capacities and enthalpies of phase transformation gives the corresponding change in molar entropy $s(p, T) - s(p, 0\,\text{K})$ [see also eqn. (10.165), and Sections 11.12 and 11.13].

According to the third law of thermodynamics (Section 11.4) the entropies of most pure substances are zero at 0 K and any pressure. Thus for such a substance $s(p, 0\,\text{K})$ is zero for all values of p, and so is the standard molar entropy s_0^0 at 0 K. Consequently the quantity $s(p, T) - s(p, 0\,\text{K})$ becomes simply $s(p, T)$.

The value of s^0 found by the method outlined above does not include the entropy of mixing isotopes and the entropy of mixing atoms having different nuclear spins. These quantities are omitted in the tabulated values of entropy. They do not affect the change in entropy attending a change in state that does not involve a nuclear reaction or a separation of isotopes (see Section 11.24).

The standard molar entropy of a perfect gas can also be computed directly from equations derived by the methods of statistical mechanics, if the internal energy levels of the molecule have been evaluated from spectroscopic data (see Section 11.14).

9.21. *Entropy changes attending chemical changes in state.* The calculation of the change in entropy ΔS of a chemical change in state from the entropies of the various substances taking part in the chemical reaction, and the calculation of the effects of pressure and temperature on this entropy change parallel the corresponding computations for enthalpy (see Sections 6.4—6.6).

Thus, for the change in state

$$\sum_i \nu_i B_i \text{ (state } i\text{)} = 0 \qquad (p, T) \tag{9.106}$$

we have

$$\Delta S = \sum_i \nu_i s_i = S(P) - S(R) \tag{9.107}$$

where s_i is the molar entropy of B_i in state i; and $S(P)$ and $S(R)$ are the sums of the entropies of all of the products and all of the reactants, respectively, of the change in state (9.106).

The variations of ΔS with temperature and pressure over a range in which no reactant or product undergoes a change in aggregation state may be found by use of equations corresponding to eqns. (6.25)—(6.29)

$$d\Delta S = \sum_i \nu_i ds_i = dS(P) - dS(R) = \Delta dS \tag{9.108}$$

Division of eqn. (9.108) first by dT at constant p, then by dp at constant T, and finally substitution from eqns. (9.13) and (9.23) give the following relations:

$$\left(\frac{\partial \Delta S}{\partial T}\right)_p = \sum_i \nu_i \left(\frac{\partial s_i}{\partial T}\right)_p = \left[\frac{\partial S(P)}{\partial T}\right]_p - \left[\frac{\partial S(R)}{\partial T}\right]_p = \Delta \left(\frac{\partial S}{\partial T}\right)_p \tag{9.109}$$

$$= \sum_i \nu_i \frac{c_{pi}}{T} = \frac{C_p(P)}{T} - \frac{C_p(R)}{T} = \frac{\Delta C_p}{T} \tag{9.110}$$

and

$$\left(\frac{\partial \Delta S}{\partial p}\right)_T = \sum_i \nu_i \left(\frac{\partial s_i}{\partial p}\right)_T = \left[\frac{\partial S(P)}{\partial p}\right]_T - \left[\frac{\partial S(R)}{\partial p}\right]_T = \Delta \left(\frac{\partial S}{\partial p}\right)_T \tag{9.111}$$

$$= -\sum_i \nu_i \left(\frac{\partial v_i}{\partial T}\right)_p = -\left[\frac{\partial V(P)}{\partial T}\right]_p + \left[\frac{\partial V(R)}{\partial T}\right]_p = -\Delta \left(\frac{\partial V}{\partial T}\right)_p \tag{9.112}$$

When both temperature and pressure change we have the analogue of eqn. (9.36)

$$\Delta S(p_2, T_2) - \Delta S(p_1, T_1) = \int_{p_1, T_1}^{T_2} \Delta C_p \frac{dT}{T} - \int_{T_2, p_1}^{p_2} \Delta \left(\frac{\partial V}{\partial T}\right)_p dp \tag{9.113}$$

If one of the reactants or products undergoes a change in aggregation state within the limits of the temperature or pressure integration, we must use the method of eqns. (6.30)—(6.32).

We recall that enthalpies of reaction may be determined calorimetrically (Section 6.3), even if the chemical reaction takes place irreversibly, provided that the *total work* of the process is zero (bomb calorimeter) or is adequately represented by the expression $p\Delta V$ or $\Sigma_i p_i \Delta V_i$ (constant-pressure or flow calorimeter) [see Table 4.4 and eqns. (4.33)].

Entropies of reaction are not determined in this way, since $T\Delta S$ is equal to a heat quantity only for a process which is essentially reversible in all respects, as for example, a chemical reaction carried out in a reversible galvanic cell. This and other methods involving equilibrium measurements have limited application. The usual method of determining ΔS for a chemical change in state by thermal measurements is to determine the entropy of each substance taking part in the chemical reaction using the method suggested by eqns. (9.104) and (9.105), followed by the application of eqn. (9.107).

REFERENCES

1 G. B. Kistiakowsky and W. W. Rice, J. Chem. Phys., 7 (1939) 281—288.
2 D. Telfair and W. H. Pielemeier, Rev. Sci. Instrum., 13 (1942) 122—126.
3 H. M. Spencer and J. L. Justice, J. Am. Chem. Soc., 56 (1934) 2311—2312; H. M. Spencer and G. N. Flannagan, ibid., 64 (1942) 2511—2513; H. M. Spencer, ibid., 67 (1945) 1859—1860; H. M. Spencer, ibid., Ind. Eng. Chem., 40 (1948) 2152—2154; see also references in Section 6.5.
4 E. Buckingham, Philos. Mag., 15 (1908) 526—538; F. G. Keyes, Temperature, Reinhold, New York, 1941, pp. 45—59; J. R. Roebuck and T. A. Murrell, ibid., pp. 60—73; J. A. Beattie, M. Benedict and J. Kaye, Proc. Am. Acad. Arts Sci., 74 (1941) 343—370.
5 See the work of Buckingham and that of Beattie et al., cited in ref. 4.
6 L. J. Gillespie, J. Am. Chem. Soc., 47 (1925) 305—312; E. Lurie and L. J. Gillespie, J. Am. Chem. Soc., 49 (1927) 1146—1157.
7 J. W. Gibbs, Scientific Papers, Vol. I, Thermodynamics, Longmans, Green and Co., New York, 1906, pp. 153—164, 165—168.
8 J. A. Beattie, Phys. Rev., 36 (1930) 132—145; Chem. Rev., 44 (1949) 141—192.
9 D. Berthelot, Trav. Mem. Bur. Int. Poids Mes., 13 (1907) 113.

Chapter 10

WORK CONTENT AND FREE ENERGY

Gibbs [1] noted that the energy and enthalpy of a system may be called, respectively, the heat function for constant volume and the heat function for constant pressure. He introduced two new extensive state properties which he distinguished by the Greek letters ψ and ζ, but never named. He pointed out that $-\psi$ is the force function for constant temperature. We may call $-\zeta$ the force function for non-expansion work at constant pressure and temperature.

Both the symbols and names used for these two functions are in disarray today. We shall use the symbol A for the ψ function and call it the work content, which is the name in general use by American chemists. This term has the disadvantage that it carries the implication that work is a state property, which it is not. Other names used for this function are equally unsatisfactory. Helmholtz called it the free energy and represented it by the symbol F. Physicists have called ψ the Helmholtz free energy, Helmholtz function, free energy, or work function, and used the symbol F for it.

We shall call ζ the free energy and represent it by the symbol G. It has been called the Gibbs function and the thermodynamic potential. It has been represented by the symbol F by American chemists, and in practically all large collections of thermodynamic data (see Section 6.5). But the use of F for the Helmholtz free energy by physicists has made it expedient to discontinue the use of this symbol for either ψ or ζ.

Unless otherwise noted, we shall consider closed systems in this chapter. Under these conditions, eqns. (8.65) and (8.66) apply

$$dE = TdS - pdV - đW_{XR} \tag{10.1}$$

$$dH = TdS + Vdp - đW_{XR} \tag{10.2}$$

We shall use the symbol W to represent all kinds of work (see Section 3.3); W_{EX} denotes expansion work, and W_X signifies all other kinds of work. Thus

$$W = W_{EX} + W_X \tag{10.3}$$

An additional subscript R denotes the work of a reversible process.

10.1. Definition of work content and free energy. The work content A and free energy G of a simple system are defined by the relations

$$A \equiv E - TS \tag{10.4}$$

$$G \equiv A + pV \equiv H - TS \equiv E - TS + pV \tag{10.5}$$

Evidently, A and G are both extensive state properties of a system. The functions A and G contain at each temperature T an integration parameter of the form

$$a - bT \tag{10.6}$$

Here a and b are, respectively, the arbitrary values assigned to the energy and the entropy of the system when in a chosen standard state (see Section 6.2). The standard state selected for a need not, and usually is not, the same as that used for b. The third law of thermodynamics assigns the value zero to the entropy constant b for each chemical substance when in a particular state at 0 K (see Chapter 11).

Let a system B undergo the following change in state

$$\text{B (state 1)} = \text{B (state 2)} \tag{10.7}$$

by any process. From eqns. (10.4) and (10.5), we find the following relations for the changes in work content ΔA and free energy ΔG of B

$$\Delta A = \Delta E - \Delta(TS) \tag{10.8}$$

$$\Delta G = \Delta A + \Delta(pV) = \Delta H - \Delta(TS) = \Delta E - \Delta(TS) + \Delta(pV) \tag{10.9}$$

When the initial and final states of a system have the same pressure, or the same temperature, or both, these relations become

$$\Delta A = \Delta E - T\Delta S \qquad (T_1 = T_2) \tag{10.10}$$

$$\Delta G = \Delta A + p\Delta V \qquad (p_1 = p_2) \tag{10.11}$$

$$= \Delta H - T\Delta S \qquad (T_1 = T_2) \tag{10.12}$$

$$= \Delta E - T\Delta S + p\Delta V \qquad (p_1 = p_2, T_1 = T_2) \tag{10.13}$$

We note that these equations apply to changes in state. Any process whatever may be employed to carry out a given change in state.

Two other forms of eqns. (10.4) and (10.5) are

$$\frac{A}{T} \equiv \frac{E}{T} - S \tag{10.14}$$

$$\frac{G}{T} \equiv \frac{H}{T} - S \tag{10.15}$$

Equations derived from these expressions have been found to be useful in chemical thermodynamics [2].

10.2. Relation of ΔA to the work of certain processes. We know that, in general, no work function exists for a system, and that the work effect of a reversible process for bringing about a given change in state is not necessarily larger than the work of an irreversible process executing the same change. However, the second-law equation (8.1) is the necessary and sufficient condition that a work function does exist for a system B connected to a single heat reservoir; and eqn. (8.3) shows that this is the work of a reversible process. From eqns. (8.3) and (8.5) we learn that the reversible work is the maximum work that the system B can produce when undergoing a given change in state. Any external system (other than the heat reservoir) may be used in the process in any way, provided that each operates in a reversible cycle and is returned to its initial state at the end of the process.

Let a closed system B which is connected to a single heat reservoir at T_0 undergo the change in state

B(state 1) = B(state 2)   (Heat reservoir at T_0)   (10.16)

by any process. The corresponding changes in energy, entropy, and work content of B are ΔE, ΔS, and ΔA, respectively. By the first-law equation (4.10) we have

$$W_B = Q_B - \Delta E = \int_1^2 đQ_B - \Delta E \tag{10.17}$$

where W_B and Q_B are the work and heat effects produced in its surroundings by the system B, Q_B being transferred between the system and the heat reservoir by one or more Carnot engines, each of which operates in a cycle and has returned to its initial state at the end of the process. Equation (8.52) applied to the system B may be written in the form

$$T_0 \int_1^2 \frac{đQ_B}{T} \leq T_0 \Delta S \tag{10.18}$$

Addition of eqns. (10.17) and (10.18) gives

$$W_B + \int_1^2 \frac{T_0 - T}{T} đQ_B \leq -(\Delta E - T_0 \Delta S) \tag{10.19}$$

By eqn. (8.40), the second term of the left-hand side of the above expression is the work W_C of the Carnot engine, or engines, which connect the system to the heat reservoir. The total work W of any process which produces the change in state (10.16) is the sum of W_B and W_C. Thus

$$W \leq -(\Delta E - T_0 \Delta S) \tag{10.20}$$

The signs of equality and inequality in this expression are the same as those in the Clausius inequality, which comes from the second-law equation (8.1). The sign of equality holds when B undergoes a reversible process (see Section 8.6); and the sign of inequality holds when B undergoes an irreversible process (see Section 8.7).

We have shown above, by reference to eqns. (8.3) and (8.5), that the work W_R produced by the system B when undergoing the change in state (10.16) by any reversible process is the maximum work that can be produced by any process which brings about this same change. Hence, from eqn. (10.20), this maximum work is

$$W_R = -(\Delta E - T_0 \Delta S) = -\Delta(E - T_0 S) \tag{10.21}$$

We note that W_R depends only on the change in state of the system and the fixed temperature T_0 of the heat reservoir.

Let us consider the following particular cases of expressions (10.20) and (10.21):

1. When the system B undergoes a cyclical process, we simply recover the second-law equations (8.1) and (8.3).

2. For a constant-volume process, the expansion work W_{EX} is zero, and hence the non-expansion work W_X and the non-expansion reversible work W_{XR} attending such a process are

$$W_X \leqslant W_{XR} = -(\Delta E - T_0 \Delta S) \qquad (V = \text{constant}) \tag{10.22}$$

3. When the temperatures of the system B in its initial and final states are uniform and equal to the fixed temperature of the heat reservoir, the work W of any process and the work W_R of any reversible process which produces the change in state (10.16) are

$$W \leqslant W_R = -(\Delta E - T \Delta S) = -\Delta A \qquad (T_1 = T_2 = T_0 \equiv T) \tag{10.23}$$

Here W_R is the maximum work that can be produced by the change in state (10.16). We note that the conditions imposed on eqn. (10.23) do not require that the process be isothermal.

Now an isothermal process fulfills the conditions imposed on eqn. (10.23). Hence, when a closed system B connected to a single heat reservoir undergoes any isothermal process subject to the restrictions imposed on eqn. (10.23), the work W which is produced in the surroundings is related to the decrease in work content $-\Delta A$ of B in the following manner

$$W < -\Delta A \quad \text{(an irreversible process)} \tag{10.24}$$

$$W_R = -\Delta A \quad \text{(a reversible process)} \tag{10.25}$$

$$W > -\Delta A \quad \text{(no process whatever)} \tag{10.26}$$

Thus, when the system B undergoes an isothermal process in which the work effect is zero, we have

$$-\Delta A \geqslant 0, \text{ or } \Delta A \leqslant 0 \qquad (W = 0) \tag{10.27}$$

Since ΔA depends only on the change in state of the system, eqns. (10.24)–(10.27) are valid even when the process is not isothermal, provided the restrictions applying to eqn. (10.23) are observed. We note that W includes the work W_C of the Carnot engine, or engines, used to connect the system to the heat reservoir.

In virtue of the foregoing discussion, we may write the following criterion of equilibrium for a system that is isolated except that it is connected to a single heat reservoir [1]: "When a system has a uniform temperature throughout, the additional conditions which are necessary and sufficient for equilibrium may be expressed by

$(\delta \psi)_t \geqslant 0"$, that is, $(\delta A)_T \geqslant 0$ \hfill (10.28)

When these conditions apply, eqn. (10.27) assures us that no "possible variation" can take place in the system. Hence the system is in a state of thermodynamic equilibrium (see Section 1.5). A "possible variation" is one that does not violate the arbitrary restrictions which we impose in defining the system, or the general laws of matter.

The subscripts t and T in eqn. (10.28) denote that the temperatures of the system in its initial and varied states are each uniform and equal to the temperature of the heat reservoir to which it is connected. When the criterion of equilibrium (10.28) is applied to a system: "Any external systems may be used in the process in any way not affecting the condition of reversibility, if restored to their original condition at the close of the process; nor does the limitation in regard to the use of heat apply to such heat as may be restored to the source from which it has been taken [1]". But the work effects of these external systems must be added to the work effect of the system under consideration in the application of the criterion.

The above criterion of equilibrium holds only for systems connected to a single heat reservoir and having a uniform temperature T, and it tests the equilibrium of the system only with respect to varied states of the same uniform temperature. All other states are not to be considered possible variations of the initial state of the system. However, the temperature of the system need not be constant during the variation, if the conditions given in the last paragraph are fulfilled.

10.3. Relation of ΔG to the work of certain processes. Let the closed system B of the last section be surrounded by a medium M (see Section 1.1) which has a uniform pressure p_0 and a uniform temperature T_0. The medium is in a state of thermodynamic equilibrium with respect both to states whose properties differ finitely as well as those whose properties differ infinitesimally from their initial values. The medium is so large that its intensive properties are not appreciably changed when it interacts with the system. All heat which passes through the boundary of the system is transferred to or from the medium by one or more Carnot engines. The outer boundary of the medium is rigid and perfectly non-conducting to heat.

Let ΔE, ΔS, ΔG, and ΔV be, respectively, the changes in energy, entropy, free energy, and volume attending the change in state

B (state 1) = B (state 2) \hfill (M at p_0, T_0) \hfill (10.29)

The work W produced in the surroundings of the system when the change in state (10.29) is brought about by any process, and the work W_R of a reversible process producing the same change are given by eqns. (10.20) and (10.21). But the concomitant expansion work W_{EX} expended in changing the volume of the medium is

$W_{EX} = p_0 \Delta V$ \hfill (10.30)

Subtraction of eqn. (10.30) from eqns. (10.20) and (10.21) gives relations which may be written

$$\left. \begin{array}{l} W_X \leqslant W_{XR} = -(\Delta E - T_0 \Delta S + p_0 \Delta V) \\ \phantom{W_X \leqslant W_{XR}} = -\Delta(E - T_0 S + p_0 V) \end{array} \right\} \quad (10.31)$$

In these equations, W_X is the net, or useful, work attending a process which brings about the change in state (10.29). It is written W_X because it consists wholly of non-expansion work. The quantity W_{XR} is the corresponding net, or useful, non-expansion work of a reversible process producing the same change. Evidently, it is the maximum useful work which can be obtained from the change in state (10.29) [see eqn. (10.21)].

From eqn. (10.31) we find that W_{XR} is the increment of the function $-(E - T_0 S + p_0 V)$ which is a joint property of the system and the medium.

When the pressures and temperatures of the system B in its initial and final states are uniform and equal to the corresponding pressure and temperature of the medium, eqn. (10.31) takes the form

$$W_X \leqslant W_{XR} = -(\Delta E - T\Delta S + p\Delta V) = -\Delta G \quad (p_1 = p_2 = p_0 \equiv p),$$
$$(T_1 = T_2 = T_0 \equiv T) \quad (10.32)$$

We note that in the derivation of this equation, the effect of gravity on the pressures of the system and the medium have been neglected.

An isopiestic, isothermal process fulfills the conditions imposed on eqn. (10.32). Hence, when a closed system B surrounded by a medium M undergoes the change in state (10.29) subject to the conditions imposed on eqn. (10.32), the non-expansion work W_X which is produced by the system in its surroundings is related to the decrease in free energy $-\Delta G$ of B in the following manner

$W_X < -\Delta G$ (an irreversible process) (10.33)

$W_{XR} = -\Delta G$ (a reversible process) (10.34)

$W_X > -\Delta G$ (no process whatever) (10.35)

Thus, when the system B undergoes an isothermal, isopiestic process in which the useful work W_X is zero, we have

$-\Delta G \geqslant 0$, or $\Delta G \leqslant 0$ $\qquad (W_X = 0)$ (10.36)

Since ΔG depends only on the change in state of the system, eqns. (10.33)–(10.36) are valid even if the process is not isothermal and isopiestic, provided the restrictions applying to eqn. (10.32) are observed. We recall that W_X includes the work W_C of the Carnot engine, or engines, used in the process (see Section 10.2).

On the basis of the above considerations we may write the following criterion of equilibrium for a system that is isolated except for heat and expansion-work interactions with a medium M surrounding the system:

When a system has a uniform pressure and temperature throughout, the additional conditions which are necessary and sufficient for equilibrium may be expressed by

$$(\delta G)_{p,T} \geqslant 0 \tag{10.37}$$

This criterion of equilibrium follows from eqn. (10.36) for a system that is isolated except for the interactions with a medium noted above.

The subscripts p, T in eqn. (10.37) denote that the pressures and temperatures of the system in its initial and final states are uniform and equal to the corresponding quantities for the medium. The process need not be isopiestic and isothermal, provided the conditions imposed on eqn. (10.32) are observed.

10.4. Available energy. Gibbs [3] defined the available energy of the system and medium discussed in Section 10.3 as the greatest amount of useful work W_x that can be produced by the system when surrounded by a medium which has a uniform pressure p_0 and uniform temperature T_0, subject to the conditions imposed on the system B and the medium M in Section 10.3. Engineers use the term availability for this quantity.

Let the system B initially in the state (E_1, S_1, V_1) undergo a reversible process whereby it reaches the state (E_0, S_0, V_0) in which it is in a state of thermodynamic equilibrium with the medium. The change in state of B and, by eqn. (10.31), the corresponding work effect ϕ are

$$B(E_1, S_1, V_1) = B(E_0, S_0, V_0) \qquad (\text{M at } p_0, T_0) \tag{10.38}$$

$$\phi = (E_1 - T_0 S_1 + p_0 V_1) - (E_0 - T_0 S_0 + p_0 V_0) \tag{10.39}$$

Since the system can undergo no further change in state, ϕ is by definition the available energy of the system and medium.

Equation (10.31) can now be written in the form

$$W_x \leqslant W_{xR} = \phi \tag{10.40}$$

10.5. Steady-flow processes. The methods developed in Section 3.11F allow us to apply the thermodynamic relations derived for systems at rest with respect to the observer to those systems undergoing a steady-flow process, for example, the Joule–Thomson experiment (Section 5.5).

Let a closed fluid system B (or one capable of flowing) be surrounded by a medium M at the fixed pressure p_0 and temperature T_0 (see Section 10.3). We recall that the outer boundary of the medium is rigid and perfectly non-conducting to heat.

When the system undergoes the change in state

$$B(\text{state } 1) = B(\text{state } 2) \qquad (\text{M at } p_0, T_0) \tag{10.41}$$

by a steady-flow process in which velocity and gravitational effects may be neglected, eqns. (10.17)–(10.21) hold without change. If the flow of the system through the inlet and outlet tubes of the vessel A (see Fig. 3.7) takes place without shear, the expansion work W_{EX} produced by the system B in the medium is given, by eqn. (3.41), as

$$W_{EX} = (p_2 V_2 - p_1 V_1) = \Delta(pV) \tag{10.42}$$

This work is expended in pushing the system through the vessel A. Subtraction of eqn. (10.42) from eqn. (10.17) gives the following first-law relation for the useful work W_{BX} of the process

$$W_{BX} = Q_B - \Delta E - \Delta(pV) = Q_B - \Delta H \tag{10.43}$$

where ΔH refers to the change in state (10.41).

Subtraction of eqn. (10.42) from eqns. (10.20) and (10.21) gives the first- and second-law equations for the useful work W_X and the useful reversible work W_{XR}, which in this case is also the maximum useful work, of the steady-flow process

$$W_X \leqslant W_{XR} = -[\Delta E + \Delta(pV) - T_0 \Delta S] = -(\Delta H - T_0 \Delta S) \tag{10.44}$$

Evidently, W_{BX}, W_X, and W_{XR} are all non-expansion work quantities. Engineers call them shaft work when they are transmitted to the surroundings by means of a piston rod or rotating shaft passing through the boundary of the vessel A as in, for example, a reciprocating engine, a turbine, or a pump. But the work quantities described above also include any non-expansion work effect of the system such as electrical work, magnetic work, etc. (see Section 3.3).

When the difference between the velocities and the elevations of the system B in the inlet and in the outlet tubes of the vessel A (see Fig. 3.7) must be taken into account, we may write eqns. (10.43) and (10.44) in the following forms

$$W_{BX} = Q_B - [\Delta H + \tfrac{1}{2} M_B \Delta(u^2) + M_B g \Delta z] \tag{10.45}$$

$$W_X \leqslant W_{XR} = -[\Delta H + \tfrac{1}{2} M_B \Delta(u^2) + M_B g \Delta z - T_0 \Delta S] \tag{10.46}$$

Here $\Delta(u^2)$ and Δz are, respectively, the differences between the square of the velocity u and the elevation z of B in its final and its initial states; ΔH refers to the change in state (10.41), in which these two effects are negligible; and M_B is the molecular weight of B.

10.6. Variations of work content and free energy with volume or pressure and temperature. Total differentiation of eqns. (10.4), (10.5), (10.14), and (10.15), and substitution from (10.1) and (10.2) with the term $đW_{XR}$ deleted, give the following relations for the work content A and free energy G of a closed, simple system

$$dA = -SdT - pdV \tag{10.47}$$

$$d\left(\frac{A}{T}\right) = -\frac{E}{T^2} dT - \frac{p}{T} dV \tag{10.48}$$

$$dG = -SdT + Vdp \tag{10.49}$$

$$d\left(\frac{G}{T}\right) = -\frac{H}{T^2}\,dT + \frac{V}{T}\,dp \tag{10.50}$$

Both A and G are functions of any two of the three variables (p, V, T), but the natural set of independent variables for work content is (V, T), and the corresponding set for free energy is (p, T).

From the above relations and eqns. (10.4) and (10.5) we find

$$\left(\frac{\partial A}{\partial T}\right)_V = -S = \frac{A-E}{T} \quad \left(\frac{\partial A}{\partial V}\right)_T = -p \tag{10.51}$$

$$\left[\frac{\partial (A/T)}{\partial T}\right]_V = -\frac{E}{T^2} \quad \left[\frac{\partial (A/T)}{\partial V}\right]_T = -\frac{p}{T} \tag{10.52}$$

$$\left(\frac{\partial G}{\partial T}\right)_p = -S = \frac{G-H}{T} \quad \left(\frac{\partial G}{\partial p}\right)_T = V \tag{10.53}$$

$$\left[\frac{\partial (G/T)}{\partial T}\right]_p = -\frac{H}{T^2} \quad \left[\frac{\partial (G/T)}{\partial p}\right]_T = \frac{V}{T} \tag{10.54}$$

The relations expressing the temperature variations of the work content and free energy given in eqns. (10.51)–(10.54) are called Gibbs–Helmholtz equations. They may also be written in the following forms

$$\left[\frac{\partial (A/T)}{\partial (1/T)}\right]_V = E \quad \left[\frac{\partial (G/T)}{\partial (1/T)}\right]_p = H \tag{10.55}$$

In Parts (a), (b), and (c) of Table 10.1, eqns. (10.47)–(10.50) are extended to hold for systems in which non-expansion work may be produced in the surroundings; and they are applied to several different processes. See eqn. (10.3) for the symbols used for the different kinds of work.

In Section (d) of Table 10.1, the quantities ΔE, ΔS, ΔH, ΔA, and ΔG are the changes in various thermodynamic properties of a system B attending an *isothermal change in state*.

B (state 1) = B (state 2) (T) (10.56)

in which a chemical reaction may occur. When B consists of two or more phases the subscript V (or p) to a partial derivative of the Gibbs–Helmholtz equations (10.1.14) and (10.1.15) denotes that the volume (or pressure) of *each phase* is held constant in the differentiation. For the modifications of these equations when non-expansion work is produced, see Section 10.23 and Table 10.6.

10.7. The chemical potential. Because of its importance in the study of equilibrium among the phases of a heterogeneous system, the partial molar free energy $\bar{g}_i$ of a component B_i of a simple, homogeneous system of c components was given a special name and symbol by Gibbs [4]. He called

TABLE 10.1

SOME EQUATIONS FOR THE WORK CONTENT AND FREE ENERGY OF A CLOSED SYSTEM

Process	$A \equiv E - TS$	$G \equiv H - TS$	
(a) General relations			
	$\Delta A = \Delta E - \Delta(TS)$	$\Delta G = \Delta H - \Delta(TS)$	(10.
	$dA = -SdT - pdV - đW_{XR}$	$dG = -SdT + Vdp - đW_{XR}$	(10.
	$\Delta\left(\dfrac{A}{T}\right) = \Delta\left(\dfrac{E}{T}\right) - \Delta S$	$\Delta\left(\dfrac{G}{T}\right) = \Delta\left(\dfrac{H}{T}\right) - \Delta S$	(10.
	$d\left(\dfrac{A}{T}\right) = -\dfrac{E}{T^2}dT - \dfrac{p}{T}dV - \dfrac{đW_{XR}}{T}$	$d\left(\dfrac{G}{T}\right) = -\dfrac{H}{T^2}dT + \dfrac{V}{T}dp - \dfrac{đW_{XR}}{T}$	(10.
(b) A reversible, isothermal process			
Constant T	$\Delta A = -W_R$	$\Delta G = \Delta(pV) - W_R$	(10.
Constant T	$dA = -pdV - đW_{XR}$	$dG = Vdp - đW_{XR}$	(10.
Constant V, T	$\Delta A = -W_{XR}$	$\Delta G = V\Delta p - W_{XR}$	(10.
Constant p, T	$\Delta A = -p\Delta V - W_{XR}$	$\Delta G = -W_{XR}$	(10.
(c) A reversible, isothermal process with $W_{XR} = 0$			
Constant T	$\Delta A = -W_{EXR}$	$\Delta G = \Delta(pV) - W_{EXR}$	(10.
Constant T	$dA = -pdV$	$dG = Vdp$	(10.1
Constant V, T	$\Delta A = 0$	$\Delta G = V\Delta p$	(10.1
Constant p, T	$\Delta A = -p\Delta V$	$\Delta G = 0$	(10.1
(d) An isothermal change in state with $W_{XR} = 0$			
	$\Delta A = \Delta E - T\Delta S$	$\Delta G = \Delta H - T\Delta S$	(10.1
	$\left(\dfrac{\partial \Delta A}{\partial T}\right)_V = -\Delta S = \dfrac{\Delta A - \Delta E}{T}$	$\left(\dfrac{\partial \Delta G}{\partial T}\right)_p = -\Delta S = \dfrac{\Delta G - \Delta H}{T}$	(10.1
	$\left[\dfrac{\partial (\Delta A/T)}{\partial T}\right]_V = -\dfrac{\Delta E}{T^2}$	$\left[\dfrac{\partial (\Delta G/T)}{\partial T}\right]_p = -\dfrac{\Delta H}{T^2}$	(10.1

When a system consists of two or more phases:
(1) Apply eqns. (4.33) to eqns. (10.1.2)–(10.1.12) if pressures of phases are different.
(2) Hold V (or p) of each phase constant in eqns. (10.1.14) and (10.1.15).

it the potential of B_i and designated it by the symbol μ_i. It is now called the chemical potential of B_i in the system. By eqn. (7.3)

$$\mu_i \equiv \bar{g}_i = \left(\dfrac{\partial G}{\partial n_i}\right)_{pTn} \qquad (i = 1, 2, \ldots, c) \qquad (10.57)$$

where G is the free energy of a solution composed of $n_1, n_2, \ldots, n_c$ moles of the components $B_1, B_2, \ldots, B_c$, respectively.

The chemical potential μ of a pure substance B is

$$\mu = g = \dfrac{G}{n} \qquad (10.58)$$

where G is the free energy of n moles of B.

All of the general relations derived in Chapter 7 for partial molar quantities hold for the chemical potential. Thus, application of eqn. (7.6) gives the following important relation for a simple, homogeneous system

$$G = \sum_i \mu_i n_i \tag{10.59}$$

Then, by the definitions (10.5), we find

$$E = TS - pV + \sum_i \mu_i n_i \tag{10.60}$$

$$H = TS + \sum_i \mu_i n_i \tag{10.61}$$

$$A = -pV + \sum_i \mu_i n_i \tag{10.62}$$

The summations extend over the c components of the system.

By eqns. (7.9) and (10.49)

$$\sum_i n_i d\mu_i = -SdT + Vdp \tag{10.63}$$

Total differentiation of eqns. (10.59)–(10.62) and use of eqn. (10.63) give

$$dE = TdS - pdV + \sum_i \mu_i dn_i \tag{10.64}$$

$$dH = TdS + Vdp + \sum_i \mu_i dn_i \tag{10.65}$$

$$dA = -SdT - pdV + \sum_i \mu_i dn_i \tag{10.66}$$

$$dG = -SdT + Vdp + \sum_i \mu_i dn_i \tag{10.67}$$

From eqns. (10.64)–(10.67) we find

$$\mu_i = \left(\frac{\partial E}{\partial n_i}\right)_{SVn} = \left(\frac{\partial H}{\partial n_i}\right)_{Spn} = \left(\frac{\partial A}{\partial n_i}\right)_{VTn} = \left(\frac{\partial G}{\partial n_i}\right)_{pTn} \tag{10.68}$$

Thus, it is only the free energy of which μ_i is the partial molar quantity.

Cross differentiation of eqns. (10.64)–(10.67) give the four Maxwell relations (see Section 9.3) and many corresponding equations involving chemical potentials. Thus, by using reciprocity relations derived from eqn. (10.67), we find

$$d\mu_i = -\bar{s}_i dT + \bar{v}_i dp \qquad (n = \text{constant}) \tag{10.69}$$

Equations (10.5) and (10.57) give

$$\mu_i = \bar{e}_i - T\bar{s}_i + p\bar{v}_i \tag{10.70}$$

In this equation, and also in eqn. (10.69), partial molar quantities become simply the corresponding molar quantities for a pure substance.

10.8. Changes in work content and free energy attending finite changes in state. Since A and G are state properties, we may integrate eqns. (10.47)–(10.50) along any path. Let a closed, simple, homogeneous system B undergo the following change in state

$$B(p_1, V_1, T_1) = B(p_2, V_2, T_2) \tag{10.71}$$

where only two of the three variables (p, V, T) are independent. We shall integrate the expressions for dA and $d(A/T)$ from T_1 to T_2 at constant V_1, and then from V_1 to V_2 at constant T_2; and we shall integrate the relations for dG and $d(G/T)$ from T_1 to T_2 at constant p_1, and then from p_1 to p_2 at constant T_2. The following relations result:

$$\Delta A = - \int_{V_1 T_1}^{T_2} S dT - \int_{T_2 V_1}^{V_2} p dV \tag{10.72}$$

$$\Delta\left(\frac{A}{T}\right) = - \int_{V_1 T_1}^{T_2} \frac{E}{T^2} dT - \frac{1}{T_2} \int_{T_2 V_1}^{V_2} p dV \tag{10.73}$$

$$\Delta G = - \int_{p_1 T_1}^{T_2} S dT + \int_{T_2 p_1}^{p_2} V dp \tag{10.74}$$

$$\Delta\left(\frac{G}{T}\right) = - \int_{p_1 T_1}^{T_2} \frac{H}{T^2} dT + \frac{1}{T^2} \int_{T_2 p_1}^{p_2} V dp \tag{10.75}$$

In the above equations, the pressure and volume integrals diverge at zero pressure for gases. The method used to overcome this difficulty is described in a later chapter.

The volume and pressure integrals of eqns. (10.72)–(10.75) can be evaluated if we are given the equations of state of the phase B in the form $p = p(V, T)$ or $V = V(p, T)$, respectively. But the temperature integrals of these equations are not expressed directly in terms of a heat capacity, which is the thermal property of a phase most frequently measured. In the next section we shall express the temperature variations of the free energy in a more tractable form.

10.9. Alternative equations for ΔA and ΔG attending finite changes in state. The methods used to transform the work content expressions (10.72) and (10.73) into forms more adaptable to numerical calculations parallel those employed for the free-energy equations (10.74) and (10.75). Hence we shall limit further discussion to a consideration of the latter two relations.

Let the system B of eqn. (10.71) undergo the change in state

$$B(T_1) = B(T) \qquad (p = p_1) \tag{10.76}$$

where T is a variable temperature. When applied to this change in state, eqns. (9.36) and (5.12) read

$$S = \int_{T_1}^{T} \frac{C_p}{T} dT + S_1 \qquad (p = p_1) \tag{10.77}$$

$$H = \int_{T_1}^{T} C_p dT + H_1 \qquad (p = p_1) \tag{10.78}$$

Here C_p is the constant pressure heat capacity of B in the state (p_1, T), and S_1 and H_1 are, respectively, the entropy and enthalpy of B in the state (p_1, T_1).

From these relations and eqns. (10.74), (10.75), and (10.5), we find the following three relations expressing G as a function of T along the isopiestic $p = p_1$

$$G = -\int_{T_1}^{T} dT' \int_{T_1}^{T'} \frac{C_p}{T''} dT'' + (G_1 + T_1 S_1) - T S_1 \qquad (p = p_1) \qquad (10.79)$$

$$\frac{G - H_1}{T} = -\int_{T_1}^{T} \frac{dT'}{T'^2} \int_{T_1}^{T'} C_p dT'' + \frac{G_1 - H_1}{T_1} \qquad (p = p_1) \qquad (10.80)$$

$$G = H - TS, \quad H = \int_{T_1}^{T} C_p dT + H_1, \quad S = \int_{T_1}^{T} \frac{C_p}{T} dT + S_1 \qquad (p = p_1) \qquad (10.81)$$

Integration by parts shows that

$$-\int_{T_1}^{T} dT' \int_{T_1}^{T'} \frac{C_p}{T''} dT'' = -T \int_{T_1}^{T} \frac{dT'}{T'^2} \int_{T_1}^{T'} C_p dT'' = \int_{T_1}^{T} C_p dT' - T \int_{T_1}^{T} \frac{C_p}{T'} dT'$$

$$(p = p_1) \qquad (10.82)$$

From eqn. (10.5) we find

$$(G_1 + T_1 S_1) - T S_1 = H_1 + \frac{G_1 - H_1}{T_1} T = H_1 - T S_1 \qquad (p = p_1) \qquad (10.83)$$

Hence, eqns. (10.79)–(10.81) are in agreement.

Returning now to the change in state (10.71), we may write eqns. (10.74) and (10.75) in the following forms

$$\Delta G = \int_{p_1, T_1}^{T_2} C_p dT - T_2 \int_{p_1, T_1}^{T_2} \frac{C_p}{T} dT - S_1 (T_2 - T_1) + \int_{T_2, p_1}^{p_2} V dp \qquad (10.84)$$

$$\Delta \left(\frac{G}{T}\right) = \frac{1}{T_2} \int_{p_1, T_1}^{T_2} C_p dT - \int_{p_1, T_1}^{T_2} \frac{C_p}{T} dT + H_1 \left(\frac{1}{T_2} - \frac{1}{T_1}\right) + \frac{1}{T_2} \int_{T_2, p_1}^{p_2} V dp \qquad (10.85)$$

Evidently, eqns. (10.72) and (10.73) may be written

$$\Delta A = \int_{V_1, T_1}^{T_2} C_V dT - T_2 \int_{V_1, T_1}^{T_2} \frac{C_V}{T} dT - S_1 (T_2 - T_1) - \int_{T_2, V_1}^{V_2} p dV \qquad (10.86)$$

$$\Delta\left(\frac{A}{T}\right) = \frac{1}{T_2}\int_{V_1,T_1}^{T_2} C_V dT - \int_{V_1,T_1}^{T_2} \frac{C_V}{T} dT + E_1\left(\frac{1}{T_2} - \frac{1}{T_1}\right) - \frac{1}{T_2}\int_{T_2,V_1}^{V_2} p\, dV \quad (10.87)$$

If the system B has the same initial and the same final states in the above four equations, S_1 has the same value in (10.84) and (10.86), while H_1 in (10.85) is equal to E_1 in (10.87) plus the product $p_1 V_1$. We note that S_1 and E_1 are the integration constants in the work content, and S_1 and H_1 are the corresponding constants in the free energy [see eqn. (10.6)].

We may evaluate the right-hand sides of eqns. (10.84)–(10.87) if we are given the equation of state of the phase B and either C_p as a function of T along some isopiestic or C_V as a function of T along some isometric.

10.10. Integrated equations for the variation of free energy with temperature at constant pressure. Let the constant-pressure heat capacity C_p of the system B of eqn. (10.76) be expressed as a function of T along the isopiestic $p = p_1$ by the equation

$$C_p = a + bT + cT^2 + dT^3 \qquad (p = p_1) \qquad (10.88)$$

Substituting from this equation into (9.13) and (5.5) and integrating, we find

$$S = a \ln T + bT + \frac{c}{2}T^2 + \frac{d}{3}T^3 + S_I \qquad (p = p_1) \qquad (10.89)$$

$$H = aT + \frac{b}{2}T^2 + \frac{c}{3}T^3 + \frac{d}{4}T^4 + H_I \qquad (p = p_1) \qquad (10.90)$$

where S_I and H_I are integration constants which have no special physical significance since eqn. (10.88) is not valid down to 0 K. Their values may be found from the above equations if we are given the values of S and H at some one temperature.

Substitution from the above eqns. into (10.53) and (10.54), integration, and use of eqn. (10.5) give

$$G = aT - aT \ln T - \frac{b}{2}T^2 - \frac{c}{6}T^3 - \frac{d}{12}T^4 + G_I - TS_I \quad (p = p_1) \quad (10.91)$$

$$G = -aT \ln T - \frac{b}{2}T^2 - \frac{c}{6}T^3 - \frac{d}{12}T^4 + H_I + IT \quad (p = p_1) \quad (10.92)$$

$$G = aT - aT \ln T - \frac{b}{2}T^2 - \frac{c}{6}T^3 - \frac{d}{12}T^4 + H_I - TS_I \quad (p = p_1) \quad (10.93)$$

In the above equations, G_I and I are integration constants. Evidently

$$G_I = H_I, \quad I = a - S_I \qquad (p = p_1) \qquad (10.94)$$

10.11. Phase diagrams. In Sections 3.16, 5.3, and 9.11 we considered some aspects of phase transformations in closed, simple, univariant systems. We recall that when a state of thermodynamic equilibrium subsists among the phases of such a system, all of the intensive properties of each phase are functions of any one intensive property of any one phase, for example, pressure, temperature, density, or the concentration of one component.

We shall continue to use a system consisting of one component in two phases in equilibrium as an example of a univariant system; but every closed, simple system of c components in $c + 1$ phases in a state of thermodynamic equilibrium is univariant. For example, the system

$$\text{NaCl(s), NaCl} \cdot n\text{H}_2\text{O (l), H}_2\text{O(g)} \tag{10.95}$$

has two components in three phases and is univariant when at equilibrium. Each of the following systems is also univariant when at equilibrium

$$\text{CaCO}_3\text{(s), CaO(s), CO}_2\text{(g)} \tag{10.96a}$$

$$\text{V}_2\text{O}_5\text{(s), V}_2\text{O}_4\text{(s), VOSO}_4\text{(s), SO}_3\text{(g)} + \text{SO}_2\text{(g)} \tag{10.96b}$$

In the first (10.96a) there is one chemical reaction among the three species, and hence there are two components in three phases [see eqn. (1.16)]. In the second system (10.96b) there are two independent chemical reactions [5], and hence there are three components in four phases.

If we choose pressure and temperature as the variables of interest in a univariant system, a graph of p against T in the pT-plane is called a phase diagram.

Phase diagrams at high and at low pressures for the one-component system water are shown in Fig. 10.1. Corresponding diagrams for sulfur are given [6] in Fig. 10.2. The solid lines divide the pT-plane into regions of pressure and temperature in which the phase indicated is stable. Along these equilibrium curves, the phases existing in the two contiguous regions are coexistent.

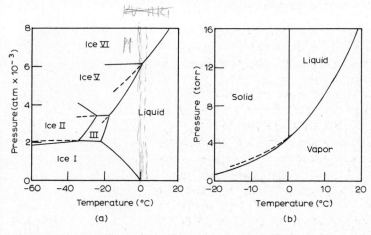

Fig. 10.1. (a) High-pressure and (b) low-pressure phase diagrams for water. There is a triple point VII—VI—liquid at 22,400 atm and 81.6°C. The liquid—vapor critical point is 218.3 atm and 374.2°C.

At a point, called a triple point, where three curves meet, the three phases in the adjoining regions are coexistent.

A broken line indicates the extension of a curve into a region where the equilibrium is metastable. For example, the curve for ice VI—liquid extends into the ice V region (Fig. 10.1a), and the fusion and sublimation curves of rhombic sulfur extend into the monoclinic region (Fig. 10.2). Along the broken line, the univariant system is stable with regard to continuous variations in state (a small change in p and T) and it is unstable with regard to discontinuous changes in state (the addition of a small amount of the phase stable in the region).

10.12. *The Clapeyron equation.* This relation expresses the slope $(dp/dT)_{SAT}$ of each of the curves $p = p(T)$ in a pT-diagram in terms of the molar (or specific) entropies (or enthalpies) and volumes of the two or more phases which are in equilibrium along the curve. In the following equations a′ and a″ are any two different aggregation states:

$$H_2O(a') = H_2O(a'') \qquad \Delta g \qquad (p, T) \qquad (10.97a)$$

$$H_2O(a') = H_2O(a'') \qquad \Delta g + d\Delta g \qquad (p + dp, T + dT) \qquad (10.97b)$$

Each change in state may be brought about by a reversible, isothermal, isopiestic process in a closed, simple system. From eqn. (10.1.12) of Table 10.1, we find that

$$\Delta g = 0 \text{ and } \Delta g + d\Delta g = 0 \qquad (10.98)$$

Hence

$$d\Delta g = d(g'' - g') = dg'' - dg' = \Delta dg = 0 \qquad (10.99)$$

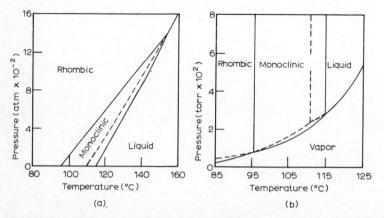

Fig. 10.2. (a) High-pressure and (b) low-pressure phase diagrams for sulfur. The liquid—vapor critical point is at approximately 115 atm and 1040°C.

Substitution from eqn. (10.49) gives

$$-s''dT + v''dp = -s'dT + v'dp \qquad (10.100)$$

whence

$$\left(\frac{dp}{dT}\right)_{SAT} = \frac{s''-s'}{v''-v'} = \frac{\Delta s}{\Delta v} = \frac{\Delta h}{T\Delta v} \qquad (10.101)$$

The last equality holds in virtue of eqn. (9.51).

A consistent set of units must be used in eqn. (10.101), for example, p in dyne cm^{-2}, Δv in cm^3 mole^{-1} (or g^{-1}), and Δh in erg mole^{-1} (or g^{-1}). In any case, Δv and Δh must be taken for the same mass of substance, and T must be on an absolute thermodynamic scale.

Equation (10.101) is an exact thermodynamic relation. It applies to every univariant system. In this equation Δv and Δh refer to a definite change in state involving a phase transformation in a univariant system at (p, T), and $(dp/dT)_{SAT}$ is the slope of the saturation curve at that same pressure and temperature.

The Clapeyron equation has many important uses:

1. It is used to correlate measurements of Δh, Δv, and the slope dp/dT of the saturation curve of a univariant system. It also furnishes a check on the temperature scale employed in the measurements, since the quantity T appearing in the equation is on the absolute thermodynamic scale (see ref. 7).

2. It is used to calculate sublimation and vapor pressures at low temperatures, where it is difficult to measure pressure, from measurements of Δh and $(dp/dT)_{SAT}$ at higher temperatures (see Section 10.14).

3. It is used to calculate Δh at high pressure, where it is relatively easy to measure Δv and $(dp/dT)_{SAT}$, but very difficult to make thermal measurements (see Section 10.15).

The Clapeyron equation imposes the following conditions on the algebraic signs of the quantities involved:

1. If Δh and Δv have the same sign, $(dp/dT)_{SAT}$ is positive. All vaporization and sublimation curves, most fusion curves, and some transition curves have positive slopes.

2. If Δh and Δv have opposite signs, $(dp/dT)_{SAT}$ is negative. A few fusion curves at lower pressures (H$_2$O, Bi, Ga, Sb), where the solid floats on the liquid phase, and some transition curves, have negative slopes.

3. If Δh is zero but Δv is not zero, then $(dp/dT)_{SAT}$ is zero. Some transition curves have a relative maximum or minimum at a point, see for example the equilibrium curve ice I—ice III in the metastable region at $-40°C$.

4. If Δv is zero but Δh not zero, $(dp/dT)_{SAT}$ is infinite. Some transition curves are vertical at a point.

5. If Δv and Δh are both zero, $(dp/dT)_{SAT}$ is indeterminate. For example, the vapor pressure curve of a pure liquid is "limited in one direction by a terminal state at which the distinction of the coexistent states vanishes" [8]. This terminal state is called a critical phase, and the corresponding point on a phase diagram is called a critical point. As this point is approached along

the vapor pressure curve, Δv and Δh approach zero. Critical states involving fluid phases have also been found in multicomponent systems. No critical points in a fusion or transition curve have been discovered.

10.13. Effect of temperature on enthalpy of phase transformation in a univariant system. Let Δh, Δs, Δg, Δv, Δc_p, and Δc_{SAT} refer to the following univariant equilibrium

$$H_2O(a') = H_2O(a'') \qquad \Delta g = 0 \qquad (p, T) \qquad (10.102)$$

From eqn. (9.14) we find

$$T\left(\frac{d\Delta s}{dT}\right)_{SAT} = \Delta c_{SAT} \qquad (10.103)$$

and from eqn. (10.2), we have

$$T\left(\frac{d\Delta s}{dT}\right)_{SAT} = \left(\frac{d\Delta h}{dT}\right)_{SAT} - \Delta v\left(\frac{dp}{dT}\right)_{SAT} \qquad (10.104)$$

From these two relations and the Clapeyron equation (10.101), we find

$$\left(\frac{d\Delta h}{dT}\right)_{SAT} = \Delta c_{SAT} + \frac{\Delta h}{T} \qquad (10.105)$$

Substitution from eqns. (9.45) and (10.101) gives

$$\left(\frac{d\Delta h}{dT}\right)_{SAT} = \Delta c_p + \frac{\Delta h}{T}\left[1 - \frac{T}{\Delta v}\left(\frac{\partial \Delta v}{\partial T}\right)_p\right] \qquad (10.106)$$

If one phase is a vapor or a gas at a moderate pressure (several atmospheres), we may write

$$\Delta v \approx v^G \approx RT/p \qquad (10.107)$$

Under these conditions, the expression enclosed in brackets in eqn. (10.106) is approximately zero, and

$$\left(\frac{d\Delta h}{dT}\right)_{SAT} = \Delta c_p \qquad (10.108)$$

Integration of this relation from a fixed temperature T_1 to a variable temperature T gives the approximate relation

$$\Delta h = \int_{T_1}^{T_2} \Delta c_p \, dT + \Delta h_1 \qquad \text{(saturation curve)} \qquad (10.109)$$

where Δh_1 is the value of Δh at T_1.

10.14. Sublimation and vaporization curves. Consider the following phase transformation in a univariant system

$$H_2O(s, \text{ or } l) = H_2O(g) \qquad (p, T) \qquad (10.110)$$

in which one of the phases is a vapor and the other is a condensed phase. If the pressure is sufficiently low, we may make the two approximations of eqn. (10.107) and obtain the approximate Clapeyron equation

$$\left(\frac{d \ln p}{dT}\right)_{SAT} = \frac{\Delta h}{RT^2} \qquad (10.111)$$

Here p may be expressed in any pressure unit, and Δh and RT must be expressed in the same energy unit. Equation (10.111) has been called the Clausius—Clapeyron equation.

Integration of eqn. (10.111) from (p_1, T_1) to (p_2, T_2) under the assumption that

$$\Delta h = \text{constant} \qquad (10.112)$$

gives

$$\ln \frac{p_2}{p_1} = \frac{\Delta h}{R}\left(\frac{T_2 - T_1}{T_1 T_2}\right) \qquad (10.113)$$

This relation rests on the assumptions of eqns. (10.107) and (10.112). Yet by a compensation of errors it gives more accurate results than we would expect, about 0.3% on the ratio (p_2/p_1) for H_2O from 90 to 100°C.

The indefinite integral of eqn. (10.111), when eqn. (10.112) applies, is

$$\ln p = -\frac{\Delta h}{RT} + \text{constant} \qquad (10.114)$$

For most substances, a graph of $\ln p$ against $1/T$ is approximated surprisingly well over a wide temperature range by a straight line. Thus we may use eqn. (10.114) to extrapolate vapor pressures to low temperature where the measurement of pressure is difficult.

One of the phase transformations occurring at a triple point is always the sum of the other two, and hence the corresponding enthalpy changes are similarly related. For example, at a solid—liquid—vapor triple point, we have

$$\Delta h \text{ (sublimation)} = \Delta h \text{ (fusion)} + \Delta h \text{ (vaporization)} \qquad (10.115)$$

Thus, if we know p, T, and the molar enthalpies of fusion and vaporization at such a triple point for a substance, we can predict its sublimation pressures at temperatures well below the triple point.

10.15. Fusion and transition curves. For the change in state

$$H_2O(s') = H_2O(l, \text{ or } s'') \qquad (p, T) \qquad (10.116)$$

we may write the exact Clapeyron equation in the form

$$\frac{1}{T}\left(\frac{dT}{dp}\right)_{SAT} = \left(\frac{d \ln T}{dp}\right)_{SAT} = \frac{\Delta v}{\Delta h} \qquad (10.117)$$

Because of the difficulty of measuring an enthalpy of fusion or transition at high pressure, this quantity is usually computed by application of eqn. (10.117) to measurements of p, T, and Δv for the appropriate equilibrium curve.

Over a pressure range in which ΔT is small, we may approximate eqn. (10.117) by the relation

$$T_2 - T_1 = T_1 \frac{\Delta v}{\Delta h}(p_2 - p_1) \tag{10.118}$$

10.16. Effect of pressure on vapor pressure. Let one part of a system contain a condensed species B and a vapor B(g), or a condensed univariant mixture which produces B(g) by a chemical reaction. In each case, a fixed amount of an inert gas may be inserted into this part of the system to increase the pressure on the condensed species.

The second part of the system contains only pure B(g) and it is connected to the first part by a rigid, thermally conducting wall permeable to B(g) alone (see Section 9.17). The pressure of this gas is called the equilibrium vapor pressure of B in part one of the system.

Consider the two changes in state

$$H_2O(l, \text{ or } s, p') = H_2O(g, p'') \qquad \Delta g \qquad (T) \tag{10.119a}$$

$$H_2O(l, \text{ or } s, p' + dp') = H_2O(g, p'' + dp'') \qquad \Delta g + d\Delta g \qquad (T) \tag{10.119b}$$

where p'' and $p'' + dp''$ are the equilibrium pressures of $H_2O(g)$ in a system containing $H_2O(l)$ at pressures of p' and $p' + dp'$, respectively. Each change may be brought about in a closed, simple system by a reversible, isothermal process with the pressure of each phase held constant.

Since the two phases of the system are under different pressures, we must modify eqn. (10.1.10) of Table 10.1 in accordance with the footnote to the Table. This does not change eqn. (10.1.12) of the same Table. Hence, eqns. (10.98) and (10.99) apply to the present experiment, and we have

$$\Delta dg = 0 \qquad (T = \text{constant}) \tag{10.120}$$

Substitution from eqn. (10.49) gives

$$v'' dp'' = v' dp' \qquad (T = \text{constant}) \tag{10.121}$$

which may be written in the form

$$dp = \frac{v^c}{v} dp^c \qquad (T = \text{constant}) \tag{10.122}$$

where p^c and v^c refer to the condensed phase, and p and v refer to the vapor. This relation is known as the Poynting equation.

If the gas phase may be considered to behave as a perfect gas, eqn. (10.122) becomes

$$d \ln p = \frac{v^c}{RT} dp^c \qquad (T = \text{constant}) \tag{10.123}$$

Integrating the right-hand side of this equation from p_0 to p^c under the assumption that v^C is constant, and the left-hand side from the corresponding vapor pressure p_0 to p, we find

$$\ln \frac{p}{p_0} = \frac{v^C}{RT} (p^C - p_0) \qquad (T = \text{constant}) \qquad (10.124)$$

The effect of pressure on the vapor pressure of a condensed phase is so small — about 0.07% per atm for water at 25°C — that the following approximate relation is sufficient for most purposes

$$\frac{\Delta p}{p} = \frac{p^C v^C}{RT} \qquad (T = \text{constant}) \qquad (10.125)$$

The above equations apply not only to the vapor pressure of a liquid or a solid, but to the pressure of any univariant system which has a gas phase. Examples of the latter are the pressure of NH_3 in the system $BaCl_2 \cdot 8NH_3(s)$, $BaCl_2(s)$, $NH_3(g)$ [9] and the gas pressures in the two systems of the expressions (10.96).

The equations derived in this section are usually employed for systems in which an excess pressure is applied to the condensed phase (or phases) by the introduction of a chemically inert gas. If the gas dissolves in the condensed phase (or phases), the value of p calculated from eqns. (10.123) and (10.125) must be corrected for this effect. The correction is appreciable only when the condensed phase is liquid.

By eqn. (9.84), which is known as Dalton's law for mixtures of perfect gases, the mole fraction y_1 of the vapor of a substance B_1 in a gas phase in which an inert gas is present is

$$y_1 = p/p^C \qquad (T = \text{constant}) \qquad (10.126)$$

where p is the equilibrium vapor pressure of B_1 at (p^C, T) and p^C is the total gas pressure applied to the condensed phase. In experimental studies of a number of systems composed of a solid, its vapor, and an inert gas, measured values of y_1 were found to be larger than those calculated from eqn. (10.126). This effect may be appreciable when the total pressure is only 1 atm (see ref. 10). When the vapor pressure of a solid is exceedingly small and the total gas pressure is several hundred atmospheres, the ratio of the measured value of y_1 to the value given by eqn. (10.126) may be greater (see ref. 11) than 1,000. This solubility of a condensed phase in an inert gas depends on the condensed phase, the inert gas, the total gas pressure, and the temperature. It can be calculated by treating the gas phase as a mixture of real gases.

10.17. Effect of temperature on the vapor pressure of a phase at constant pressure. Consider the two changes in state

$$H_2O(l, \text{ or } s, p') = H_2O(g, p'') \qquad \Delta g \qquad (T) \qquad (10.127a)$$
$$H_2O(l, \text{ or } s, p') = H_2O(g, p'' + dp'') \qquad \Delta g + d\Delta g \qquad (T + dT) \qquad (10.127b)$$

Each may be brought about by a reversible, isothermal process, with the pressure on each phase constant. Hence eqn. (10.120) applies, and

$$\Delta dg = 0 \tag{10.128}$$

Substitution from eqn. (10.49) gives

$$-s'dT = -s''dT + v''dp'' \qquad (p' = \text{constant}) \tag{10.129}$$

This equation may be written in the form

$$\left(\frac{\partial p}{\partial T}\right)_{p^C} = \frac{\Delta s}{v} = \frac{\Delta h}{Tv} \tag{10.130}$$

In this equation [12] p^C is the constant pressure of the condensed phase, p is the pressure and v the volume of the vapor of the change in state (10.127), while Δs and Δh refer to this same change in state.

If the vapor acts as a perfect gas, eqn. (10.130) may be written

$$\left(\frac{\partial \ln p}{\partial T}\right)_{p^C} = \frac{\Delta h}{RT^2} \qquad (p^C = \text{constant}) \tag{10.131}$$

This equation may be compared with the approximate Clapeyron equation (10.111).

Equation (10.131) applies to vapor pressures determined by the gas current method, in which an inert gas under atmospheric pressure is bubbled through a liquid or allowed to flow over a solid. In computing the vapor pressure of the liquid or solid substance from the composition of the saturated inert gas, we must consider the possibility that the effect discussed in Section 10.16 may be appreciable.

The results of this section may be applied to any univariant system having a gas phase.

10.18. Perfect gases. Let n moles of a perfect gas B undergo the change in state

$$nB(p_1, V_1, T_1) = nB(p_2, V_2, T_2) \tag{10.132}$$

where only two of the three variables (p, V, T) are independent. Substitution from the perfect-gas equation (9.11) into eqns. (10.84) and (10.85) gives

$$\Delta G = \int_{p_1 T_1}^{T_2} nc_p dT - T_2 \int_{p_1 T_1}^{T_2} \frac{nc_p}{T} dT - S_1(T_2 - T_1) + nRT_2 \ln \frac{p_2}{p_1} \tag{10.133}$$

$$\Delta \left(\frac{G}{T}\right) = \frac{1}{T_2} \int_{p_1 T_1}^{T_2} nc_p dT - \int_{p_1 T_1}^{T_2} \frac{nc_p}{T} dT + H_1 \left(\frac{1}{T_2} - \frac{1}{T_1}\right) + nR \ln \frac{p_2}{p_1} \tag{10.134}$$

Here c_p is the molar constant pressure heat capacity of the gas.

Corresponding relations derived from eqns. (10.86) and (10.87) hold for work content.

10.19. Standard molar enthalpy and free energy of a pure substance. From eqns. (9.39) and (10.53) we find that the molar enthalpy $h(p, T)$ and free energy $g(p, T)$ of a pure solid or liquid in the state (p, T) vary slowly with pressure; and hence when p is approximately 1 atm they are virtually equal to the respective standard molar enthalpy h^0 and free energy g^0 of the substance at T.

Application of the general method used for entropy (see Section 9.20) to the above equations gives the following pairs of relations, corresponding to eqns. (9.103) and (9.105), for calculating the values of h^0 and g^0 of a pure gas at T from the respective values $h(p, T)$ and $g(p, T)$ derived from calorimetric measurements

$$h^0 = h(p) - \left(B' - T\frac{dB'}{dT}\right)p - \left(C' - T\frac{dC'}{dT}\right)\frac{p^2}{2} - \cdots \quad \text{(at } T\text{)} \quad (10.135a)$$

$$h^0 = h(p) - \frac{9}{128}\frac{RT_c}{p_c}\left(1 - 18\frac{T_c^2}{T^2}\right)p \quad \text{(at } T\text{)} \quad (10.135b)$$

$$g^0 = g(p) - RT \ln p - B'p - C'\frac{p^2}{2} - \cdots \quad \text{(at } T\text{)} \quad (10.136a)$$

$$g^0 = g(p) - RT \ln p - \frac{9}{128}\frac{RT_c}{p_c}\left(1 - 6\frac{T_c^2}{T^2}\right)p \quad \text{(at } T\text{)} \quad (10.136b)$$

Here B', C', etc. are the coefficients in the pressure virial equation of state (2.44), and p_c and T_c are the critical pressure and temperature of the gas.

Let one mole of a pure substance undergo a change in temperature from $(p, 0\,\text{K})$ where it is in a condensed aggregation state to (p, T) by a constant-pressure process at about 1 atm. From calorimetric measurements we can calculate $h(p, T) - h_0^0$ by use of eqns. (5.12) and (5.28), and $g(p, T) - h_0^0$ from eqn. (10.79), (10.80), or (10.81) (see also Section 10.22). In the free-energy expression, the standard molar free energy g_0^0 of the substance at 0 K has been replaced by the corresponding enthalpy h_0^0 since, in accordance with the third law of thermodynamics, they approach the same value as the Kelvin temperature approaches zero (see Section 11.1). For each pure substance we may assign to h_0^0 any value consistent with the change in state under consideration.

The expression $(g^0 - h_0^0)/T$ for a pure substance is called its free-energy function. In some tables of thermodynamic properties this quantity, together with $h^0 - h_0^0$ and s^0, are listed for equally spaced temperature intervals. In other tables, the fiducial temperature 298 K is used. Since $g = h - Ts$ at any temperature, we have

$$\frac{g^0 - h_0^0}{T} = \frac{h^0 - h_0^0}{T} - s^0 \quad \text{(at } T\text{)} \quad (10.137a)$$

$$\frac{g^0 - h^0_{298}}{T} = \frac{h^0 - h^0_{298}}{T} - s^0 \qquad \text{(at } T\text{)} \qquad (10.137\text{b})$$

The addition of the term $(h^0_0 - h^0_{298})/T$ to each side of eqn. (10.137a) gives eqn. (10.137b). Note that each term of each equation can be evaluated from calorimetric data. Actually we need to compute only two terms in either equation to evaluate the third.

Let eqn. (9.106) be a standard change in state

$$\sum_i \nu_i B_i \text{ (standard state } i\text{)} = 0 \qquad (T) \qquad (9.106)$$

If we are given the value of the free energy function $[(g^0 - h^0_0)/T]_i$ at any temperature T, and the standard molar enthalpy of formation h^0_{0i} (Section 6.4) at 0 K for each reactant and product B_i we can calculate from eqn. (10.137a) the change in free energy ΔG^0 attending the change in state (9.106)

$$\Delta G^0 = T \sum_i \nu_i \left(\frac{g^0 - h^0_0}{T} \right)_i + \sum_i \nu_i h^0_{0i} \qquad \text{(at } T\text{)} \qquad (10.138)$$

We may replace h^0_0 in the above equation by h^0_{298} [see eqn. (10.137b)].

Table 10.2 gives the values at 800 K of certain properties of the substances (listed in Column 1 of the Table) which enter into the Deacon process for the production of chlorine (see ref. 13). The bottom line gives the increment Δ in each quantity attending the standard change in state appearing at the head of the Table.

In Columns 6 and 5 of Table 10.2, h^0 and g^0 are, respectively, the standard molar enthalpies and free energies of formation at 800 K of the substances listed in Column 1. On this basis, h^0 and g^0 are zero for each pure element in its stable aggregation state at every temperature. We could also determine the integration constant in s^0 in the same manner.

The entropy constants in the standard molar entropies s^0 of Column 2 of the Table are determined by assigning the value zero to the standard molar entropy s^0_0 of each pure element and compound at 0 K, in accordance with the third law. On this basis, the free energies $h^0 - Ts^0$ of Column 7 do not have the same values as the corresponding free energies g^0 of Column 5 because of the difference between the integration constants assigned to s^0. We find from the last line of the Table that the free energies of Columns 5 and 7 give the same value of ΔG^0_{800} attending the standard change in state at the head of the Table.

From the data of Table 10.2 we may calculate ΔG^0_{800} by four different routes

Column 5: $\sum_i \nu_i g^0_i = -1098$ cal \qquad (10.139a)

Column 7: $\sum_i \nu_i (h^0 - Ts^0)_i = -1098$ cal \qquad (10.139b)

Columns 2, 6: $\sum_i \nu_i h^0_i - T \sum_i \nu_i s^0_i = -1098$ cal \qquad (10.139c)

TABLE 10.2
CALCULATION OF $\Delta G°$ FOR THE DEACON REACTION AT 800 K
$2HCl(g) + \tfrac{1}{2}O_2(g) = H_2O(g) + Cl_2(g)$

$\Delta G°_{800}$

Gas	$s°$ (cal deg^{-1} mole^{-1})	$-(g° - h°_{298})/T$ (cal deg^{-1} mole^{-1})	$h° - h°_{298}$ (cal mole^{-1})	$g°$ (cal mole^{-1})	$h°$ (cal mole^{-1})	$h° - Ts°$ (cal mole^{-1})	$h°_{298}$ (cal mole^{-1})
$H_2O(g)$	53.464	48.089	4,300	-48,646	-58,905	-101,676	-57,798
$Cl_2(g)$	61.747	56.334	4,331	0	0	-49,398	0
$HCl(g)$	51.595	47.163	3,546	-23,774	-22,440	-63,716	-22,063
$O_2(g)$	56.361	51.629	3,786	0	0	-45,089	0
Δ	-16.159[a]	-15.717[a]	-354[b]	-1,098[b]	-14,025[b]	-1,098[b]	-13,672[b]

[a] Units cal deg^{-1}. [b] Units cal.

Columns 3,8: $T\sum_i \nu_i \left(\dfrac{g^0 - h^0_{298}}{T}\right)_i + \sum_i \nu_i h^0_{298i} = -1098$ cal (10.139d)

10.20. Free energy changes attending chemical changes in state. In order to calculate ΔG for the chemical change in state

$\sum_i \nu_i B_i$ (state i) = 0 (p, T) (10.140)

and determine the effects of pressure and temperature on ΔG, we employ the methods of Sections 6.4–6.6, together with eqn. (10.58), and use the relations developed in Sections 10.8–10.10. Thus

$$\Delta G = \sum_i \nu_i g_i = \sum_i \nu_i \mu_i = G(\mathrm{P}) - G(\mathrm{R}) \qquad (10.141)$$

Here g_i and μ_i are the molar free energy of B_i in the state i, and $G(\mathrm{P})$ and $G(\mathrm{R})$ are the sums of the free energies of all of the products and of all of the reactants, respectively, involved in the change in state.

When no reactant or product of the chemical reaction undergoes a change in aggregation state over the pressure and temperature range of interest, we have

$$d\Delta G = \sum_i \nu_i dg_i = dG(\mathrm{P}) - dG(\mathrm{R}) = \Delta dG \qquad (10.142)$$

From eqns. (10.53) and (10.54), we find

$$\left(\dfrac{\partial \Delta G}{\partial T}\right)_p = -\Delta S \qquad \left(\dfrac{\partial \Delta G}{\partial p}\right)_T = \Delta V \qquad (10.143)$$

$$\left[\dfrac{\partial \Delta(G/T)}{\partial T}\right]_p = -\dfrac{\Delta H}{T^2} \qquad \left[\dfrac{\partial \Delta(G/T)}{\partial p}\right]_T = \dfrac{\Delta V}{T} \qquad (10.144)$$

Application of eqns. (10.84) and (10.85) gives the following relations for the differences between the changes in free energy ΔG_2 and $(\Delta G_2)/T_2$ attending the change in state (10.140) at (p_2, T_2) and the corresponding values ΔG_1 and $(\Delta G_1)/T_1$ attending the change in state at (p_1, T_1)

$$\Delta G_2 = \int_{p_1 T_1}^{T_2} \Delta C_p\, dT - T_2 \int_{p_1 T_1}^{T_2} \dfrac{\Delta C_p}{T}\, dT - \Delta S_1 (T_2 - T_1) + \int_{T_2 p_1}^{p_2} \Delta V dp + \Delta G_1$$

(10.145)

$$\dfrac{\Delta G_2}{T_2} = \dfrac{1}{T_2} \int_{p_1 T_1}^{T_2} \Delta C_p\, dT - \int_{p_1 T_1}^{T_2} \dfrac{\Delta C_p}{T}\, dT + \Delta H_1 \left(\dfrac{1}{T_2} - \dfrac{1}{T_1}\right) + \dfrac{1}{T_2} \int_{T_2 p_1}^{p_2} \Delta V dp$$

$$+ \dfrac{\Delta G_1}{T_1} \qquad (10.146)$$

Here ΔS_1 and ΔH_1 are the changes in entropy and enthalpy, respectively, attending the change in state (10.140) at (p_1, T_1).

We may express ΔG as a function of temperature along the isopiestic p_1

by use of any one of the eqns. (10.91)–(10.93). For example, from (10.92) we find

$$\Delta G(T) = -\Delta a\, T \ln T - \frac{\Delta b}{2} T^2 - \frac{\Delta c}{6} T^3 - \frac{\Delta d}{12} T^4 + \Delta H_\mathrm{I} + \Delta I T$$

$(p = p_1)$ (10.147)

where a, b, c, d are the coefficients of eqn. (10.88) for C_p. From eqn. (10.90) we find

$$\Delta H(T) = \Delta a\, T + \frac{\Delta b}{2} T^2 + \frac{\Delta c}{3} T^3 + \frac{\Delta d}{4} T^4 + \Delta H_\mathrm{I} \qquad (p = p_1) \qquad (10.148)$$

If we know $\Delta H(T)$ at one temperature and also $\Delta G(T)$ at one temperature we can calculate the values of the integration constants ΔH_I and ΔI in eqn. (10.147).

If one of the reactants or products of a chemical reaction undergoes a change in aggregation state within the limits of the temperature or pressure integration, we must use the methods of eqns. (6.30)–(6.32).

10.21. Relation of ΔG and ΔA to chemical changes in state. The free-energy and work-content functions have particular importance in chemical thermodynamics since the natural set of independent variables for these quantities are those most commonly used in experimental investigations; the set (p, T) for free energy and the set (V, T) for work content. In this section we shall consider changes in state which involve chemical reactions. The results so obtained are valid whether or not chemical changes occur.

In Section 10.3 we derived equations for the work effect attending certain processes in a closed system B surrounded by a medium M of uniform and constant pressure p and temperature T. The medium acts as a single heat reservoir connected to the system, and all expansion-work effects of the system are produced in M. Let ΔG be the change in free energy of B and W_X the non-expansion work produced in the surroundings when B undergoes the proposed change in state

$$\mathrm{B_1 = B_2} \qquad (p, T) \qquad (10.149)$$

in which a chemical reaction occurs. Application of eqns. (10.33)–(10.36) gives the following results.

The necessary and sufficient condition that the system $\mathrm{B_1}$ *may* undergo the change in state (10.149) at a finite rate is

$$\Delta G < -W_\mathrm{X} \text{ or } -\Delta G > W_\mathrm{X} \qquad (10.150)$$

The corresponding condition that the system $\mathrm{B_1}$ *cannot* undergo the change in state (10.149) is

$$\Delta G > -W_\mathrm{X} \text{ or } -\Delta G < W_\mathrm{X} \qquad (10.151)$$

The corresponding condition that the system $\mathrm{B_1}$ is in a state of thermodynamic equilibrium with respect to the change in state (10.149) is

$$\Delta G = W_X \tag{10.152}$$

When the non-expansion work W_X of the process producing the change in state (10.149) is zero, we find the following results.

The necessary and sufficient condition that the system B_1 *may* undergo the change in state (10.149) at a finite rate is

$$\Delta G < 0 \qquad (W_X = 0) \tag{10.153}$$

The corresponding condition that the system B_1 *cannot* undergo the change in state (10.149) is

$$\Delta G > 0 \qquad (W_X = 0) \tag{10.154}$$

The corresponding condition that the system B_1 is in a state of thermodynamic equilibrium with respect to the change in state (10.149) is

$$\Delta G = 0 \qquad (W_X = 0) \tag{10.155}$$

Equations (10.150)–(10.155) were derived for a system which is surrounded by a medium and undergoes the change in state (10.149). They are also valid when the change in state (10.149) is brought about by an isopiestic, isothermal process. Such a process fulfills the conditions imposed on eqn. (10.32), from which the above equations were derived.

The various phases of a heterogeneous system may be under different pressures. Equations (10.150)–(10.155) hold when such a system undergoes an isothermal process in which a constant pressure is applied to each phase [see eqns. (4.33) and Table 10.1].

Although thermodynamics tells us whether the change in state (10.149) may or cannot take place in a system surrounded by a medium, it does not give us any information concerning the rate of the process. If the decrease in free energy of the system is greater than the non-expansion work produced in the surroundings, but the rate of the process is considered to be too slow, it would be worthwhile to seek a catalyst to increase the rate. But no catalyst could be found for this change in state if the decrease in free energy of the system is less than the non-expansion work produced in the surroundings.

Let us now apply the methods used above to the work-content function. In Section 10.2 we derived equations for the work effect attending certain processes in a closed system B' connected to a single heat reservoir at a uniform and constant temperature T. Let ΔA be the change in work content of B' and W the total work produced in the surroundings when B' undergoes the proposed change in state

$$B'_1 = B'_2 \qquad (T) \tag{10.156}$$

in which a chemical reaction occurs. Application of eqns. (10.24)–(10.27) gives the following results.

The necessary and sufficient condition that the system B'_1 *may* undergo the change in state (10.156) at a finite rate is

$$\Delta A < -W \text{ or } -\Delta A > W \tag{10.157}$$

The corresponding condition that the system B'_1 *cannot* undergo the change in state (10.156) is

$$\Delta A > -W \text{ or } -\Delta A < W \tag{10.158}$$

The corresponding condition that the system B'_1 is in a state of thermodynamic equilibrium with respect to the change in state (10.156) is

$$\Delta A = W \tag{10.159}$$

When the total work W of the process producing the change in state (10.156) is zero, we find the following results.

The necessary and sufficient condition that the system B'_1 *may* undergo the change in state (10.156) at a finite rate is

$$\Delta A < 0 \qquad (W = 0) \tag{10.160}$$

The corresponding condition that the system B'_1 *cannot* undergo the change in state (10.156) is

$$\Delta A > 0 \qquad (W = 0) \tag{10.161}$$

The corresponding condition that the system B'_1 is in a state of thermodynamic equilibrium with respect to the change in state (10.156) is

$$\Delta A = 0 \qquad (W = 0) \tag{10.162}$$

Equations (10.157)–(10.162) were derived for a system connected to a single heat reservoir. They are also valid when we replace this condition by the requirement that the change in state (10.156) be brought about by an isothermal process. Such a process fulfills the condition imposed on eqn. (10.23), from which the above equations were derived.

10.22. Equations for h^0, s^0, and g^0 referred to their values at 1 atmosphere and 0 K. Tables of standard thermodynamic properties (see Section 10.19) usually list for each substance the values of the quantities $(h^0 - h_0^0)$, $(s^0 - s_0^0)$, and $-(g^0 - h_0^0)/T$ for a series of equally spaced temperatures T. Here h^0, s^0, and g^0 refer to T, and h_0^0 and s_0^0 to a fiducial temperature T_0, which may or may not have the same value for both quantities. In this Section we derive equations for the above properties when $T_0 = 0$ K; and we discuss the evaluation of the integration constants h_0^0 and s_0^0.

Let the substance B undergo the change in state

$$B(s_1, 0\,K) = B(g, T) \qquad (1\text{ atm}) \tag{10.163}$$

from a fixed temperature 0 K where it has the standard molar enthalpy h_0^0, entropy s_0^0, and free energy g_0^0 to a variable temperature T where it has the corresponding values h^0, s^0, and g^0. The change in state is brought about by a reversible, isopiestic process in which the system passes through a series of standard states (see Section 6.2). The substance B undergoes a transition from one crystalline form s_1 to another form s_2 at T_{tr}, fusion at

T_f, and vaporization to a perfect gas at T_v, all at 1 atm. The corresponding standard molar enthalpies of phase transformation are Δh_{tr}^0, Δh_f^0, and Δh_v^0.

The following equations are obtained by application of relations derived from the first and second laws [see eqns. (5.12), (5.28), (9.36), and (9.51)]

$$h^0 - h_0^0 = \int_0^{T_{tr}} c_p^0(s_1) dT + \Delta h_{tr}^0 + \int_{T_{tr}}^{T_f} c_p^0(s_2) dT + \Delta h_f^0$$

$$+ \int_{T_f}^{T_v} c_p^0(l) dT + \Delta h_v^0 + \int_{T_v}^{T} c_p^0(g) dT \equiv \phi(T) \tag{10.164}$$

$$s^0 - s_0^0 = \int_0^{T_{tr}} c_p^0(s_1) \frac{dT}{T} + \frac{\Delta h_{tr}^0}{T_{tr}} + \int_{T_{tr}}^{T_f} c_p^0(s_2) \frac{dT}{T} + \frac{\Delta h_f^0}{T_f}$$

$$+ \int_{T_f}^{T_v} c_p^0(l) \frac{dT}{T} + \frac{\Delta h_v^0}{T_v} + \int_{T_v}^{T} c_p^0(g) \frac{dT}{T} \equiv \psi(T) \tag{10.165}$$

Thus, for the change in state (10.163), we have

$$\left. \begin{array}{l} g^0 = h^0 - Ts^0 \\ = (h^0 - h_0^0) - T(s^0 - s_0^0) + h_0^0 - Ts_0^0 \\ = \phi(T) - T\psi(T) + h_0^0 - Ts_0^0 \end{array} \right\} \tag{10.166}$$

These equations may be written in the form

$$-\frac{g^0 - h_0^0}{T} = -\frac{\phi(T)}{T} + \psi(T) + s_0^0 \tag{10.167}$$

[compare eqns. (10.79)–(10.83) and eqn. (10.137)].

Both experimental and theoretical results indicate that the first integrals of the right-hand sides of eqns. (10.164) and (10.165) converge for all substances as T approaches zero.

We may assign any values we please to the integration constants h_0^0 and s_0^0 for any substance B, subject only to the condition that they are consistent with the changes in state in which B is to take part. Although the above equations are written for a fiducial temperature $T_0 = 0$ K, they hold for any value of T_0. Let us consider the following two cases.

1. If B acts as a system of one component throughout an investigation, we may assign any value, for example zero, to both h_0^0 and s_0^0 at 0 K (or at any fiducial temperature T_0). In general, the standard molar enthalpy h^0, entropy s^0, and hence free energy g^0 do not appear in the equation for the change in free energy ΔG attending an isothermal change in state of any substance [see for example eqn. (10.74)].

2. Application of eqn. (10.166) to the standard chemical change in state

$$\sum_i \nu_i B_i \text{ (standard state } i) = 0 \qquad (T) \qquad (10.168)$$

gives

$$\Delta G^0 = \sum_i \nu_i (\phi_i - T\psi_i) + \sum_i \nu_i h^0_{0i} - T\sum_i \nu_i s^0_{0i} \qquad (10.169)$$

The substances B_i taking part in the change of state (10.168) are not conserved, but the elements composing them are. In principle, we may assign any values we please to the enthalpy and entropy constants h^0_{0i} and s^0_{0i} of all elementary substances at 0 K (or at the fiducial temperature T_0). The values of the corresponding quantities for a compound can then be computed from these numbers and the standard molar enthalpy and entropy of formation of B_i at 0 K (or at T_0).

In practice, we equate h^0_{0i} to the standard molar enthalpy of formation of B_i at the fiducial temperature T_0, which is equivalent to giving this constant the value zero for all elementary substances at T_0. The temperatures commonly used for T_0 are 0 K, 298 K, or the temperature T of the given isothermal change in state. This procedure is practicable from an experimental viewpoint since it involves the application of the first-law equations (5.1) and (5.2), which are the equations used in calorimetry (Sections 3.18—3.24). These relations do not require that a process be reversible in all respects, but only that the total work effect shall consist of reversible, or essentially reversible, expansion work.

But the determination of the entropy of formation of a substance at any temperature by a thermodynamic method involves the use of a second-law equation. This requires either the measurement of the heat effect attending a chemical process that is essentially reversible in all respects, or of the properties of a system in a state of thermodynamic equilibrium. Even when practicable, these methods involve difficult experimental procedures. For many substances these methods are not feasible. Here, in practice, the value of the entropy constant s^0_{0i} at 0 K is determined by application of the third law (see Chapter 11).

For example, only ordinary calorimetric measurements are required to determine the values of the standard molar enthalpies of the three isomeric pentanes at any temperature, and to reduce these values to 0 K. But the determination of the entropy of formation of an isomer, or the change in entropy attending the conversion of one isomer into another by a thermal or an equilibrium measurement, is not practicable. Hence, from these kinds of experiments we cannot calculate the composition of the equilibrium mixture of the three isomers at any temperature. The third law makes such a calculation possible by fixing the value of the entropy constant s^0_0 at 0 K for each elementary substance and compound.

In the JANAF Thermochemical Tables (see Section 10.19) the fiducial temperature for enthalpy is 298 K, and that for entropy is 0 K, where s^0_0 is placed equal to zero for all substances in accordance with the third law.

From Table 10.2 we can find several entirely different values for the standard molar free energy of a substance at a given temperature. We have shown in the four equations (10.138) that these different sets of numbers give the same value for the standard change in free energy ΔG^0 attending the Deacon reaction at 800 K.

The quantities listed at any temperature in the JANAF Tables are connected by the following relations

$$(h^0 - Ts^0) - g^0 = T\sum_j \nu_j s_j^0 \tag{10.170}$$

$$(h^0 - Ts^0) - \left[T\left(\frac{g^0 - h_{298}^0}{T}\right) + h_{298}^0\right] = \sum_j \nu_j (h^0 - h_{298}^0)_j \tag{10.171}$$

Here a quantity without an identifying subscript refers to a substance B, the subscript j denotes an elementary substance, ν is its stoichiometric coefficient (negative for a reactant) in the chemical equation for the formation of 1 mole of B, and the summation extends over all of the *reactants* of this relation. Since the summations on the right-hand sides of eqns. (10.170) and (10.171) depend only on the properties of elementary substances, each cancels in the equation for the change in free energy ΔG attending an isothermal change in state based on a balanced chemical reaction. Thus, for the change in state (10.168), we find

$$\left.\begin{aligned}
\sum_i \nu_i g_i^0 &= \sum_i \nu_i (h^0 - Ts^0)_i \\
&= \sum_i \nu_i h_i^0 - T\sum_i \nu_i s_i^0 \\
&= T\sum_i \nu_i \left(\frac{g^0 - h_{298}^0}{T}\right)_i + \sum_i \nu_i h_{298i}^0
\end{aligned}\right\} \tag{10.172}$$

Here the subscript i refers to the substance B_i in eqn. (10.168). The above relations may be compared with eqns. (10.138).

10.23. Thermodynamic relations of systems producing non-expansion work. Let a closed system B undergo the change in state

$$B \text{ (state 1)} = B \text{ (state 2)} \tag{10.173}$$

by a reversible process in which expansion work and one kind of non-expansion work W_{XR} are produced in the surroundings. Then

$$dW_{XR} = ydX, \quad y = -\left(\frac{\partial E}{\partial X}\right)_{SV} \tag{10.174}$$

where y is a generalized force exerted by the system on its surroundings, and X is the conjugate generalized displacement coordinate.

The total differential of the energy E of the above system is

$$dE = TdS - pdV - ydX \tag{10.175}$$

We can define seven Legendre transforms (see ref. 14) of the energy of the system B by subtracting from E all combinations of the three products

$$(TS), (-pV), (-yX) \tag{10.176}$$

taken one at a time, two at a time, and three at a time.

The energy equation and its seven transformations, together with their total differentials, are listed in Table 10.3. Here E, H, A, and G have their usual meanings [see eqns. (4.13), (10.4), and (10.5)]; and E_1, H_1, A_1, and G_1 are derived by subtracting $-yX$ from the corresponding quantities without subscripts.

The twenty-four reciprocity relations obtained by cross-differentiation of the right-hand sides of the eight total differentials of Table 10.3 are given in Table 10.4. Equations in Column A may be compared with the Maxwell relations (9.20)—(9.23).

Four heat capacities of the system B are defined in Table 10.5. Eight Gibbs—Helmholtz equations are given in Table 10.6 [see eqns. (10.1.14) and (10.1.15) of Table 10.1].

Now let the system B of eqn. (10.173) produce ω different kinds of reversible non-expansion work in its surroundings, each defined by equations similar to (10.174) and (10.175). The total differential of the energy E of the system is

$$dE = TdS - pdV - \sum_i y_i dX_i, \quad y_i = -\left(\frac{\partial E}{\partial X_i}\right)_{SVX} \tag{10.177}$$

Here the summation extends over the ω kinds of non-expansion work; and the subscript X to a partial derivative denotes that all of the X's are held constant except the one appearing in the denominator.

We can define $2^{\omega+2} - 1$ Legendre transforms of the energy of the system B, corresponding to the seven given in Table 10.3, by subtracting from E all combinations of the $\omega + 2$ products

$$(TS), (-pV), (-y_1 X_1), (-y_2 X_2), \ldots, (-y_\omega X_\omega) \tag{10.178}$$

TABLE 10.3

THERMODYNAMIC FUNCTIONS OF A SYSTEM PRODUCING NON-EXPANSION WORK

Function	Total differential	
E	$dE = TdS - pdV - ydX$	(10.3.1)
$E_1 = E + yX$	$dE_1 = TdS - pdV + Xdy$	(10.3.2)
$H = E + pV$	$dH = TdS + Vdp - ydX$	(10.3.3)
$H_1 = E + pV + yX$	$dH_1 = TdS + Vdp + Xdy$	(10.3.4)
$A = E - TS$	$dA = -SdT - pdV - ydX$	(10.3.5)
$A_1 = E - TS + yX$	$dA_1 = -SdT - pdV + Xdy$	(10.3.6)
$G = E - TS + pV$	$dG = -SdT + Vdp - ydX$	(10.3.7)
$G_1 = E - TS + pV + yX$	$dG_1 = -SdT + Vdp + Xdy$	(10.3.8)

When the system consists of two or more phases:
   Application of eqns. (4.33) for the pV-product and corresponding relations for the yX-product may be necessary.

TABLE 10.4
RECIPROCITY RELATIONS OF A SYSTEM PRODUCING NON-EXPANSION WORK

A	B	C	
$\left(\frac{\partial T}{\partial V}\right)_{SX} = -\left(\frac{\partial p}{\partial S}\right)_{VX}$	$\left(\frac{\partial T}{\partial X}\right)_{SV} = -\left(\frac{\partial y}{\partial S}\right)_{XV}$	$\left(\frac{\partial p}{\partial X}\right)_{VS} = \left(\frac{\partial y}{\partial V}\right)_{XS}$	(10.4.1)
$\left(\frac{\partial T}{\partial V}\right)_{Sy} = -\left(\frac{\partial p}{\partial S}\right)_{Vy}$	$\left(\frac{\partial T}{\partial y}\right)_{SV} = \left(\frac{\partial X}{\partial S}\right)_{yV}$	$\left(\frac{\partial p}{\partial y}\right)_{VS} = -\left(\frac{\partial X}{\partial V}\right)_{yS}$	(10.4.2)
$\left(\frac{\partial T}{\partial p}\right)_{SX} = \left(\frac{\partial V}{\partial S}\right)_{pX}$	$\left(\frac{\partial T}{\partial X}\right)_{Sp} = -\left(\frac{\partial y}{\partial S}\right)_{Xp}$	$\left(\frac{\partial V}{\partial X}\right)_{pS} = -\left(\frac{\partial y}{\partial p}\right)_{XS}$	(10.4.3)
$\left(\frac{\partial T}{\partial p}\right)_{Sy} = \left(\frac{\partial V}{\partial S}\right)_{py}$	$\left(\frac{\partial T}{\partial y}\right)_{Sp} = \left(\frac{\partial X}{\partial S}\right)_{yp}$	$\left(\frac{\partial V}{\partial y}\right)_{pS} = \left(\frac{\partial X}{\partial p}\right)_{yS}$	(10.4.4)
$\left(\frac{\partial S}{\partial V}\right)_{TX} = \left(\frac{\partial p}{\partial T}\right)_{VX}$	$\left(\frac{\partial S}{\partial X}\right)_{TV} = \left(\frac{\partial y}{\partial T}\right)_{XV}$	$\left(\frac{\partial p}{\partial X}\right)_{VT} = \left(\frac{\partial y}{\partial V}\right)_{XT}$	(10.4.5)
$\left(\frac{\partial S}{\partial V}\right)_{Ty} = \left(\frac{\partial p}{\partial T}\right)_{Vy}$	$\left(\frac{\partial S}{\partial y}\right)_{TV} = -\left(\frac{\partial X}{\partial T}\right)_{yV}$	$\left(\frac{\partial p}{\partial y}\right)_{VT} = -\left(\frac{\partial X}{\partial V}\right)_{yT}$	(10.4.6)
$\left(\frac{\partial S}{\partial p}\right)_{TX} = -\left(\frac{\partial V}{\partial T}\right)_{pX}$	$\left(\frac{\partial S}{\partial X}\right)_{Tp} = \left(\frac{\partial y}{\partial T}\right)_{Xp}$	$\left(\frac{\partial V}{\partial X}\right)_{pT} = -\left(\frac{\partial y}{\partial p}\right)_{XT}$	(10.4.7)
$\left(\frac{\partial S}{\partial p}\right)_{Ty} = -\left(\frac{\partial V}{\partial T}\right)_{py}$	$\left(\frac{\partial S}{\partial y}\right)_{Tp} = -\left(\frac{\partial X}{\partial T}\right)_{yp}$	$\left(\frac{\partial V}{\partial y}\right)_{pT} = \left(\frac{\partial X}{\partial p}\right)_{yT}$	(10.4.8)

TABLE 10.5
HEAT CAPACITIES OF A SYSTEM PRODUCING NON-EXPANSION WORK

$C_{VX} = \left(\frac{\partial E}{\partial T}\right)_{VX} = T\left(\frac{\partial S}{\partial T}\right)_{VX}$	$C_{pX} = \left(\frac{\partial H}{\partial T}\right)_{pX} = T\left(\frac{\partial S}{\partial T}\right)_{pX}$	(10.5.1)
$C_{Vy} = \left(\frac{\partial E_1}{\partial T}\right)_{Vy} = T\left(\frac{\partial S}{\partial T}\right)_{Vy}$	$C_{py} = \left(\frac{\partial H_1}{\partial T}\right)_{py} = T\left(\frac{\partial S}{\partial T}\right)_{py}$	(10.5.2)

TABLE 10.6
GIBBS—HELMHOLTZ EQUATIONS OF A SYSTEM PRODUCING NON-EXPANSION WORK

$\left(\frac{\partial \Delta A}{\partial T}\right)_{VX} = -\Delta S = \frac{\Delta A - \Delta E}{T}$	$\left(\frac{\partial \Delta G}{\partial T}\right)_{pX} = -\Delta S = \frac{\Delta G - \Delta H}{T}$	(10.6.1)
$\left(\frac{\partial \Delta A_1}{\partial T}\right)_{Vy} = -\Delta S = \frac{\Delta A_1 - \Delta E_1}{T}$	$\left(\frac{\partial \Delta G_1}{\partial T}\right)_{py} = -\Delta S = \frac{\Delta G_1 - \Delta H_1}{T}$	(10.6.2)
$\left[\frac{\partial (\Delta A/T)}{\partial T}\right]_{VX} = -\frac{\Delta E}{T^2}$	$\left[\frac{\partial (\Delta G/T)}{\partial T}\right]_{pX} = -\frac{\Delta H}{T^2}$	(10.6.3)
$\left[\frac{\partial (\Delta A_1/T)}{\partial T}\right]_{Vy} = -\frac{\Delta E_1}{T^2}$	$\left[\frac{\partial (\Delta G_1/T)}{\partial T}\right]_{py} = -\frac{\Delta H_1}{T^2}$	(10.6.4)

When the system consists of two or more phases:
  Hold the quantities represented by the subscripts to the partial derivatives constant for each phase.

taken one at a time, two at a time,..., $\omega + 2$ at a time. Differentiation of these equations and substitution from (10.177) give, together with the energy relation itself, $2^{\omega+2}$ differential equations (compare ref. 15).

We can derive $2^{\omega+2}a_{\omega+1}$ reciprocity relations similar to those of Table 10.4 by cross-differentiation of the right-hand sides of the above $2^{\omega+2}$ equations. Here

$$a_{\omega+1} = \frac{(\omega+1)(\omega+2)}{2} \qquad (10.179)$$

is the $(\omega + 1)$th triangular number. Similarly, there are 2^ω equations for each of the heat capacities at constant volume and at constant pressure, and 2^ω equations for each of the four kinds of Gibbs—Helmholtz relations corresponding to those appearing in Tables 10.5 and 10.6.

REFERENCES

1 J. W. Gibbs, Scientific Papers, Longmans, Green and Co., New York, 1906, Vol. I, Thermodynamics, pp. 89—92.
2 M. F. Massieu, C. R. Acad. Sci., 69 (1869) 858—862, 1057—1061; Massieu used the negative of these functions which he denoted by the symbols ψ and ψ', respectively.
3 J. W. Gibbs, ref. 1, pp. 49—54; see also J. H. Keenan, Thermodynamics, Wiley, New York, 1941, Chapter XVII, pp. 402—403, 426.
4 J. W. Gibbs, ref. 1, pp. 62—417.
5 H. Flood and O. J. Kleppa, J. Am. Chem. Soc., 69 (1947) 998—1002.
6 P. W. Bridgman, in E. W. Washburn (Ed.), International Critical Tables, McGraw-Hill, New York, 1928, Vol. IV, p. 11; E. W. Washburn, ibid., Vol. III, pp. 210—212; A. W. C. Menzies, ibid., Vol. III, p. 201; K. A. Kobe and R. E. Lynn, Jr., Chem. Rev., 52 (1953) 117—236; G. Tammann, The States of Aggregation, Van Nostrand, New York, 1925, pp. 177—178.
7 F. G. Keyes, J. Chem. Phys., 15 (1947) 602—612, 17 (1949) 923—934.
8 J. W. Gibbs, ref. 1, p. 129.
9 E. Lurie and L. J. Gillespie, J. Am. Chem. Soc., 49 (1927) 1146—1157.
10 H. T. Gerry and L. J. Gillespie, Phys. Rev., 40 (1932) 269—280.
11 R. B. Hinckley and R. C. Reid, Am. Inst. Chem. Eng. J., 10 (1964) 416—417.
12 G. N. Lewis and M. Randall, Thermodynamics, McGraw-Hill, New York, 1923, pp. 183—184.
13 D. R. Stull et al., JANAF Thermochemical Tables, Dow Chemical Company, Midland, Michigan, 1964.
14 H. B. Callen, Thermodynamics, Wiley, New York, 1960, pp. 90—102.
15 A. H. Wilson, Thermodynamics and Statistical Mechanics, Cambridge University Press, Cambridge, 1957, pp. 44—49, 192.

Chapter 11

THE THIRD LAW OF THERMODYNAMICS

The first law of thermodynamics defines the difference between the energies of two states of a system, and the second law defines the difference between the entropies of two states. Thus each law defines a new state property except for an additive arbitrary constant of integration. These constants appear in equations for the work content A and free energy G of a system as an additive expression of the form $a - bT$, where a is the energy constant of the system and b is its entropy constant (see Section 10.1).

In contrast to the first two laws of thermodynamics, the third law does not define a new state property of a system. It enables us to assign a value to the standard molar entropy s_0^0 of every element and compound at 0 K. Thus, it fixes the value of the entropy constants s_0^0 appearing in eqns. (10.165)–(10.167) which express the standard molar entropy s^0 and free energy g^0 of any pure substance as a function of temperature. From the values of s_0^0 we can then compute the entropy constant $\Sigma_i \nu_i s_{0i}^0$ in eqn. (10.169) which gives the temperature variation at constant pressure of the change in free energy ΔG^0 attending the standard change in state (10.168). Thus, from purely calorimetric measurements, we can calculate values of g^0 and ΔG^0 at any temperature. Corresponding relations can be derived for the work content.

The application of the third law, unlike those of the first two laws, relies on a knowledge of the molecular constitution of matter. Moreover, the third law is concerned only with the temperature region in the vicinity of 0 K, whereas the other two laws are not so delimited.

In Sections 10.8–10.10 and 10.22, we derived equations expressing the free energy G (or work content A) of a closed, simple, homogeneous system composed of a fixed mass of a pure substance as a function of temperature at constant pressure (or volume) by integration of the Gibbs–Helmholtz equations (10.51)–(10.55) with respect to T at constant p (or V). In the present chapter we shall use the following forms of these equations

$$\left(\frac{\partial G}{\partial T}\right)_p = -S = \frac{G-H}{T} \tag{11.1}$$

$$\left(\frac{\partial A}{\partial T}\right)_V = -S = \frac{A-E}{T} \tag{11.2}$$

In Sections 10.20 and 10.22, we derived corresponding relations expressing the change in free energy ΔG (or change in work content ΔA) attending an isothermal physical or chemical change in state at constant pressure (or volume) by integration of the Gibbs–Helmholtz equations (10.1.14) and

(10.1.15) of Table 10.1 with respect to T at constant pressure (or volume) for each substance taking part in the change. The forms of these equations used in the present chapter are as follows

$$\left(\frac{\partial \Delta G}{\partial T}\right)_p = -\Delta S = \frac{\Delta G - \Delta H}{T} \tag{11.3}$$

$$\left(\frac{\partial \Delta A}{\partial T}\right)_V = -\Delta S = \frac{\Delta A - \Delta E}{T} \tag{11.4}$$

In the following review of the history of the third law we shall use the notations of eqns. (11.1)—(11.4), rather than those used by the original investigators.

11.1. Historical introduction. Both Thomsen [1] and M. Berthelot [2] carried out extensive programs of thermochemical measurements. After consideration of their results, Berthelot proposed his *Principle of Maximum Work* often called the *Principle of Thomsen and Berthelot*:

"Every chemical change produced without the intervention of external energy tends towards the production of the substance or system of substances which evolve the most heat".

The principle predicts that an isothermal process, in which the total work $W = 0$, tends to go in the direction in which $\Delta E < 0$; and an isopiestic, isothermal process, in which the non-expansion work $W_X = 0$, tends to go in the direction in which $\Delta H < 0$.

These conditions continued to be applied long after Gibbs had shown that the corresponding correct conditions are $\Delta A < 0$ and $\Delta G < 0$, respectively (see Section 10.21).

The Principle of Thomsen and Berthelot neglects the term $T\Delta S$ in the Gibbs—Helmholtz equations (10.10) and (10.12) as applied to isothermal chemical changes in state. This implies that at all temperatures

$$\Delta A = \Delta E, \Delta G = \Delta H \tag{11.5}$$

for such changes. We note from eqns. (11.3) and (11.4) that the above relations are correct at 0 K, if the corresponding change in entropy ΔS remains finite down to the absolute zero.

Equations (11.5) are surprisingly good approximations for chemical changes in condensed systems at room temperature, and they become more accurate as the temperature decreases. But their application to chemical reactions involving gases may lead to significant errors.

With regard to the integration of the Gibbs—Helmholtz equations, Le Chatelier [3] wrote: "It is very probable that the constant of integration, like the coefficients of the differential equation, should be a definite function of certain physical properties of the substances present. The determination of the nature of this function would give us a complete knowledge of the laws of equilibrium. ... So far we have not determined the exact nature of this constant; ... ."

Using the available measurements of the electromotive forces of ten galvanic cells (of the Daniel-cell type) at 1 atm and 291 K, and the available values of ΔH and ΔC_p for the changes in state produced by the passage of electricity through these cells, Richards [4] calculated ΔG, $(\partial \Delta G/\partial T)_p$, ΔH, and $(\partial \Delta H/\Delta T)_p$ at 1 atm and 291 K for each relevant change in state. He then drew curves of ΔG against T and of ΔH against T for six of these cells: the segment of each curve from 273 K to 323 K was drawn as a straight line passing through the correct (ΔG or ΔH) point at 291 K and having the calculated slope. From a consideration of these graphs he wrote: "it is evident that each pair [of lines] tends to converge at a point not far from absolute zero" and "in order to converge at the absolute zero these lines must be not exactly straight, but slightly curved".

Among a number of other conclusions, Richards stated: "The fact that in many reactions the change of heat capacity is small and the concentration effect about balanced [as in the reactions of the above cells] affords an explanation of the frequent fulfilment of Berthelot's approximate 'rule of maximum work' ". But Richards did not relate his results to the determination of the entropy constant, but rather to his "hypothesis of compressible atoms".

Starting from the Gibbs—Helmholtz equations (11.4), Haber [5] (1905) expressed the change in work content ΔA attending an isothermal chemical change in state involving only gases as a function of temperature at constant volume. He assumed that ΔC_V for this change in state is constant, and he supposed that the resulting equation for ΔA applied down to the lowest temperatures, all of the reactants and products remaining gaseous. He realized that if ΔC_V, that is $(\partial \Delta E/\partial T)_V$, for an isothermal change in state remains finite as T approaches the absolute zero, then $(\partial \Delta A/\partial T)_V$ and hence ΔS tend towards positive or negative infinite values. In 1905 there was no reason to believe that the heat capacities of gases decreased to zero at 0 K.

Haber chose a fiducial temperature 1 K to get rid of the troublesome infinity, and investigated various methods of evaluating the "indeterminate thermodynamic constant" in the equation for ΔA. He presented evidence that the entropy constant is small in general, and that it may be zero when $\Delta C_V = 0$. Haber based his conclusions, in part, on the knowledge that the Principle of Thomsen and Berthelot increases in accuracy as the temperature decreases. With regard to Richard's statement, quoted above, concerning this principle, Haber wrote: "Gas reactions do not contradict this conclusion, ... although our knowledge may not be sufficient to demonstrate that they support it".

Influenced by the very good approximation of the Principle of Thomsen and Berthelot, and aware of the work of Richards and Haber, Nernst [6] (1906) proposed [7—9] that the Gibbs—Helmholtz equation (11.4) for the change in work content ΔA attending an isothermal physical or chemical change in state in a condensed system be integrated, subject to the conditions

$$\lim_{T \to 0} \left(\frac{\partial \Delta E}{\partial T}\right)_V = 0, \quad \lim_{T \to 0} \left(\frac{\partial \Delta A}{\partial T}\right)_V = 0 \tag{11.6}$$

The corresponding conditions for the integration of the Gibbs–Helmholtz equation (11.3) for the change in free energy ΔG attending a similar change in state are

$$\lim_{T \to 0} \left(\frac{\partial \Delta H}{\partial T}\right)_p = 0, \quad \lim_{T \to 0} \left(\frac{\partial \Delta G}{\partial T}\right)_p = 0 \tag{11.7}$$

By virtue of Kirchhoff's formula (6.26) and eqns. (11.3) and (11.4), we may write the above equations in the following forms

$$\lim_{T \to 0} \Delta C_V = 0, \quad \lim_{T \to 0} \Delta C_p = 0 \tag{11.8}$$

$$\lim_{T \to 0} \Delta S = 0 \tag{11.9}$$

Equation (11.9), when not limited to condensed systems, is the mathematical statement of the third law. Equations (11.8) are necessary but not sufficient conditions for the validity of this law. Thus (see ref. 10), from eqns. (11.4) and (11.6), we find that

$$\lim_{T \to 0} \frac{\Delta A - \Delta E}{T} = -\lim_{T \to 0} \Delta S = \lim_{T \to 0} \left(\frac{\partial \Delta A}{\partial T}\right)_V = 0 \tag{11.10}$$

Application of L'Hôpital's rule to the first section of these equations gives

$$\lim_{T \to 0} \left(\frac{\partial \Delta A}{\partial T}\right)_V - \lim_{T \to 0} \left(\frac{\partial \Delta E}{\partial T}\right)_V = \lim_{T \to 0} \left(\frac{\partial \Delta A}{\partial T}\right)_V = 0 \tag{11.11}$$

Whence

$$\lim_{T \to 0} \left(\frac{\partial \Delta E}{\partial T}\right)_V = \lim_{T \to 0} \Delta C_V = 0 \tag{11.12}$$

Similarly, we can prove from eqns. (11.3) and (11.7) that

$$\lim_{T \to 0} \left(\frac{\partial \Delta H}{\partial T}\right)_p = \lim_{T \to 0} \Delta C_p = 0 \tag{11.13}$$

But we cannot derive eqn. (11.9) from either of eqns. (11.8) and the second law of thermodynamics.

The third law requires that: (1) when the volume of each phase of an isothermal change in state is not allowed to change with temperature, plots of the corresponding values of ΔA and ΔE against T have a common horizontal tangent at 0 K; (2) when the pressure of each phase of an isothermal change in state is not allowed to change with temperature, plots of the corresponding values of ΔG and ΔH against T also have a common horizontal tangent at 0 K (see Fig. 11.1); and (3) ΔC_V and ΔC_p for an

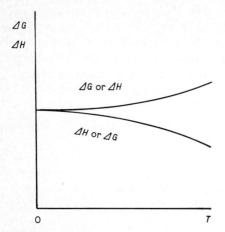

Fig. 11.1. Changes in free energy and enthalpy attending an isopiestic, isothermal change in state plotted against Kelvin temperature, according to Richards [4] and the Nernst Heat Theorem [6].

isothermal physical or chemical change in state approach zero as the Kelvin temperature tends to zero.

We recall that Richards' graphs of ΔG and ΔH plotted against temperature exhibited the same behavior as those of Fig. 11.1.

In 1906 very little was known about heat capacities of solids at low temperatures. Only a few measurements extended as far down as the boiling point of liquid air, and these were mostly average values over fairly large temperature ranges. The heat capacities showed some decrease with temperature, but the dramatic drop in the value of this property that takes place below about 100 K was not suspected. Two empirical laws were universally accepted. The law of Dulong and Petit [11] states that the heat capacities of one gram-atom of all solid elements have the same value; Boltzmann [12] showed that the value of this constant is 3R. The law of Neumann [13] and Kopp [14] states that the molar heat capacity of a solid compound is equal to the sum of the heat capacities of the constituent elements. At first Nernst accepted these laws; but he argued that although the heat capacities of the substances entering into a change in state may not be zero at 0 K, the *change* in heat capacity attending this change in state is zero at this temperature.

Nernst [6] treated isothermal changes in state in systems having gaseous constituents by employing his "chemical constants". This procedure has only historical interest in chemical thermodynamics today.

The application of quantum mechanics has shown, as Nernst [15] surmised, that the molar heat capacity c_V of a perfect gas, when in the state of lowest energy, is zero at 0 K. But Nernst's belief that the third law is a consequence of the second law and the disappearance of the heat capacity at the absolute zero of temperature proved unfounded.

Planck [16] (1912) went one step further than Nernst and stated the third law in the following form:

"The entropy of a solid or liquid chemically homogeneous substance has the value zero at the absolute zero of temperature".

Later Planck revised his statement of the law to read that, as the temperature decreases indefinitely, the entropy of a chemically homogeneous body of finite density approaches indefinitely near to a definite value which is independent of the pressure, the aggregation state, and the special chemical modification. He proposed to assign the value zero to this limit.

Planck's proposal is equivalent to assigning a fixed value, for example zero, to the entropy of each element when in the state described above at 0 K, and then determining the entropy of each compound when in a similar state at 0 K by application of eqn. (11.9) to the change in state in which the compound is formed from its elements.

According to Planck's proposal

$$\lim_{T \to 0} S = 0 \tag{11.14}$$

for a fixed mass of a pure element or compound when in the condition described above. By application of the methods employed to derive eqns. (11.12) and (11.13) from eqns. (11.1) and (11.2), we can show that the relations

$$\lim_{T \to 0} C_V = 0 \text{ and } \lim_{T \to 0} C_p = 0 \tag{11.15}$$

are consequences of eqn. (11.14).

Thus, for a substance meeting the conditions given above, the entropy constants s_0^0 may be given the value zero, and all of the temperature integrals are convergent in eqns. (10.164)–(10.167). Also, the entropy constant $\Sigma_i \nu_i s_{0i}^0$ is zero, and all of the temperature integrals in the term $\Sigma_i \nu_i (\phi_i - T\psi_i)$ are convergent in eqn. (10.169), which expresses the change in free energy ΔG attending the isothermal chemical change in state (10.168) as a function of temperature at constant pressure of each substance entering into the change.

In Fig. 11.2, curves [8] of entropy plotted against temperature for two values of the pressure p are drawn for a fixed mass of a classical perfect gas which has a finite heat capacity C_p at the absolute zero. The entropy of such a substance is negatively infinite at 0 K.

In Fig. 11.3, curves of entropy plotted against temperature are drawn for two values of a parameter P for each of two substances. In each case, the ratio C_P/T approaches zero as the Kelvin temperature tends to zero. The third law is valid for the substance of graph (a), where the entropy curves for different values of P meet in a common point at 0 K. In accordance with Planck's proposal, this point has been placed at $S = 0$. The third law is invalid for the substance of graph (b), where the entropy curves are not concurrent at the absolute zero.

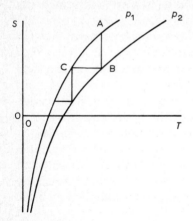

Fig. 11.2. Entropy of a perfect gas as a function of Kelvin temperature for two values of the pressure p (after Simon [8]).

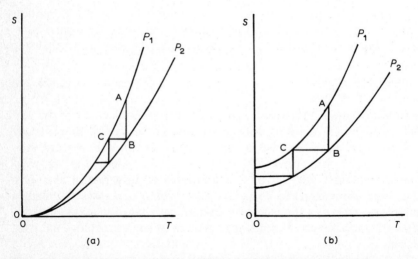

Fig. 11.3. Entropy as a function of Kelvin temperature for two values of a parameter P under assumptions of (a) validity and (b) non-validity of the third law (after Simon [8]).

Other statements of the third law have been proposed by Lewis and Gibson [17], Eastman and Milner [9], and Simon [8]. Simon has given two statements of this law which, he feels, express the same facts in slightly different forms:

"At absolute zero the entropy differences disappear between all of those states of a system between which reversible transitions are possible at least in principle".

"[At absolute zero] the entropy differences disappear between all those states of a system which are in internal thermodynamic equilibrium".

11.2. Principle of the unattainability of the absolute zero. In order to express his Heat Theorem in the form of an empirical declaration of an impossibility similar to those in which the first and second laws may be phrased, Nernst [6b] offered, in 1912, a proof that a law which he called the Principle of the Unattainability of the Absolute Zero can be deduced from the Heat Theorem. We may state this law in the following form:

It is impossible to reduce the temperature of any system to the absolute zero in a finite number of processes.

Let each of the systems of Figs. 11.2 and 11.3 undergo a reversible, isothermal process, from a state A to a state B, in which its entropy decreases, followed by a reversible, adiabatic process to a state C, whereby its temperature decreases. It is evident from the curves of Figs. 11.2 and 11.3(a) that this procedure, or a multistage process composed of a finite number of these procedures, cannot reduce the temperature of a classical perfect gas or of a system for which the third law is valid to zero. But we can reduce the temperature of the substance whose entropy curves are shown in Fig. 11.3(b) to zero in a finite number of processes. Although the third law is not valid for this substance, its heat capacity C_p approaches zero as the Kelvin temperature tends to zero.

Nernst's derivation of the Unattainability Principle was based on a (reversible) Carnot cycle operating between a finite temperature and the 0 K isotherm which, by the third law, is also a reversible adiabatic for the given system. Criticisms of this proof by Einstein, Planck, and Bennewitz are discussed by Eastman [9] and by Simon [7] (1951). Einstein's objection is that the 0 K isotherm of the above cycle cannot be carried out even in principle, since the slightest heat leak or the least irreversibility will raise the temperature of the system above the absolute zero. Thus the cycle will be destroyed.

Proofs of the equivalence of the Unattainability Principle and the statement of the third law based on eqn. (11.9), which do not employ a Carnot cycle, have been given. One proof [18] makes use of an adiabatic process; another [19] does not require that the process be adiabatic.

Physicists seem to prefer the Unattainability statement of the third law; chemists prefer the more directly applicable entropy statement of the law given in eqn. (11.9) or (11.14).

11.3. Entropy of mixing. The molar entropy of mixing, usually called simply *the entropy of mixing*, of the various substances $B_1, B_2, \ldots, B_m$ at the pressure p and temperature T is the change in entropy ΔS attending the change in state for the formation of one mole of solution from these constituents, each reactant and the product having the same pressure p and temperature T, and being in the same aggregation state a'. The entropy of mixing is ΔS for the change in state

$$x_1 B_1 + x_2 B_2 + \ldots + x_m B_m = [x_1 B_1 \cdot x_2 B_2 \cdots x_m B_m] \qquad (p, T, a') \quad (11.16)$$

Here x_i is the mole fraction of the constituent B_i in the solution which may be solid, liquid, or gaseous.

In Section 9.18 we found that the entropy of mixing m perfect gases at p and T is

$$\Delta S^M = -R \sum_i x_i \ln x_i \qquad (p, T = \text{constant}) \tag{11.17}$$

where the summation extends over the m constituents.

In a later chapter we shall define perfect solid and liquid solutions from a thermodynamic standpoint, and prove that eqn. (11.17) also applies for condensed perfect solutions.

We note that ΔS in eqn. (11.17) is always positive, and is independent of pressure, temperature, and the substances mixed. The entropy of mixing the constituents of a real solution does depend upon all three of these variables. It is expressed by an equation in which the leading term is the right-hand section of eqn. (11.17); and, with a very few exceptions, it is positive.

11.4. The third law of thermodynamics. We shall state the third law in the following form:

The change in entropy attending an isothermal change in state that may be brought about by a reversible process approaches the limit zero as the Kelvin temperature of the change in state tends to zero.

Let a system B undergo the isothermal physical or chemical change in state

$$B(\text{state } 1) = B(\text{state } 2) \qquad (T) \tag{11.18}$$

where the initial and final states of B have the same temperature, but not necessarily the same values of pressure, volume, or any other state property. If, as the temperature T approaches 0 K, the above change in state may be brought about by any reversible process, which need not itself be isothermal, the third law requires that

$$\lim_{T \to 0} \Delta S = 0 \tag{11.19}$$

The following statement is a sufficient but not a necessary condition for the validity of the third law:

The entropy of a system in a state of thermodynamic equilibrium, with respect to all reversible processes of interest that reach this state, approaches the limit zero as the Kelvin temperature tends to zero. Thus, for each such equilibrium state of a system (or of a substance), we have

$$\lim_{T \to 0} S = 0 \tag{11.20}$$

In the above statement, the phrase "all reversible processes of interest" includes all of the reversible processes we may choose to employ in bringing about all of the changes in state which we intend to produce in the system during a given investigation.

We note that eqn. (11.20) does not specify the process whereby the temperature of a system is reduced to the absolute zero, except that it be essentially reversible.

The above two statements of the third law may be compared with those of Simon [8] (see Section 11.1).

In statistical mechanics, the entropy S of a system in a given state is defined by the equation

$$S = k \ln W \tag{11.21}$$

where $k = 1.38054 \times 10^{-16}$ erg deg^{-1} molecule^{-1} is Boltzmann's constant, and W is the total number of different quantum states (complexions) corresponding to the given thermodynamic state. In the above equation, the arbitrary, additive constant which occurs in the thermodynamic definition of entropy has been determined by placing $S = 0$ when $W = 1$.

A system in a thermodynamic state having a large number of complexions W has a large entropy S, and is said to be in a more disordered state than one in which W, and consequently S, are smaller. A system in a state for which $W = 1$ and hence $S = 0$ is said to be in a completely ordered state. Thus we may regard entropy as the measure of the extent of disorder of the state of a system.

For all known systems, the number of quantum states with energies less than kT, for T sufficiently small, is less than a^N for any value of a greater than one, N being the number of molecules in the system. Hence S at these temperatures is essentially zero. This is the statistical-mechanical basis for the third law.

It has been found experimentally that eqn. (11.20) is not valid for substances which, on being cooled down to 0 K, do not continue to pass through states of thermodynamic equilibrium with respect to all reversible processes of interest that reach each of these states. As the temperature decreases, the rate of some physical or chemical reaction in such a substance becomes so small that it does not attain a completely ordered state at 0 K in any reasonable length of time (see Section 1.5). Simon [20] proposed the expression "frozen-in" state or phase for such a condition. We shall consider these substances in a later section. In every case, the system can be considered to be a mixture of two or more different species and hence, by eqn. (11.17), it has a positive entropy.

11.5. Some consequences of the third law. In the present section, we shall limit our discussion to systems for which eqn. (11.20) is valid. We recall that this relation holds no matter what state properties are held constant as 0 K is approached. Hence, the partial derivative of S at constant T with respect to any other independent state variable approaches zero as the Kelvin temperature tends to zero.

The left-hand sides, and hence the right-hand sides, of the Maxwell relations (9.22) and (9.23) for a closed, simple, homogeneous system approach the limit zero as the Kelvin temperature tends to zero

$$\lim_{T \to 0} \left(\frac{\partial S}{\partial V} \right)_T = \lim_{T \to 0} \left(\frac{\partial p}{\partial T} \right)_V = 0 \qquad (11.22)$$

$$\lim_{T \to 0} \left(\frac{\partial S}{\partial p} \right)_T = -\lim_{T \to 0} \left(\frac{\partial V}{\partial T} \right)_p = 0 \qquad (11.23)$$

In terms of the thermometric coefficients of the system, eqns. (11.22) and (11.23) become

$$\lim_{T \to 0} \alpha_V = 0, \ \lim_{T \to 0} \alpha_p = 0 \qquad (11.24)$$

Here α_V is the coefficient of thermal pressure increase of the system at constant volume, and α_p is its coefficient of thermal expansion at constant pressure [see eqns. (3.102) and (3.98)]. Similar relations hold for the corresponding coefficients α'_V and α'_p defined by eqns. (3.103) and (3.99), respectively.

From the foregoing discussion and from eqns. (10.4.6) and (10.4.7) of Column A and eqns. (10.4.5) and (10.4.6) of Column B of Table 10.4, we can derive the following third-law equations for a system producing expansion and also one kind of non-expansion work in its surroundings (see Section 10.23).

$$\lim_{T \to 0} \left(\frac{\partial S}{\partial V} \right)_{Ty} = \lim_{T \to 0} \left(\frac{\partial p}{\partial T} \right)_{Vy} = 0 \qquad (11.25)$$

$$\lim_{T \to 0} \left(\frac{\partial S}{\partial p} \right)_{TX} = -\lim_{T \to 0} \left(\frac{\partial V}{\partial T} \right)_{pX} = 0 \qquad (11.26)$$

$$\lim_{T \to 0} \left(\frac{\partial S}{\partial X} \right)_{TV} = \lim_{T \to 0} \left(\frac{\partial y}{\partial T} \right)_{XV} = 0 \qquad (11.27)$$

$$\lim_{T \to 0} \left(\frac{\partial S}{\partial y} \right)_{TV} = -\lim_{T \to 0} \left(\frac{\partial X}{\partial T} \right)_{yV} = 0 \qquad (11.28)$$

The above procedure may be applied to a system which produces more than one kind of non-expansion work in its surroundings.

We shall now consider two examples of the application of the above equations [21].

The reversible non-expansion work attending an infinitesimal increase da in the surface area a of a liquid of surface tension σ is $đW_X = -\sigma \, da$. Application of eqn. (11.27) gives

$$\lim_{T \to 0} \left(\frac{\partial S}{\partial a} \right)_{Tp} = -\lim_{T \to 0} \left(\frac{\partial \sigma}{\partial T} \right)_{ap} = 0 \qquad (11.29)$$

The subscript p may be replaced by V.

The measured surface tensions of both liquid ^{4}He and liquid ^{3}He are in agreement with the above relation.

The reversible non-expansion work effect attending an infinitesimal increase dI in the total magnetization of a homogeneous, isotropic system composed of a fixed mass of a magnetic substance in a uniform, static magnetic field of magnetic field strength H is $\mathrm{d}W_X = -\mathrm{H}\mathrm{d}I$. Application of eqn. (11.28) gives

$$\lim_{T\to 0}\left(\frac{\partial S}{\partial \mathrm{H}}\right)_{Tp} = \lim_{T\to 0}\left(\frac{\partial I}{\partial T}\right)_{\mathrm{H}p} = 0 \qquad (11.30)$$

The subscript p may be replaced by V. The magnetic susceptibility per unit volume χ_m of the system is

$$\chi_m = \frac{I}{V\mathrm{H}} \qquad (11.31)$$

Differentiation of this relation with respect to T at constant p and H, and substitution from eqn. (11.26), gives

$$\lim_{T\to 0}\left(\frac{\partial \chi_m}{\partial T}\right)_{\mathrm{H}p} = 0 \qquad (11.32)$$

The subscript p may be replaced by V.

Measured values of χ_m are in agreement with the above relation.

11.6. Phase diagrams at low temperatures. In Section 10.12 we found that the Clapeyron equation

$$\left(\frac{\mathrm{d}p}{\mathrm{d}T}\right)_{\mathrm{SAT}} = \frac{\Delta s}{\Delta v} \qquad (11.33)$$

gives the slope of the pressure—temperature curve on a phase diagram for a univariant system which, for a pure substance, consists of two phases in equilibrium (Figs. 10.1 and 10.2). According to the third-law equation (11.20) the molar entropy of each phase of such a system approaches zero as T tends to zero. Hence, in eqn. (11.33), we have

$$\lim_{T\to 0} \Delta s = 0 \qquad (11.34)$$

If Δv does not approach zero as T tends to zero, then eqns. (11.33) and (11.34) give

$$\lim_{T\to 0}\left(\frac{\mathrm{d}p}{\mathrm{d}T}\right)_{\mathrm{SAT}} = 0 \qquad (\lim_{T\to 0} \Delta v \neq 0) \qquad (11.35)$$

The vapor pressure curves of liquid ^{4}He and ^{3}He are in accord with eqn. (11.35). The slopes of the sublimation-pressure curves of all other substances are essentially zero at temperatures well above 0 K.

For the application of eqn. (11.35) to phase transformations in condensed systems, we turn to phase diagrams [22] of ^{4}He and ^{3}He. No solid—liquid—vapor triple point has been discovered for either isotope, and presumably each remains liquid when under its vapor pressure down to absolute zero.

The fusion curve of ^{4}He above 1 K is shown in Fig. 11.4 [23]. The liquid undergoes a phase transformation under its vapor pressure at 2.17 K from a superfluid, designated He(II), to a normal fluid, He(I), attended by a λ-anomaly (Fig. 3.10) in its heat capacity. Simon and Swenson [23] found that the slope of the fusion curve between 1 and 1.4 K is proportional to T^7, which suggests that it approaches zero rapidly as T tends to zero.

However, measurements [24] from about 0.5 to 0.8 K show that the fusion curve (Fig. 11.5) has a shallow minimum at about 0.8 K. Since the change in volume Δv attending fusion approaches a positive value as T tends to zero, Δs of fusion passes through zero at the minimum of the curve,

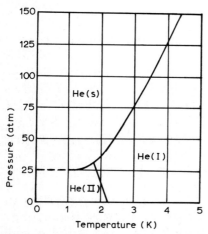

Fig. 11.4. The fusion curve above 1 K and the λ line of ^{4}He (after Simon and Swenson [23]).

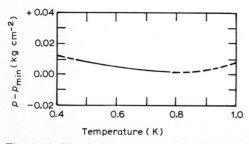

Fig. 11.5. The fusion curve of ^{4}He below 1 K (after Le Pair et al. [24]). The value of p_{min} is about 25.9 kg cm^{-2}.

and becomes negative at lower temperatures. Thus, in the latter region, the solid has a larger entropy than the liquid. This effect has been explained on the basis that phonon excitations, which account for practically all of the entropies of both phases, are more numerous in the solid phase (which supports three modes of vibration) than in the liquid (which supports only one) [24b]. Ultimately the fusion curve must turn so that its tangent, and hence Δs of fusion, approach zero as T tends to zero, in accordance with eqns. (11.34) and (11.35). Thus both solid and liquid ^{4}He have zero entropy at 0 K. This conclusion is substantiated by the following calculations.

Wilks [25] found agreement, within experimental error, between the standard molar entropies of liquid ^{4}He under its vapor pressure at 1.75 K computed in two entirely different ways: (1) integration of c_{SAT}/T with respect to T for the liquid along the vapor pressure curve from 0 to 1.75 K, with its entropy at 0 K equated to zero; and (2) subtraction of the entropy of vaporization of the liquid at 1.75 K (calculated from the Clapeyron equation) from the entropy of the coexistent vapor (computed from the Sackur—Tetrode equation explained in Section 11.14). In the latter calculation, a correction was applied for the deviation of the vapor from a perfect gas.

Only one liquid phase of ^{3}He has been found. The fusion curve [26] is shown in Fig. 11.6: it has a well-defined minimum at about 0.3 K. The volume change Δv attending fusion approaches a positive limit as T tends to zero. Hence, as in the case of ^{4}He, the solid has a larger entropy than the liquid in the temperature region where the coexistence curve has a negative slope. This effect has been explained on the basis that the ordering of the nuclear spins (see Section 11.24) of ^{3}He atoms has proceeded to a greater extent in the liquid than in the solid phase [26d].

Calculations [25] of the entropies of ^{3}He under its vapor pressure at 1.5 K similar to those made for ^{4}He (account being taken of the nuclear-spin entropy in the gas phase) indicate that the fusion curve of ^{3}He finally turns so that its slope approaches zero as T tends to zero, as required by the third law. Both liquid and solid ^{3}He have zero entropy at the absolute zero.

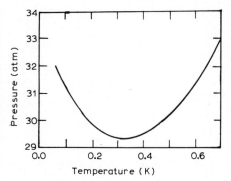

Fig. 11.6. The fusion curve of ^{3}He below 0.7 K (after Edwards et al. [26]).

Three solid phases have been found [27] for each of the helium isotopes ^{4}He and ^{3}He. The only transition curve that can be extrapolated to 0 K in either diagram is the curve along which hexagonal close-packed and body-centered cubic ^{3}He are in equilibrium: apparently its slope approaches zero as T tends to zero, in accord with the third-law equation (11.35).

11.7. Heat capacities at low temperatures. We have shown that the third law requires that C_V and C_p of a closed, simple, homogeneous system composed of a pure substance approach zero as T tends to zero [see eqns. (11.15)]. In a similar manner, we can prove from the eqns. of Table 10.5 that the following relations hold for a closed, homogeneous system consisting of a pure substance

$$\lim_{T \to 0} C_{VX} = 0 \qquad \left[\lim_{T \to 0} C_{pX} = 0\right] \tag{11.36}$$

From eqns. (9.41), (11.23), and (11.15), or corresponding relations expressed in terms of the heat capacities of eqns. (11.36), we find that the following relation holds for the heat capacity C_C of the systems considered in the last paragraphs along any path C, if $(dp/dT)_C$ remains finite down to the absolute zero

$$\lim_{T \to 0} C_C = 0 \qquad \left[\lim_{T \to 0} \left(\frac{dp}{dT}\right)_C \neq \infty\right] \tag{11.37}$$

In particular, if eqn. (11.35) is valid, the above expression gives

$$\lim_{T \to 0} C_{SAT} = 0 \tag{11.38}$$

for each of the phases that are in equilibrium along any saturation curve of a univariant one-component system.

11.8. Einstein's heat-capacity equation. The introduction of Planck's quantum theory (1901) into statistical mechanics by Einstein (1907) led to an explanation [28] of the disappearance of the heat capacities of solids at 0 K.

Einstein [29] supposed that the lattice energy E of a solid crystalline element consisting of N atoms, each having three degrees of freedom, could be approximated by the energy of vibration of $3N$ independent, one-dimensional, harmonic oscillators, each obeying quantum mechanics and each vibrating with the same frequency ν_E. For this model of an atomic solid, and setting $N = N_A$ (where N_A = Avogadro's number), he found

$$\frac{e - e_0}{3RT} = \frac{x}{e^x - 1} \qquad \left(x = \frac{h\nu_E}{kT} \equiv \frac{\Theta_E}{T}\right) \tag{11.39}$$

Here e_0 is the energy of one gram-atom of the solid at 0 K, h is Planck's constant, k is Boltzmann's constant, and Θ_E is a parameter called the Einstein temperature of the given solid.

From eqn. (11.39) and general thermodynamic relations we find [30] the following expressions for the lattice heat capacity c_V and entropy s of one gram-atom of the solid

$$\frac{c_v}{3R} = \frac{x^2 e^x}{(e^x - 1)^2} \tag{11.40}$$

$$\frac{s}{3R} = \frac{x}{e^x - 1} - \ln(1 - e^{-x}) \tag{11.41}$$

The quantity c_V/R of the Einstein heat-capacity equation (11.40) is sometimes represented by the symbol $3E(\Theta_E)$, where 3 is the number of degrees of freedom of the particle at each of the N lattice points of the crystal, $E(\Theta_E)$ denotes the right-hand section of eqn. (11.40), called the Einstein heat capacity function, and Θ_E is the Einstein temperature of the solid. When each particle consists of an assembly of two or more atoms, the above notation for c_V/R becomes $nE(\Theta_E)$, where n is the number of degrees of freedom of each particle. In this case, the number 3 appearing in the denominators of the left-hand sections of eqns. (11.39)–(11.41) must be replaced [31] by n.

When the above changes are made, the Einstein equation (11.40) represents the lattice heat capacities c_V of solid elements and compounds qualitatively, but not always quantitatively, over a wide range of temperature. But below about $T = \Theta_E/10$ the calculated values of c_V approach zero much too rapidly [28].

Einstein energy, entropy, and heat-capacity functions are tabulated in several places [28, 32, 33].

Equation (11.40) gives

$$\lim_{T \to 0} c_V = 0, \quad \lim_{T \to \infty} c_V = 3R \tag{11.42}$$

in agreement with the third law and the law of Dulong and Petit, respectively.

11.9. Debye's heat-capacity equation. Debye [34] assumed that the $3N$ oscillators of Einstein's model of an atomic solid, consisting of N atoms each having 3 degrees of freedom, are coupled, and that they have the same frequency distribution up to a maximum frequency ν_D as the distribution of the allowed frequencies of standing acoustic waves in a continuous, isotropic, elastic medium of volume V (the gram-atomic volume of the solid). He further assumed that the longitudinal and transverse waves in the medium have the same velocity, which is independent of frequency. Debye chose the value of ν_D so that the number of normal modes of vibration of the medium is equal to $3N$, the number of degrees of freedom of the solid. From these considerations Debye derived the following equation for the lattice energy e of one gram-atom of the solid

$$\frac{e-e_0}{3RT} = \frac{3}{x^3}\int_0^x \frac{y^3 dy}{e^y - 1} \equiv D_e(x) \qquad \left(x = \frac{h\nu_D}{kT} \equiv \frac{\Theta_D}{T}, y = \frac{h\nu}{kT}\right) \quad (11.43)$$

The quantity Θ_D is called the Debye temperature of the crystal.

From eqn. (11.43) and general thermodynamic relations we find [30] the following expressions for the lattice heat capacity c_v and the corresponding entropy s of one gram-atom of the solid

$$\frac{c_V}{3R} = 4D_e(x) - \frac{3x}{e^x - 1} \qquad (11.44)$$

$$\frac{s}{3R} = \frac{4}{3}D_e(x) - \ln(1 - e^{-x}) \qquad (11.45)$$

A notation similar to that used in Section 11.8 has been employed for the Debye heat-capacity equation (11.44). Thus c_V/R in this relation is represented by the symbol $3D(\Theta_D)$, where $D(\Theta_D)$ denotes the right-hand section of eqn. (11.44), called the Debye heat-capacity function, and Θ_D is the Debye temperature of the solid. When each particle in the lattice has n degrees of freedom instead of 3, the notation $nD(\Theta_D)$ is used, and the number 3 in the denominators of the left-hand sections of eqns. (11.43)–(11.45) are replaced by n.

When the above changes are made, the Debye equation (11.44) gives a fairly quantitative representation of the lattice heat capacities c_v of solid elements and compounds over a wide range of temperature. It is particularly successful in the temperature region below $\Theta_D/10$ where the Einstein equation fails badly.

Debye energy, entropy, and heat-capacity functions are tabulated in several places [28, 32, 35].

Equations (11.42) hold for the Debye heat-capacity equation (11.44) at zero and at infinite Kelvin temperatures.

Below about $T = \Theta_D/15$, eqn. (11.44) is approximated by the well-known Debye T^3 law

$$\frac{c_V}{3R} = \frac{4\pi^4}{5x^3} = \frac{77.92727}{x^3} \qquad (11.46)$$

or

$$c_V = \alpha T^3, \alpha = \frac{1943.73}{\Theta_D^3} \text{ J deg}^{-4} \text{ mole}^{-1} = \frac{464.56}{\Theta_D^3} \text{ cal}_{th} \text{ deg}^{-4} \text{ mole}^{-1} \qquad (11.47)$$

The error of this approximation is 2.8% for $T = \Theta_D/10$, 0.71% for $T = \Theta_D/12$, 0.079% for $T = \Theta_D/15$, and 0.0015% for $T = \Theta_D/20$.

The values of Θ_D for solids extend from less than 100 K to more than 1000 K. Many lie in the range 100–300 K.

The outstanding success of the Debye equation is its prediction of the T^3 law for the *lattice heat capacity* of crystalline solids in the region below about 15 K. The lattice heat capacities of graphite, black phosphorus (Stephenson, Potter, Maple, and Morrow [31]), and other layer crystals obey a T^3 law at temperatures below a few degrees Kelvin.

The heat capacity of ^{4}He obeys the T^3 law below about 0.6 K; but that of ^{3}He approaches a linear variation with temperature below about 0.1 K.

11.10. Electronic heat capacity of a metal. Application of quantum statistics to the free electrons in a metal gives the following approximate relation for the electronic contribution c_{V_e} to the constant-volume heat capacity of the solid

$$c_{V_e} = \gamma T \tag{11.48}$$

This relation is valid [36] for temperatures small in comparison with 30,000 K. The value of the parameter γ can be calculated from theoretical considerations; alternatively, it can be determined by the application of a graphical method, described in Section 11.12, to measured values of c_V extending from about 15 K down to about 1 K. The calculated values of γ are of the order of 10^{-4} cal deg^{-2} mole^{-1}. The measured values are often larger than this.

The electronic heat capacity of a metal is of the order of 1% of the lattice heat capacity at 300 K, but it becomes larger than the latter at temperatures below a few degrees Kelvin.

11.11. Difference between c_p and c_V of a solid. The Einstein and Debye equations for the heat capacity of a solid express c_V as a function of temperature. But the quantity determined experimentally is c_p. The thermodynamic relation (9.43) gives the difference $c_p - c_V$ in terms of quantities whose values have not been measured over a wide temperature range for most solids. It has been found [37] that if we write eqn. (9.43) in the form

$$c_p - c_V = \left(\frac{v\alpha_p^2}{\beta_T c_p^2}\right) c_p^2 \, T \equiv a \, c_p^2 \, T \tag{11.49}$$

the parameter a is practically independent of temperature, and hence may be evaluated at any convenient temperature. It varies somewhat from solid to solid, but has a value of the order of 10^{-5} cal^{-1} mole.

11.12. Heat capacity c_p of a solid at low temperatures. In this section we shall assume that all heat capacities c_p of condensed phases are measured at a pressure of 1 atm. The effect of the variations in pressure from this value encountered in ordinary calorimetric determination of c_p can be neglected in the calculation of entropies.

Some measurements of c_p extend to temperatures well below 1 K. But in general the measurements of this quantity for the purpose of determining entropies do not extend below 10 or 15 K, although those for metals often go down to about 1 K where the electronic heat capacity becomes important.

Up to about 15 K, the difference between c_p and c_V is much less than the error in the experimental determination of either quantity. Thus, in this temperature range we may use the Debye T^3 law for c_p with negligible error. From eqns. (11.47) and (11.48) we find the following relations

$$c_p = \alpha T^3 \qquad \text{(dielectric, } T \leqslant 15 \text{ K)} \qquad (11.50)$$

$$c_p = \alpha T^3 + \gamma T \qquad \text{(metal, } T \leqslant 15 \text{ K)} \qquad (11.51)$$

According to eqn. (11.51), the graph of c_p/T against T^2 for a metal below about 15 K is a straight line. This has been confirmed experimentally [38]. The slope of this line is α and the intercept on the c_p/T axis is γ.

The above equations hold for the heat capacities of layer crystals, but only at temperatures below a few degrees Kelvin. Anomalies in the heat capacities of paramagnetic solids at low temperatures are considered in Section 11.23.

When the thermal data do not extend below about 15 K, the parameter α in eqns. (11.50) and (11.51) has been evaluated by fitting the Debye heat-capacity equation (11.44) to the measured values of c_p over the range 15 K to about 50 K; the effect of the difference between c_p and c_V on the value of α may be neglected in the computation of entropies at 15 K by the T^3 law.

A good representation of the measured values of c_p up to about 100 K can be obtained by the use of a combination of one or more Debye and one or more Einstein heat-capacity functions, each of the type described in Sections 11.8 and 11.9. Thus, for NH_4NO_3 Stephenson, Bentz, and Stevenson [31] used the equation

$$c_p/R = 6D(192) + 3E(340) + 3E(400) \qquad (11.52)$$

which represents the measured values of c_p from 15 to 100 K within a few hundredths of a cal deg^{-1} mole^{-1}. Here the term $6D(192)$ represents the contributions to the molar heat capacity c_p of NH_4NO_3 from excitation of $3N$ degrees of freedom of lattice vibrations of the NO_3^- ions and $3N$ degrees of freedom of lattice vibrations of the NH_4^+ ions, each with $\Theta_D = 192$ K; $3E(340)$ represents the contributions to c_p from excitation of $3N$ degrees of freedom of torsional oscillations of the NO_3^- ions with $\Theta_E = 340$ K; and $3E(400)$ represents the corresponding contributions of the NH_4^+ ions with $\Theta_E = 400$ K. The contributions to c_p from excitation of the remaining $6N$ degrees of freedom of internal vibrations of the NO_3^- ions and the corresponding $9N$ degrees of freedom of the NH_3^+ ions are negligible below 100 K.

11.13. Calorimetric entropies. According to the third-law equation (11.20) we may equate the integration constant s_0^0 in eqn. (10.165) to zero for a substance for which this law is valid. Hence we can calculate from this relation the standard molar entropy s_T^0 of the substance at any temperature T if we are given its standard molar heat capacities c_p^0 from 0 to T K and the relevant standard molar enthalpies of phase transformation Δh^0 under equilibrium conditions.

We recall that calorimetric measurements usually give the molar quantities c_p and Δh of a substance at 1 atm rather than the standard values. Introduction of these quantities into eqn. (10.165) gives the molar entropy of the substance in the state (s, l, or g, 1 atm, T).

In Section 6.2 we found that the standard state of a solid or liquid is simply the state (s, or l, 1 atm, T). Hence the measured molar entropy at (1 atm, T) is the corresponding standard value

$$s_T^0 = s(1 \text{ atm}, T) \qquad \text{(solid or liquid)} \qquad (11.53)$$

The standard state of a pure gas is the state (PG, 1 atm, T) (see Section 6.2). The molar entropy of the gas in the state (g, 1 atm, T), or in the state (g, p, T) if the liquid (or solid) is vaporized under a vapor pressure p rather than its normal boiling (or sublimation) pressure of 1 atm, must be corrected for the imperfection of the gas. Usually the reduction of the molar entropy $s(p, T)$ to the standard molar value s_T^0 is made by application of the approximate eqn. (9.105). For many gases, this correction is of the order of 0.1 cal deg^{-1} mole^{-1} when the pressure of the real gas is 1 atm, and hence the error of the approximation is not serious.

The actual computation of the molar entropy of a substance in the state (1 atm, T) from calorimetric measurements is usually carried out in two steps. The first step is to calculate its molar entropy at (1 atm, T^*) where T^* is about 15 K. Substitution from eqn. (11.51) into (10.165) leads to the following relation for the standard molar entropy s_T^0 of the solid at T^*

$$s_T^0 = \tfrac{1}{3}\alpha T^{*3} + \gamma T^* \qquad (T^* \approx 15 \text{ K}) \qquad (11.54)$$

where $\gamma = 0$ for a dielectric. The parameters α and γ are evaluated by the methods considered in Sections 11.10 and 11.12.

The second step is to calculate the change in molar entropy of the substance from (1 atm, T^*) to (1 atm, T) by eqn. (10.165), evaluating the definite integrals graphically from a graph of c_p/T against T.

As an example of the application of these procedures, we give in Table 11.1 the results of Giauque and Clayton [39] on the determination of $s_{298.1}^0$ for $N_2(g)$. This element exists in two allotropic forms which we shall label α and β, the latter being the stable form at 0 K. Nitrogen undergoes the following phase transformations

$N_2(\beta) = N_2(\alpha)$ $\qquad$ (1 atm, 35.61 K) $\qquad\qquad$ (11.55)

$N_2(\alpha) = N_2(l)$ $\qquad$ (1 atm, 63.14 K) $\qquad\qquad$ (11.56)

$N_2(l) = N_2(g)$ $\qquad$ (1 atm, 77.32 K) $\qquad\qquad$ (11.57)

The change in the standard molar entropy of $N_2(g)$ from 77.32 K to 298.1 K was computed from spectroscopic data by the methods of statistical mechanics, which we shall consider briefly in the next section. The value of

TABLE 11.1

STANDARD MOLAR ENTROPY OF $N_2(g)$ AT 298.1 K AND $p = 1$ atm

Temperature range (K) or phase transition	s	Method	Molar entropy (E.U. mole^{-1})
0—10		T^3 law	0.458
10—35.61		graphical	6.034
↓ β = α			1.536
35.61—63.14		graphical	5.589
↓ α = l			2.729
63.14—77.32		graphical	2.728
↓ l = g			17.239
	$sN_2(g, 1$ atm, 77.32 K)		36.31 ± 0.1
Correction for gas imperfection			0.22
	$s^0_{77.32} N_2(g)$		36.53
77.32—298.1		spectroscopic	9.37
	$s^0_{298.1} N_2(g)$	calorimetric	45.9
	$s^0_{298.1} N_2(g)$	spectroscopic	45.788

$s^0_{298.1}$ of $N_2(g)$ calculated in the manner indicated in Table 11.1 is labeled *calorimetric*. The value derived entirely from spectroscopic data is labeled *spectroscopic*. In this Table, E. U. denotes the usual entropy unit, namely cal deg^{-1}.

11.14. Spectroscopic entropies. In quantum statistical mechanics [40—42] the energy of a simple system composed of a pure perfect gas is treated, in the first approximation, as the sum of the contributions of the translational energy and the internal energy (rotational, vibrational, internal rotational, and electronic) of its molecules, or, when relevant, of its atoms. The energy levels of the internal energy of a molecule or atom can be evaluated from spectroscopic data; those of the translational energy cannot.

Entropy, heat capacity, and other extensive thermodynamic properties can be treated in a similar manner.

The translational entropy of a perfect gas is expressed as a function of pressure and temperature by the Sackur—Tetrode equation [43], which can be derived by the methods of quantum statistical mechanics. This equation may be written in the following forms for the molar entropy $s(p, T)$ of a perfect gas in the state (p, T) and the corresponding standard molar entropy s^0_T

$$s(p, T) = R \left[2.5 \ln T - \ln p + 1.5 \ln M + \ln \frac{k}{a}\left(\frac{2\pi k}{N_A h^2}\right)^{1.5} + 2.5 \right] \quad (11.58)$$

$$s^0_T = R(2.5 \ln T + 1.5 \ln M - 1.16495) = 4.575617 \ (2.5 \log_{10} T$$
$$+ 1.5 \log_{10} M) - 2.31495 \text{ cal deg}^{-1} \text{ mole}^{-1} \quad (11.59)$$

Here k = 1.38054 × 10^{-16} erg deg^{-1}, h = 6.6256 × 10^{-27} erg sec, N_A = 6.02252 × 10^{23} mole^{-1}, M is the molecular weight of the gas, p is its pressure in atm, and a = 1.01325 × 10^6 dyne cm^{-2} atm^{-1}.

The only property of the gas occurring in eqns. (11.58) and (11.59) is its molecular (or atomic) weight. The relation is valid down to temperatures at which quantum effects become appreciable.

Equation (11.59) expresses the total standard molar entropy of a monatomic gas as a function of temperature up to temperatures at which the effect of electron excitation becomes appreciable.

The internal standard molar entropy of a perfect gas can be expressed as a function of temperature by equations which contain parameters that depend on the energy levels of the constituent molecules or atoms. When the values of these energy levels relative to the energy of the ground state of the gas can be determined from an interpretation of the spectroscopic measurements, we can compute its internal standard molar entropy at any temperature T, whether or not the solid or liquid phase of the substance has a measurable vapor pressure at T.

The sum of the translational and internal standard molar entropies of a perfect gas at a temperature T is its corresponding *spectroscopic entropy* s_T^0. From eqn. (11.21) we find that s_0^0 is zero at 0 K if the ground state of the molecule is non-degenerate. Thus spectroscopic and calorimetric entropies have the same zero.

No thermal data are employed in the above computations. The calculation of the standard molar entropy of the solid or liquid phase of a substance at a temperature T'' from the spectroscopic entropy of its gas phase at T' does require the use of calorimetric data, namely, heat capacities of condensed phases and the relevant enthalpies of phase transformation in the range $T' - T''$. But no thermal measurements outside this range are needed.

In general, calorimetric and spectroscopic values of standard molar entropies agree within the experimental errors of the two methods (see, for example, Table 11.1). The few substances for which the two methods do not give concordant results are considered below.

11.15. Entropies of polymorphic forms at 0 K. When a solid undergoes a transition from one polymorphic form to another at temperature T, the high-temperature form, which we shall label α, can in some cases be cooled down to the lowest attainable temperature and remain there for an indefinitely long time with no perceptible reversion to the low-temperature modification, which we shall denote by β. The α form is said to be in a state of metastable equilibrium, the β form in a state of stable equilibrium (see Section 10.11). The isothermal change in state

$$\alpha = \beta \qquad (1 \text{ atm}, T) \tag{11.60}$$

can be brought about in either direction by a three-step reversible process, which is not itself isothermal. Hence by the third law the change in entropy Δs attending the change in state (11.60) approaches zero as T tends to zero [see eqn. (11.19)]. Thus, in accordance with eqn. (11.20), the entropy of each polymorphic form approaches zero as the Kelvin temperature decreases to zero.

This conclusion has been verified experimentally for white and gray tin [44], the two crystalline forms of cyclohexanol [44], rhombic and monoclinic sulfur [45], and β and γ phosphine [46].

As an example of these results we shall consider the measurements of Stephenson and Giauque [46] on phosphine. This substance has four polymorphs at 1 atm, which we shall label α, β, γ, δ, which undergo the following transitions

$PH_3(\beta) = PH_3(\alpha)$ (1 atm, 49.43 K) (11.61)

$PH_3(\alpha) = PH_3(\delta)$ (1 atm, 88.10 K) (11.62)

$PH_3(\gamma) = PH_3(\alpha)$ (1 atm, 30.29 K) (11.63)

At 1 atm the β form is stable from 0 to 49.43 K, the α form is stable from 49.43 to 88.10 K, and the δ form is stable from 88.10 K to the melting point, 139.35 K. At 1 atm the γ form is always metastable, and it undergoes a transition at 30.29 K to the α form, which is also metastable at this transition point.

We can calculate the standard molar entropy s_T^0 of δ-PH_3 at 88.10 K along each of the two paths given in Table 11.2, one starting with stable β-PH_3 at 0 K, and the other starting with metastable γ-PH_3 at the same temperature. In accord with the third-law equation (11.20), s_0^0 of each form was assigned the value zero at 0 K. The two values of $s_{88.10}^0$ for δ-PH_3 agree well within the experimental errors of the two sets of measurements.

11.16. Practical difficulties in determining spectroscopic entropies: internal rotation. In general, values of the standard molar entropies of gases calculated from spectroscopic data agree with those derived from calorimetric measurements within the experimental errors of the two methods. A discrepancy which depends on temperature denotes use of an incorrect partition function. This may be due to errors in the determination of the vibrational or the internal rotational energy levels of the gas from spectroscopic data, or to both effects. We shall now consider briefly the internal rotations of polyatomic molecules.

TABLE 11.2

STANDARD MOLAR ENTROPY OF $PH_3(\delta)$ AT 88.10 K AND p = 1 atm

Path $\beta = \alpha = \delta$			Path $\gamma = \alpha = \delta$		
Temperature range (K) or phase transition	Method	Molar entropy (E.U. mole^{-1})	Temperature range (K) or phase transition	Method	Molar entropy (E.U. mole^{-1})
0—15	T^3 law	0.338	0—15	T^3 law	0.495
15—49.43	graphical	4.041	15—30.29	graphical	2.185
↓ $\beta = \alpha$		3.757	↓ $\gamma = \alpha$		0.647
49.43—88.10	graphical	6.705	30.29—88.10	graphical	11.505
↓ $\alpha = \delta$		1.314	↓ $\alpha = \delta$		1.314
$s_{88.10}^0 PH_3(\delta)$		16.155	$s_{88.10}^0 PH_3(\delta)$		16.146

Formerly it was supposed that [47] "a single bond, such as carbon to carbon in ethane (CH_3—CH_3), acted like a well-greased axle about which the two groups rotate freely". In the period 1935—1937, a discrepancy was found between the value of the entropy of ethane derived from spectroscopic data on the basis of free rotation of the methyl groups, and the corresponding value derived either from equilibrium data on the hydrogenation of ethylene [48] or from calorimetric measurements [49]. This and similar discrepancies for a number of organic compounds were explained on the basis of the existence of potential barriers restricting free rotation in molecules [50].

If the potential barrier to internal rotation per mole of gas is very small compared with RT, such as for the CH_3 groups in H_3C—$C\equiv C$—CH_3 at room temperature, essentially free rotation takes place. But if this barrier is large compared with RT, such as for the CH_2 groups in H_2C=CH_2 at room temperature, torsional oscillation of the two groups occurs. In each case, the internal rotational energy levels are given by quantum mechanics. The corresponding partition function of the gas, and the contribution to the equation expressing its standard molar entropy as a function of temperature, are then given by statistical mechanics.

In some gases, the potential barrier is of intermediate height and the rotation is said to be hindered. The internal rotational energy levels of such a molecule are determined from the wave equation into which an appropriate expression for the potential barrier has been introduced.

When the rotating groups are a pair of symmetrical coaxial tops, for example the CH_3 groups in ethane, the potential barrier is well represented by a periodic function of the form $0.5\ V_0\ (1 - \cos n\phi)$, where V_0 is the maximum value of the potential energy, ϕ is the angle of rotation, and n is the number of times the molecule returns to an equivalent position in each revolution [40—42]. The values of V_0 for a given gas may be derived from spectroscopic data in the infrared and microwave regions of the spectrum. The use of the above expression for the potential barriers of the type considered here leads to values of spectroscopic entropies that are in good agreement with the calorimetric values.

The problem of determining the internal rotational energy levels of molecules with rotating tops that are not symmetrical and coaxial is much more difficult, because of the coupling between the internal and the overall rotation.

11.17. Practical difficulties in determining calorimetric entropies. The spectroscopic values of the standard molar entropies of some gases are larger than the corresponding calorimetric values by significant amounts which are independent of temperature: only the former (i.e. spectroscopic) values are found to be correct. The final criterion of the validity of either value is the corresponding entropy derived from the measured equilibrium conditions of a chemical reaction (or physical change) together with appropriate thermal data [see eqn. (11.3)].

We shall now consider two sources of error in entropies calculated from purely thermal measurements.

11.17A. Frozen-in disorder at 0 K. Some substances form condensed phases which have certain elements of configurational disorder that persist substantially unchanged during the time required to cool a system down to a low temperature and carry out a series of normal constant-pressure heat-capacity measurements, each run being completed in a matter of an hour or so. From statistical mechanics we find that the configurational contribution to the heat capacity of such a "frozen-in" phase [8] is virtually zero, and hence the corresponding contribution to its entropy is practically independent of temperature.

Straightforward application of the method described in Section 11.13 to the determination of entropies of such a substance at temperatures above the "freezing-in" temperature of the configurational disorder gives values which are smaller than the correct entropies by a constant amount. The constant difference is termed the residual entropy of the substance, or its frozen-in entropy at 0 K. The expression zero-point entropy has also been used for this quantity.

We recall (Section 1.5) that Gibbs [51] clearly distinguished between the equilibrium "caused by the balance of the active tendencies of the system" and that due to "passive forces or resistances to change, [potential barriers in the present case] in so far, at least, as they are capable of *preventing* change". He pointed out that, in the application of his criteria of equilibrium and stability to a system, those variations which involve changes prevented by passive forces or resistances to change are to be rejected as impossible.

According to the second law, we cannot calculate by thermodynamic methods the entropy of a system in a state which cannot be reached from a standard state by a reversible process (see Sections 8.14 and 8.18). Now the state of a system with frozen-in configurational disorder at a given pressure and temperature cannot be reached from any completely ordered state of the system at 0 K by a reversible process without the use of variations which are not allowed because they are prevented by passive forces. Hence the third law (Section 11.4) is clearly not applicable to systems with frozen-in configurational disorder.

Evidently, the state of a system with a certain frozen-in configurational disorder at a given pressure and temperature can be reached from a state with virtually the same frozen-in configurational disorder at 0 K by a process which is essentially reversible. Hence the change in entropy attending this change in state can be evaluated by the methods of Section 11.13.

So-called entropies of phases with frozen-in configurational disorder are never used in a purely thermodynamic calculation of the equilibrium conditions of a system, whether homogeneous or heterogeneous. They do find an application in the study of the arrangements or orientations of the molecules, atoms, or ions in the crystal lattice of a solid.

Examples of frozen-in phases are condensed solutions, glasses, and certain pure crystalline solids that retain elements of configurational disorder at low temperatures (see Sections 11.18, 11.20, and 11.21).

11.17B. Incorrect extrapolation to 0 K. Entropies derived from calorimetric measurements are computed by using the Debye T^3 law for extrapolation of the lattice heat capacity of the solid from the lowest temperature of measurement to 0 K (see Section 11.13). The true heat-capacity curve may follow quite a different course at low temperatures, because at the lowest temperature of *measurement* there may be elements of disorder in the crystal which disappear at still lower temperatures. The Debye equation may represent the lattice heat capacity satisfactorily, but the actual heat capacity may have one or more λ-anomalies (Fig. 3.10) or Schottky anomalies (Fig. 3.11), occasioned by a phase transformation (see Section 3.15) in which a more or less rapid decrease in entropy takes place in a small temperature interval. A Schottky anomaly may occur at temperatures well below 1 K. When a substance is cooled to below the temperature of its lowest heat-capacity anomaly, its heat capacity approaches zero as the Kelvin temperature tends to zero, and so does its entropy if it has no frozen-in disorder.

Systems presenting the kind of difficulty mentioned here are treated in Sections 11.22 and 11.23.

In regard to the entropy effects considered above, we recall that the molar entropy of mixing at constant pressure and temperature of m distinguishably different species *which form perfect solutions* is given by the thermodynamic equation (11.17). The statistical mechanical relation (11.21) gives the same expression [52]. When each of the m species has the same mole fraction $1/m$, eqn. (11.17) gives the following relation for the molar entropy of mixing

$$\Delta S^M = -Rm \frac{1}{m} \ln \frac{1}{m} = R \ln m \tag{11.64}$$

11.18. Entropies of solid solutions at 0 K. Eastman and Milner [53] calculated the changes in free energy ΔG and enthalpy ΔH attending the change in state

$$0.728 \, \text{AgBr(s)} + 0.272 \, \text{AgCl(s)} = [0.728 \, \text{AgBr} \cdot 0.272 \, \text{AgCl}] \, \text{(s)}$$

(1 atm, 298 K) \hfill (11.65)

from measurements of the electromotive forces of relevant galvanic cells, and pertinent enthalpies of solution at 298 K. They also measured the constant-pressure heat capacities of the reactants and products of the reaction from about 15 to 298 K. From these results they found the following values for the changes in entropy attending the change in state (11.65) at 298 K, and at 0 K

$$\Delta S^M_{298} = 1.12 \pm 0.10 \text{ cal deg}^{-1}, \Delta S^M_0 = 1.03 \pm 0.10 \text{ cal deg}^{-1} \tag{11.66}$$

These values agree within their estimated experimental errors.

From the standpoint of statistical mechanics, the above solid solution is a frozen-in phase from room temperatures down to 0 K. Thus, in this temperature range, the configurational disorder of the phase is independent of temperature, the configurational contribution to its heat capacity is zero, and the corresponding contribution to its entropy is constant. The solid solution has residual entropy.

The entropy of a solution can decrease to zero on being cooled to 0 K by: (1) total separation into two or more phases each in a completely ordered state; or (2) total transformation into a single ordered state.

Below a temperature called the critical solution temperature, some solid solutions separate into two or more coexistent solutions each of whose compositions approach purity in one of the constituent substances as the temperature decreases. In general, the rates of diffusion of the constituent molecules, atoms, and ions become so small at lower temperatures that configurational disorder is frozen-in well above the absolute zero, and the system has residual entropy.

When an alloy having essentially the same composition as a compound, such as CuZn or Cu_3Au, is cooled, it may undergo a transformation from a disordered to an ordered arrangement of the constituent atoms, termed a superlattice [54]. This change is attended by a λ-anomaly in the heat capacity, and hence a relatively large decrease in the entropy of the system. In general, the entropy of the alloy approaches zero as the Kelvin temperature tends to zero.

11.19. Entropies of liquid solutions of ^4He and ^3He below 1 K. Figure 11.7 is the phase diagram for liquid mixtures of ^4He and ^3He at 1 atm and below 1 K [35]. The critical mixing temperature is about 0.8 K. The coexistent liquids at 0 K are pure ^3He and a liquid mixture containing about 6 mole percent ^3He. Solutions with from 0 to 6 mole percent ^3He are homogeneous down to 0 K. We recall (see Section 11.6) that the entropies of the pure liquid isotopes of He approach zero as the Kelvin temperature tends to zero. The statistical mechanical theory of degenerate quantum liquids indicates [56] that the entropies of mixtures containing 6 or less mole percent ^3He also tend to zero with T. Thus the entropy of a system composed of any liquid mixture of ^4He and ^3He approaches zero as it is cooled to the absolute zero.

11.20. Entropies of glasses at 0 K. In general, a liquid may be cooled a number of degrees below its freezing point and remain for an indefinite time in a state of metastable thermodynamic equilibrium, called a supercooled liquid, before nucleation starts (see Section 10.11). Graphs of viscosity, vapor pressure, heat capacity, and other properties of the liquid plotted against temperature pass through the freezing point smoothly, without an abrupt or even a relatively sudden change in slope or curvature (see, for example, Figs. 10.1 and 10.2).

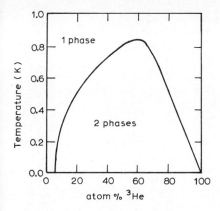

Fig. 11.7. The phase diagram for liquid solutions of ^{3}He and ^{4}He (after Wilks [55]).

When care is taken to prevent nucleation, some liquids, particularly those having a rather high rate of increase in viscosity with decrease in temperature, may be cooled far below their normal freezing points without crystallization taking place. Long before the absolute zero is reached, the mobile supercooled liquid undergoes an enormous increase in viscosity in a finite temperature interval, called the transformation region, to a quite rigid, vitreous substance, termed a glass, which has a viscosity of at least 10^{13} poise. Other properties of the substance also undergo rapid change in this temperature region.

Many liquids form glasses. The one most thoroughly investigated from a thermodynamic standpoint is glycerol, $C_3H_8O_3$, which may be cooled far below its freezing point (291.00 K) without crystallization occurring. In fact, there is some difficulty in initiating nucleation [57]. The metastable supercooled liquid undergoes a transformation into a glass in a temperature interval extending from about 190 to 180 K.

Graphs of the molar heat capacity c_p of crystalline and non-crystalline phases of glycerol plotted against temperature at 1 atm are given in Fig. 11.8 [57—62]. We note that, in the transformation region, the heat-capacity curve (a) of the supercooled liquid drops rather suddenly, but without discontinuity in its slope or curvature, along curve (b) to the heat-capacity curve (c) of the glass. Thereafter it runs approximately parallel to the heat-capacity curve (d) of the crystalline solid, which is the stable phase below 291.00 K.

From the entropy of fusion of crystalline glycerol and the heat-capacity curves of Fig. 11.8, we find that the "residual molar entropy" of the glass is about 5 cal deg^{-1} mole^{-1} [57, 61]. This value depends to some extent on the rate of cooling of the substance through the transformation region.

Oblad and Newton [62] determined the heat capacities of glycerol from

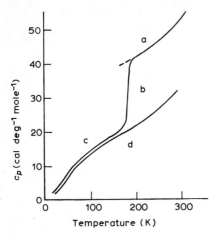

Fig. 11.8. The molar heat capacity c_p of glycerol at 1 atm plotted against Kelvin temperature (after Jones and Simon [59], and Oblad and Newton [62]).

thermal measurements on samples that had been held for various periods at each of a number of fixed temperatures below 190 K. By extending this treatment from "a few hours" at 183 K to "about a week" at 175 K, they found that c_p increased from the values given by curves (b) and (c) of Fig. 11.8 to those given by the broken portion of curve (a). The latter is the extension of the heat-capacity curve of the metastable supercooled liquid. The experiment was not continued to temperatures below 175 K because of the time required to complete a single run and the frequency of spontaneous crystallization of liquid glycerol when held for a long time in this temperature region.

It is apparent from these results that the physical condition of amorphous glycerol is time-dependent in the temperature range 190—175 K. Below this region each sample of the glass has a chance configurational disorder determined, in part, by its rate of cooling through and somewhat below the transformation region [59]. When the glass is cooled a little further, this disorder becomes essentially frozen-in and persists virtually unchanged down to the absolute zero.

Clearly, amorphous glycerol is not in a state of thermodynamic equilibrium when in the temperature region where its properties are changing with time at an observable rate (see Section 1.5). Therefore, the third law is not applicable to glycerol when in this condition.

At lower temperatures, where each sample of the glass has a random frozen-in configurational disorder, the condition or "state" of glycerol is not determined solely by its pressure and temperature [8]. Such a "state" of the glass cannot be reached from any ordered state of the substance at 0 K by a reversible process (see Section 11.17A). Hence the third law is also not applicable to the glass.

Thus the third law is not valid for glycerol when in a condition indicated by any point in the curves (b) and (c) of Fig. 11.8.

From the standpoint of statistical mechanics, the large increase in the viscosity of supercooled liquid glycerol as it is cooled through the transformation region at any reasonable rate produces a large reduction in the mobility of its molecules. This greatly retards the rate of rearrangement of the molecules to a greater degree of short-range order, and consequently produces a rapid decrease in the rate of change of configurational disorder with temperature [58, 59]. Hence there is a large decrease in the configurational contribution to the heat capacity of the substance.

Somewhat below 175 K the configurational disorder remaining in the glass becomes frozen-in for all practical purposes, the configurational contribution to its constant-pressure heat capacity becomes essentially zero, and the corresponding contribution to its entropy becomes virtually independent of temperature. There remain, however, the vibrational contributions to the heat capacity and entropy, which are somewhat larger than those of the crystalline solid.

Gibson et al. [63] measured the heat capacities C_p of the liquid, supercooled liquid, and glass phases of ethyl alcohol, propyl alcohol, and an equimolar mixture of these two liquids over various temperature ranges extending from about 77 to 275 K. The transformation regions of these liquids are approximately 95 to 105 K. The investigators concluded that the change in heat capacity ΔC_p attending the mixing of the two alcohols is essentially zero over the temperature range of the measurements. Hence the corresponding entropy of mixing ΔS^M persists substantially unchanged from about 275 to 77 K and presumably down to the absolute zero.

This conclusion is substantiated by thermal measurements extending down to 9 K on an approximately equimolar mixture of water and glycerol [64].

11.21. Entropies of solids with configurational disorder at 0 K. At the freezing points T of some liquids, the change in energy attending a change in the orientation of a constituent molecule, atom, or ion in the crystal lattice of the solid is very small compared with kT. Such substances crystallize with certain elements of configurational disorder due to thermal agitation. If the rate of reorientation of the particles into an ordered array is sufficiently small at T, the disordered condition is essentially frozen-in; and this condition persists practically unchanged as the substance is cooled at any reasonable rate to the absolute zero. The solid has residual entropy.

Under the conditions described above, the constituent molecules, atoms, or ions of the given solid are distributed almost equally among the allowed states. Thus, the difference between the molar entropies of the solid with random and with completely ordered arrangements of the particles in the crystal lattice is given by eqn. (11.64)

$$\Delta S^M \equiv s_0 = R \ln m \qquad (11.67)$$

where s_0 is the molar residual entropy of the solid and m is the number of species (allowed orientations of a particle) in this phase.

Consider, for example, nitrous oxide. It has a linear molecule with the unsymmetrical structure NNO. It freezes at 182.26 K to a crystalline solid having a lattice with random end-for-end orientation of molecules

---NNO—ONN—NNO—NNO—ONN—NNO—ONN--- (11.68)

rather than a lattice with the completely ordered orientation

---NNO—NNO—NNO—NNO—NNO—NNO—NNO--- (11.69)

Since each molecule has two allowed orientations in the crystal lattice, the molar residual entropy s_0 of the solid is given by

$$s_0 = R \ln 2 = 1.377 \text{ cal deg}^{-1} \text{ mole}^{-1} \qquad (11.70)$$

Table 11.3 gives, at the temperature T of Column 2, the values of the molar entropies calculated from spectroscopic (or from chemical-equilibrium and thermal) data and those derived from calorimetric measurements for the inorganic substances which have been proved experimentally to have residual entropy. In each case, the molar value of the latter quantity is listed, together with the appropriate form of eqn. (11.67) selected by the investigators.

The residual entropy of nitric oxide per formula weight NO is in good agreement with the value derived on the assumption that in the solid phase it has the molecular formula N_2O_2 which has a rectangular structure with two allowed orientations in the crystal lattice [65].

The molar residual entropies of nitrous oxide N_2O [66], carbon monoxide CO [67], and sulfuryl fluoride SO_2F_2 [68] are explained on the assumptions that the molecules of each substance have two allowed orientations in the crystal lattice. The N_2O and CO molecules are linear. The SO_2F_2 molecule is a slightly asymmetrical top. It was suggested [66] that a negative

TABLE 11.3

MOLAR RESIDUAL ENTROPIES OF SOME INORGANIC COMPOUNDS

Substance	Temperature (K)	Molar entropy (E.U. mole^{-1})			
		Spectroscopic	Calorimetric	Residual	Theoretical
NO(g)	121.36	43.75	43.03	0.72	$\tfrac{1}{2} R \ln 2 = 0.689$
N_2O(g)	184.59	48.50	47.36	1.14	$R \ln 2 = 1.377$
CO(g)	81.61	38.32	37.22	1.10	$R \ln 2 = 1.377$
SO_2F_2(g)	217.78	64.14	62.66	1.48	$R \ln 2 = 1.377$
H_2O(g)	298.10	45.10	44.28	0.82	$R \ln \tfrac{6}{4} = 0.806$
D_2O(g)	273.10	46.66	45.89	0.77	$R \ln \tfrac{6}{4} = 0.806$
ClO_3F(g)	226.48	62.59	60.17	2.42	$R \ln 4 = 2.75\overline{5}$
$Na_2SO_4 \cdot 10H_2O$(s)	298.15	141.46[a]	139.95	1.51	$R \ln 2 = 1.377$[b]

[a] From chemical-equilibrium and thermal measurements.
[b] See ref. 74.

difference between the experimental and the theoretical values of residual entropy, such as is the case for N_2O and CO, "indicates that some approach to orderly arrangement in the crystal has been made". The positive value of this difference for SO_2F_2 was attributed to experimental error [68].

The molar residual entropies of water H_2O [69] and deuterium oxide D_2O [70] are attributed by Pauling [71] to random orientation of hydrogen and deuterium bonds in the crystal lattices. For each substance, the experimental and theoretical values of the molar residual entropy are in excellent agreement.

The perchloryl fluoride molecule ClO_3F has one fluorine and three oxygen atoms at the vertices of an approximately regular tetrahedron, with a centrally located chlorine atom. There are thus four almost equivalent positions, and hence four possible orientations of a molecule in the crystal lattice with almost the same energy. The experimental value of the molar residual entropy of this substance is noticeably smaller than the theoretical value [72].

The molar residual entropy of $Na_2SO_4 \cdot 10H_2O$ [73] has been explained [74] on the basis that in the crystal lattice there are rings consisting of four water molecules (two from each of two $Na_2SO_4 \cdot 10H_2O$ formula units). Each such water molecule is so oriented that one of its hydrogen atoms is in the bonds in the ring and the other is in a bond to a sulfate oxygen. The former atom may be oriented in the ring in two ways, and hence each ring has two possible configurations in the crystal lattice.

It was noted [74] that compounds similar to $Na_2SO_4 \cdot 10H_2O$, but with S replaced by Cr, Se, Mo, or W, have the same ring structure. We would expect that they would also have a molar residual entropy of $R \ln 2$.

The organic substances considered below have been proved experimentally to have residual entropies.

Comparison of the molar entropies of the 1-olefins with those of the corresponding n-paraffins at the same temperature indicated that the 1-olefins with more than ten carbon atoms have end-for-end orientational disorder which increases with the number of carbon atoms until the molar residual entropy of 1-hexadecene has become $R \ln 2$ [75].

Benzothiophene, C_8H_6S, undergoes a phase transition from a low-temperature crystalline form I to a high-temperature crystalline form II in the region 250—261.6 K, attended by a λ-anomaly in its molar heat capacity c_p [76]. Phase II supercools readily down to the lowest temperature of measurement and presumably to the absolute zero. It is not a metastable phase, such as monoclinic sulfur or γ-PH_3 (see Section 11.15), but has frozen-in configurational disorder with a residual entropy of 1.46 ± 0.14 cal deg^{-1} mole^{-1}. This is equal, within the estimated error, to $R \ln 2$.

Measurements of the relevant thermal properties of methane and the deuterated methanes down to about 2.5 K [77] gave the following values of molar residual entropies in units of cal deg^{-1} mole^{-1}, each with an estimated error of 0.10 unit:

CH$_4$ −0.58 CH$_3$D 2.83 CH$_2$D$_2$ 3.53
CHD$_3$ 2.75 CD$_4$ ∼0

Although the molar residual entropies of CH$_3$D, CH$_2$D$_2$, and CHD$_3$ are in good agreement with the values R ln 4 = 2.75$\bar{5}$ or R ln 6 = 3.561 cal deg^{-1} mole^{-1}, Colwell et al. do not explain them on the basis of frozen-in configurational disorder. They write "... the properties of the solid methanes at the lowest temperatures are interpreted as resulting from separate contributions of lattice-vibrational and perturbed-molecular (rotational) energy levels". Apparently the situation in the solid methanes is quite complicated.

11.22. Rotational heat capacity of hydrogen. Because of the small moment of inertia of the H$_2$ molecule, the rotational energy levels of the gas are so widely spaced that differences between consecutive values are not small in comparison with kT at temperatures below about 300 K. Thus the classical rotational partition function is not valid for gaseous H$_2$, and the molar rotational heat capacity of the gas does not approximate the classical value R below this temperature. In fact, the molar value of c_V of H$_2$ gas at 1 atm decreases steadily from almost the classical value $\tfrac{5}{2}$R for a diatomic gas at 300 K to almost the classical value $\tfrac{3}{2}$R for a monatomic gas at 40 K. Thus in this temperature range the rotational contribution to c_V has decreased from about R to about zero. Below about 40 K practically every molecule in gaseous H$_2$ is in its lowest allowed rotational level.

Prior to 1927, the value of the standard molar entropy of hydrogen at any temperature (or its chemical constant), when determined by the calorimetric method (see Section 11.13) with a Debye T^3-law extrapolation of the heat capacity from about 14 to 0 K [78], was incorrect. The error was due to a λ-anomaly in the molar heat capacity of solid H$_2$ at about 1.6 K, well below the lowest temperature of the thermal measurements (11 K) [79].

Also, before 1927, the standard molar constant-pressure heat capacity of gaseous H$_2$ could not be represented quantitatively below about 300 K by the methods of statistical mechanics, because of the use of an erroneous rotational partition function [80].

On the basis of wave mechanics, Heisenberg [81] showed in 1927 that a substance composed of molecules having two or more identical isotopes with non-zero nuclear spins consists, in fact, of a mixture of two kinds of molecules: one with symmetric nuclear-spin wave functions, and the other with antisymmetric nuclear-spin wave functions. These two modifications came to be termed ortho and para molecules, respectively.

In the same year, Dennison [82] found, independently of the above work, that the measured molar rotational heat capacity c_{rot} of gaseous H$_2$ from 70 to 286 K could be quantitatively represented as the sum of two temperature functions

$$c_{rot} = (1-x)\, c_{rot}(\text{ortho}) + x\, c_{rot}(\text{para}) \tag{11.71}$$

Here x is the mole fraction of para H_2 in the gas, and $c_{rot}(\text{ortho})$ and $c_{rot}(\text{para})$ are the spectroscopic molar rotational heat capacities of ortho and para H_2, respectively.

Dennison assumed that the time of transition between ortho and para states of H_2 is very long compared with the time required to complete a series of heat-capacity measurements, so that x in eqn. (11.71) is essentially independent of temperature. On this basis he found that the best representation of the existing heat-capacity data on gaseous H_2 was obtained when $x = 1/4$.

Giauque and Johnston [83] calculated from quantum statistics that the nuclear-spin and rotational contributions to the standard molar entropy of H_2 gas with 25 mole percent of para H_2 at a temperature below about 40 K is 4.39 cal deg^{-1}. In effect, this is the difference between the spectroscopic value of the standard molar entropy of H_2 at any temperature above the region of the λ-anomaly in the heat capacity of solid H_2, and the value determined by the calorimetric method with a Debye T^3-law extrapolation of the heat capacity below 14 K.

From the existing data, Giauque [84] computed the values of the above two standard molar entropies of H_2 gas at 298.1 K as 33.98 and 29.74 cal deg^{-1} mole^{-1}, respectively. The difference, 4.24 cal deg^{-1} mole^{-1}, is in satisfactory agreement with the theoretical value. We shall consider these quantities further below.

Also, Giauque derived from quantum statistics the relation expressing the mole (or weight) percent of para H_2 in the equilibrium mixture as a function of temperature. Some values are: 25% at infinite T, 25.07% at 298.1, 99.99% at 15, and 100% at 0 K. In general, "ordinary hydrogen" is considered to contain 25% para H_2.

Bonhoeffer and Harteck [85] found that ordinary hydrogen gas adsorbed on charcoal at liquid hydrogen temperatures is rapidly converted into practically pure para H_2. Hydrogen containing more than 99% ortho H_2 has been prepared by a three-stage adsorption—desorption of ordinary H_2 gas on alumina at 50 torr and 20.4 K [86].

Let us now enumerate the nuclear and rotational quantum states of the molecules of H_2 gas [87].

The hydrogen nucleus has a spin quantum number $I = \frac{1}{2}$. Thus the H atom has $2I + 1 = 2$ nuclear spin states (see Section 11.24B).

The diatomic, homonuclear molecule H_2 has $(2I + 1)^2 = 4$ nuclear spin states. Ortho H_2, with a total nuclear-spin quantum number $T = 1$ (atomic nuclear spins aligned parallel) has $(I + 1)(2I + 1) = 3$ symmetric nuclear-spin states of virtually the same energy, corresponding to the three possible orientations of the nuclear spin with respect to any fixed axis. Para H_2, with a total nuclear-spin quantum number $T = 0$ (atomic nuclear spins aligned antiparallel) has $I(2I + 1) = 1$ antisymmetric nuclear spin state [88].

Since the hydrogen nucleus has an odd number of nucleons (a proton), the total wavefunctions of H_2 are antisymmetric with respect to nuclear exchange. Hence ortho H_2, with symmetric nuclear-spin states, has antisymmetric rotational levels for which the rotational quantum number J has only the odd values $J = 1,3,5,\ldots$. Para H_2, with an antisymmetric nuclear-spin state, has symmetric rotational levels for which J has only the even values $J = 0,2,4,\ldots$.

The lowest rotational level of ortho H_2 has the quantum number $J = 1$, and it has $2J + 1 = 3$ quantum states, corresponding to the three possible orientations of the axis of rotation with respect to any fixed axis. In the absence of an applied magnetic field these states have almost the same energy. The lowest rotational level of para H_2 has the quantum number $J = 0$ and it has 1 rotational state, which is the ground rotational level of the H_2 molecule.

At about 14 K, liquid ordinary H_2 crystallizes to a frozen-in solid solution having approximately 25 mole percent para H_2. This composition persists essentially unchanged down to 0 K for long periods of time. Ortho H_2 molecules rotate in both the liquid and solid phases [89].

Since the energy levels $T = 1$ and $J = 1$ are each 3-fold degenerate, the lowest rotational level of ortho H_2 consists of $(2T + 1)(2J + 1) = 9$ different quantum states of practically the same energy, no two of which have the same orientations of both nuclear spins and rotational axes. In the lowest rotational level of para H_2 $T = 0$ and $J = 0$, and there is just one such quantum state.

In ordinary H_2 the mole fraction of ortho H_2 is $\frac{3}{4}$ and that of para H_2 is $\frac{1}{4}$. Below about 40 K ortho H_2 consists of 9 species, each having almost the same mole fraction which is $\frac{1}{9} \times \frac{3}{4} = \frac{1}{12}$. The one kind of para H_2 has a mole fraction of $\frac{1}{4}$. From eqn. (11.17) we find that the molar entropy of mixing these 10 species is

$$\Delta S^M = -\tfrac{9}{12} R \ln \tfrac{1}{12} - \tfrac{1}{4} R \ln \tfrac{1}{4} \qquad (T < 40 \text{ K})$$
$$= R \ln 4 + \tfrac{3}{4} R \ln 3 \qquad\qquad\qquad\qquad (11.72)$$
$$= 2.75\bar{5} + 1.637 = 4.392 \text{ cal deg}^{-1} \text{ mole}^{-1}$$

The above equation gives the nuclear-spin and rotational contributions to the standard molar entropy of H_2 gas at temperatures below about 40 K. It is also valid for the condensed phases of H_2 down to the temperature at which the λ-anomaly in the heat capacity [79] begins to affect the entropy.

Let us first consider the term $R \ln 4$ in eqn. (11.72). Above about 300 K, both ortho and para states of H_2 have the same classical molar rotational partition function, and hence ordinary H_2 gas is a mixture of 4 identifiable species: the three nuclear-spin states of ortho H_2 and the one such state of para H_2. Each state has the mole fraction $\frac{1}{4}$. Thus, at temperatures above about 300 K, the contribution to the standard molar entropy of H_2 gas, due to the mixing of the nuclear-spin states, called the *nuclear-spin entropy* (see Section 11.24B), is just $R \ln 4$. By convention, the nuclear-spin entropy of a substance is not included in tables of entropy. Subject to this convention, the spectroscopic value of the standard molar entropy of H_2 gas at 298.1 K found by Giauque [84] is $33.98 - 2.75 = 31.23$ cal deg^{-1} mole^{-1}.

This procedure for a gas with ortho and para forms yields values of standard molar entropies of gases which can be used to calculate correct values of ΔS^0 attending a standard physical or chemical change in state only at temperatures where the classical rotational partition function of the gas is valid.

We now turn to the term $\frac{3}{4} R \ln 3$ of eqn. (11.72), which represents the change in the standard molar entropy of solid ordinary H_2 attending the λ-anomaly in c_p at about 1.6 K [79]; it is caused by the removal of the 3-fold degeneracy of the $J = 1$ rotational level of ortho H_2. Addition of this quantity to the calorimetric standard molar entropy of H_2 gas at 298.1 K [84] gives $29.74 + 1.64 = 31.38$ cal deg^{-1} mole^{-1} (see Section 11.17B). This value is in satisfactory agreement with the corresponding spectroscopic entropy of 31.23 cal deg^{-1} mole^{-1}.

The deuterium nucleus has an even number of nucleons, and a spin quantum number $I = 1$. The total wave functions of D_2 are symmetric with respect to nuclear exchange.

The quantum states of the D_2 molecule, together with those of H_2 (and T_2), are enumerated in Table 11.4 [90].

The spectroscopic value of the standard molar entropy (nuclear-spin contribution included) of D_2 gas at 298.1 K, and the calorimetric value, determined with a Debye T^3-law extrapolation of the heat capacity below 12 K, are 38.98 and 33.92 cal deg^{-1} mole^{-1}, respectively [91]. From the data of Table 11.4 we find the following relations for the nuclear-spin and rotational contributions to the standard molar entropy of D_2 when its molecules are in their lowest rotational levels

$$\begin{aligned}\Delta S^M &= \tfrac{2}{3} R \ln 9 + \tfrac{1}{3} R \ln 27 \\ &= R \ln 9 + \tfrac{1}{3} R \ln 3 \\ &= 4.366 + 0.728 = 5.094 \text{ cal deg}^{-1} \text{ mole}^{-1}\end{aligned} \quad (11.73)$$

Compare eqn. (11.72).

TABLE 11.4

QUANTUM STATES OF H_2, D_2, AND T_2

	H_2 and T_2		D_2	
I	$\tfrac{1}{2}$		1	
Total wavefunctions	antisymmetric		symmetric	
Designation	ortho	para	para	ortho
T	1	0	1	2,0
Nuclear-spin state	symmetric	antisymmetric	antisymmetric	symmetric
No. of nuclear states	3	1	3	6
Rotational states	antisymmetric	symmetric	antisymmetric	symmetric
J	1,3,5,...	0,2,4,...	1,3,5,...	0,2,4,...

The spectroscopic value of the standard molar entropy (nuclear-spin contribution deleted) of D_2 gas at 298.1 K is $38.98 - 4.37 = 34.61$ cal deg^{-1} mole^{-1}; and the corresponding calorimetric value, corrected for the removal of the 3-fold degeneracy of the $J = 1$ rotational level of para D_2, is $33.92 + 0.73 = 34.65$.

A λ-anomaly in the heat capacity of solid D_2 in which the 3-fold degeneracy mentioned above is removed takes place at about 1.3 K [92].

The tritium nucleus has an odd number of nucleons, and a nuclear spin $I = \frac{1}{2}$. Thus the enumeration of the nuclear-spin and rotation states of T_2 is the same as that of H_2 (see Table 11.4). Its entropy and other thermodynamic properties have been computed from spectroscopic data [93].

Substances composed of molecules having two or more isotopes symmetrically placed have ortho and para forms. However, with the possible exception [77] of CH_4, the molecules of all such gaseous substances, except H_2 and its homonuclear isotopes, have such large moments of inertia, and consequently such small spacings of their rotational energy levels, that the classical rotational partition function is valid at all temperatures above the normal boiling points of the respective liquids. Also, the molecules of these substances do not rotate in the solid phases. Hence the values of their spectroscopic and calorimetric standard molar entropies can be calculated in the usual manner.

11.23. Paramagnetic substances. A substance which has a positive magnetic susceptibility is termed paramagnetic. One having an extremely large positive magnetic susceptibility is called ferromagnetic. Paramagnetic materials contain molecules, atoms, or ions that have permanent magnetic moments arising from the spin of one or more unpaired electrons. This magnetic moment is a vector which can take any one of $2K + 1$ orientations with respect to a fixed axis, where K is the total angular momentum quantum number of the particle.

At a sufficiently low temperature, the magnetic moments of the constituent molecules, atoms, or ions of ferromagnetic materials are aligned parallel. Those of antiferromagnetic and ferrimagnetic materials are aligned antiparallel [94]. Thus, each of these substances is in an ordered state, and hence the magnetic contribution to its standard molar entropy is zero.

Above a characteristic temperature, called the Curie temperature for a ferromagnetic and the Néel temperature for an antiferromagnetic or a ferrimagnetic substance, the ordered alignment of the magnetic dipoles of the constituent particles is destroyed by thermal agitation. The substance is then paramagnetic. The magnetic dipoles are randomly oriented, and for each value of K there are $2K + 1$ quantum states which, in the absence of an applied magnetic field, have almost the same energy. The corresponding magnetic contribution s_M to the standard molar entropy of the substance, given by eqn. (11.67), is

$$s_M = R \ln(2K+1) \tag{11.74}$$

The transformation of the magnetic moments of the constituent particles of a magnetic substance from an ordered to a random orientation is attended by an anomaly in its susceptibility, and either a λ or a Schottky anomaly in its heat capacity.

The Curie or Néel temperatures of a number of paramagnetic salts are below the temperature to which heat-capacity measurements usually extend. For example, $MnBr_2 \cdot 4H_2O$ has a λ-anomaly in its heat capacity at about 2.2 K. The heat capacities of salts that have paramagnetic ions which are separated by a large number of non-magnetic particles may have an anomaly well below 1 K, for example, $KCr(SO_4)_2 \cdot 12H_2O$ has a Schottky anomaly at about 0.1 K. The standard molar entropies of these substances, when determined by the usual calorimetric method with a Debye T^3-law extrapolation of the heat capacity below about 15 K, are too small by the amount $R \ln(2K+1)$.

11.24. Contributions to entropy not included in tabulated values. In this section, we shall consider two contributions to the standard molar entropy of an element or compound which are not observed in the calorimetric method (Section 11.13) nor included in the spectroscopic method (Section 11.14) of evaluating this quantity. Since each is primarily an atomic property, its consistent omission introduces no error in the calculation of the change in entropy attending any ordinary physical or chemical change in state.

11.24A. Entropy of isotope mixing. Many elements occur in nature with appreciable proportions of two or more naturally occurring isotopes. Thus many "pure" elements and compounds are in fact solutions composed of species of different atomic weights. The molar entropy of mixing these species is given by eqn. (11.17). According to the third law, the contribution of the entropy of isotope mixing to the standard molar entropy of a substance should approach zero as the Kelvin temperature tends to zero. Thus the mixture should approach an ordered state as T approaches zero.

The isotopes of a substance form nearly perfect solutions, and hence the change in energy attending their mixing at constant pressure and temperature is so small that separation into the constituent isotopes would tend to take place at a very low temperature. Long before this temperature is reached, all such solutions, except liquid mixtures of ^3He and ^4He, are frozen-in solid solutions (see Section 11.18). This condition persists essentially unchanged as the solution is cooled to 0 K, and the substance has residual entropy (see Section 11.17A) equal to the entropy of isotope mixing.

The entropy of isotope mixing cancels in the calculation of the change in entropy attending a physical or chemical change in state that does not include a nuclear reaction or a fractionation of isotopes.

11.24B. Nuclear spin entropy. Atomic nuclei have small magnetic moments (revealed by high-resolution spectroscopic or other experimental methods) proportional in magnitude to a nuclear spin quantum number I, which is an integer $(0,1,2,...)$ for a nucleus with an even number of nucleons and a half odd-integer $(1/2, 3/2,...)$ for a nucleus with an odd number of nucleons.

The nuclear magnetic moment of an atom is a vector which can take any one of $2I + 1$ orientations with respect to a fixed axis. There are thus $2I + 1$ nuclear-spin quantum states which, on account of the very small magnitude of the magnetic moment, do not differ appreciably in energy.

By convention, the nuclear spin contribution s_N to the standard molar entropy of a substance is given by

$$s_N = \Sigma_i m_i R \ln(2I_i + 1) \tag{11.75}$$

where the sum is over the different types of atoms composing the substance, m_i is the number of atoms of species i in a molecule of the substance, and I_i is the nuclear-spin quantum number of an atom of species i. The above relation follows from eqn. (11.67) and the hypothesis that there are $2I_i + 1$ nuclear-spin states of essentially the same energy for each nucleus of species i.

All of the nuclear spins of the constituent atoms of a substance are in an ordered array at 0 K, and hence the nuclear-spin contribution to its standard molar entropy is zero at this temperature, in accord with the third law.

Subtraction of s_N from the spectroscopic value of the standard molar entropy (nuclear-spin contribution included) of a gas at a temperature at which the classical rotational partition function is valid gives the quantity listed in tables of thermodynamic properties.

The nuclear-spin quantum number of an atom is not changed in any ordinary physical or chemical process. Hence the nuclear-spin entropy defined by eqn. (11.75) cancels in the calculation of the change in entropy attending a physical or chemical change in state which does not involve a nuclear reaction or a fractionation of isotopes [95].

REFERENCES

1 J. Thomsen, Ann. Phys. Chem., 92 (1854) 34—57; J. Thomsen, Thermochemische Untersuchungen, J. A. Barth, Leipzig, 1882, Vol. 1, pp. 9—18: see particularly pp. 15—16.
2 M. Berthelot, Essai de Méchanique Chimique, Dunod, Paris, 1879, Vol. 2, pp. 417—472: see particularly p. 421.
3 H. M. Le Chatelier, Ann. Mines, 13 (1888) 157—382: see particularly p. 336.
4 T. W. Richards, Proc. Am. Acad. Arts Sci., 38 (1902) 291—317.
5 F. Haber, Thermodynamics of Technical Gas—Reactions, translated by A. B. Lamb, Longmans, Green and Co., London, 1908, pp. 41—46. The German edition is dated 1905.
6 (a) W. Nernst, Experimental and Theoretical Applications of Thermodynamics to Chemistry, C. Scribner's Sons, New York, 1907; (b) W. Nernst, The New Heat Theorem, translated by G. Barr, E. P. Dutton, New York, 1926. Note the bibliography on pp. 256—263.

7 F. Simon, Physica (Utrecht), 4 (1937) 1089—1096; Z. Naturforsch. Teil A, 6 (1951) 397—400.
8 For an account of the inception of the Nernst Heat Theorem, see F. Simon, Year Book of the Physical Society, pp. 1—22 (1956).
9 For an account of the status of the third law in 1936, see E. D. Eastman, Chem. Rev., 18 (1936) 257—275.
10 See P. S. Epstein, Textbook of Thermodynamics, Wiley, New York, 1937, p. 227.
11 A. T. Petit and P. L. Dulong, Ann. Chim. Phys., 10 (1819) 395—413.
12 L. Boltzmann, Sitzungsber. K. Akad. Wiss. Wien, 63 (1871) 679—711, 712—732.
13 F. E. Neumann, Ann. Phys. Chem., 99 (1831) 1—39.
14 H. Kopp, Ann. Chem. Pharm., Suppl. III, Part 2, (1864) 289—342.
15 W. Nernst, Z. Elektrochem., 20 (1914) 357—360.
16 M. Planck, Ber. Dtsch. Chem. Ges., 45 (1912) 5—23; M. Planck, Treatise on Thermodynamics, translated by A. Ogg, Longmans, Green and Co., London, 1927, p. 273.
17 See G. N. Lewis and M. Randall, Thermodynamics, McGraw-Hill, New York, 1923, p. 448.
18 See, for example, R. H. Fowler and E. A. Guggenheim, Statistical Thermodynamics, University Press, Cambridge, 1939, pp. 224—227.
19 D. Chandler and I. Oppenheim, J. Chem. Educ., 43 (1966) 525—527.
20 F. Simon, Ergeb. Exakten Naturwiss., 9 (1930) 222—274.
21 See, for example, J. Wilks, The Third Law of Thermodynamics, in W. Jost (Ed.), Physical Chemistry: An Advanced Treatise, Vol. I, Thermodynamics, Academic Press, New York, 1971, Chapter 6, pp. 441, 445.
22 For a full discussion of the He phase diagrams, see J. Wilks, The Properties of Liquid and Solid Helium, Clarendon Press, Oxford, 1967.
23 W. H. Keesom, Helium, Elsevier, Amsterdam, 1942, pp. 202, 203; F. E. Simon and C. A. Swenson, Nature (London), 165 (1950) 829—831; C. A. Swenson, Phys. Rev., 79 (1950) 626—631, 86 (1952) 870—876, 89 (1953) 538—544; E. R. Grilly and R. L. Mills, Ann. Phys. (N.Y.), 18 (1962) 250—263; J. Wilks, ref. 22, p. 29.
24 (a) C. Le Pair, K. W. Taconis, R. De Bruyn Ouboter and P. Das, Physica (Utrecht), 29 (1963) 755—756; (b) J. Wilks, ref. 22, p. 477.
25 J. Wilks, The Third Law of Thermodynamics, Oxford University Press, London, 1961, pp. 130, 131.
26 (a) B. Weinstock, B. M. Abraham and D. W. Osborne, Phys. Rev., 85 (1952) 158—159; (b) E. R. Grilly and R. L. Mills, Ann. Phys. (N.Y.), 8 (1959) 1—23; (c) D. O. Edwards, J. L. Baum, D. F. Brewer, J. G. Daunt and A. McWilliams, in G. M. Graham and A. C. Hollis Hallet (Eds.), Proc. Int. Conf. Low Temp. Phys., VIIth, University of Toronto Press, 1961, pp. 610—613; (d) J. Wilks, ref. 22, p. 475.
27 J. H. Vignos and H. A. Fairbank, Phys. Rev. Lett., 6 (1961) 265—267, 646; E. R. Grilly and R. L. Mills, refs. 23 and 26; J. S. Dugdale and J. P. Franck, Philos. Trans. R. Soc. London, 257 (1964) 1—29; J. Wilks, ref. 22, pp. 590, 592, 632, and 633.
28 For a full discussion of heat capacities at low temperatures see E. S. R. Gopal, Specific Heats at Low Temperatures, Plenum Press, New York, 1966.
29 A. Einstein, Ann. Phys., 22 (1907) 180—190.
30 See, for example, F. H. MacDougall, Thermodynamics and Chemistry, Wiley, New York, 1939, pp. 423—431.
31 See, for example, C. C. Stephenson, D. R. Bentz and D. A. Stevenson, J. Am. Chem. Soc., 77 (1955) 2161—2164; C. C. Stephenson and A. M. Karo, J. Chem. Phys., 48 (1968) 104—108; C. C. Stephenson, R. L. Potter, T. G. Maple and J. C. Morrow, J. Chem. Thermodyn., 1 (1969) 59—76.
32 K. Schäfer and E. Lax (Eds.), Landolt—Börnstein, Zahlenwerte und Funktionen, Springer Verlag, Berlin, 1961, 6th edn., Vol. II, Part 4, pp. 736—749.

33 J. Hilsenrath and G. G. Ziegler, Natl. Bur. Stand. U.S., Monogr., 49 (1962) 258 pp.
34 P. Debye, Ann. Phys., 39 (1912) 789—839.
35 J. A. Beattie, J. Math. Phys. (Cambridge, Mass.), 6 (1926) 1—32.
36 C. Kittel, Introduction to Solid State Physics, 2nd edn., Wiley, New York, 1956, pp. 134—136, 257—259; E. S. R. Gopal, ref. 28, pp. 55—64.
37 W. Nernst and F. A. Lindemann, Z. Elektrochem., 17 (1911) 817—827.
38 W. S. Corak, M. P. Garfunkel, C. B. Satterthwaite and A. Wexler, Phys. Rev., 98 (1955) 1699—1707.
39 W. F. Giauque and J. O. Clayton, J. Am. Chem. Soc., 55 (1933) 4875—4889.
40 J. E. Mayer and M. G. Mayer, Statistical Mechanics, Wiley, New York, 1940.
41 G. N. Lewis and M. Randall, Thermodynamics, revised by K. S. Pitzer and L. Brewer, McGraw-Hill, New York, 1961, Chapter 27.
42 N. Davidson, Statistical Mechanics, McGraw-Hill, New York, 1962.
43 O. Sackur, Ann. Phys., 36 (1911) 958—980, 40 (1913) 67—86; H. Tetrode, Ann. Phys., 38 (1912) 434—442.
44 K. K. Kelley, J. Am. Chem. Soc., 51 (1929) 1400—1406. The thermal measurements on tin used here were made by F. Lange, Z. Phys. Chem., 110 (1924) 343—362, and by J. N. Brönsted, Z. Phys. Chem., 88 (1914) 479—489.
45 E. D. Eastman and W. C. McGavock, J. Am. Chem. Soc., 59 (1937) 145—151; E. D. West, J. Am. Chem. Soc., 81 (1959) 29—37.
46 C. C. Stephenson and W. F. Giauque, J. Chem. Phys., 5 (1937) 149—158.
47 E. B. Wilson, Jr., Science, 162 (1968) 59—66.
48 E. Teller and B. Topley, J. Chem. Soc., (1935) 876—885.
49 J. D. Kemp and K. S. Pitzer, J. Chem. Phys., 4 (1936) 749; R. K. Witt and J. D. Kemp, J. Am. Chem. Soc., 59 (1937) 273—276.
50 J. D. Kemp and K. S. Pitzer, J. Am. Chem. Soc., 59 (1937) 276—279; K. S. Pitzer, J. Chem. Phys., 5 (1937) 469—472, 473—479; K. S. Pitzer and L. Brewer, ref. 41, pp. 438—443; N. Davidson, ref. 42, pp. 194—202.
51 J. W. Gibbs, Scientific Papers, Longmans, Green and Co., New York, 1906, Vol. I, Thermodynamics, pp. 56—58.
52 J. E. Mayer and M. G. Mayer, ref. 40, pp. 138—140, 209—213.
53 E. D. Eastman and R. T. Milner, J. Chem. Phys., 1 (1933) 444—456.
54 A. H. Wilson, Thermodynamics and Statistical Mechanics, University Press, Cambridge, 1957, pp. 457—476.
55 G. K. Walters and W. M. Fairbank, Phys. Rev., 103 (1956) 262—263; W. M. Fairbank and G. K. Walters, Nuovo Cimento Suppl. 1, 9 (1958) 297—305; D. O. Edwards, D. F. Brewer, P. Seligman, M. Skertic and M. Yaqub, Phys. Rev. Lett., 15 (1965) 773—775; J. Wilks, ref. 21, p. 449.
56 E. G. D. Cohen and J. M. J. van Leeuwen, Physica (Utrecht), 26 (1960) 1171—1173; R. De Bruyn Ouboter and J. J. M. Beenakker, Physica (Utrecht), 27 (1961) 219—229.
57 G. E. Gibson and W. F. Giauque, J. Am. Chem. Soc., 45 (1923) 93—104.
58 F. Simon, Ann. Phys., 68 (1922) 241—280; F. Simon and F. Lange, Z. Phys., 38 (1926) 227—236; F. Simon, Z. Anorg. Allg. Chem., 203 (1932) 219—227; F. Simon, refs. 8 and 20.
59 G. O. Jones and F. E. Simon, Endeavour, 8 (1949) 175—181; G. O. Jones, Rep. Prog. Phys., 12 (1948/1949) 133—162.
60 J. Wilks, ref. 21, pp. 460—463.
61 J. E. Ahlberg, E. R. Blanchard and W. O. Lundberg, J. Chem. Phys., 5 (1937) 539—551.
62 A. G. Oblad and R. F. Newton, J. Am. Chem. Soc., 59 (1937) 2495—2499.
63 G. E. Gibson, G. S. Parks and W. M. Latimer, J. Am. Chem. Soc., 42 (1920) 1542—1550.
64 F. Simon, in F. Henning (Ed.), Handbuch der Physik, Vol. X, J. S. Springer, Berlin, 1926, pp. 392—394.

65 H. L. Johnston and W. F. Giauque, J. Am. Chem. Soc., 51 (1929) 3194—3214; W. J. Dulmage, E. A. Meyers and W. N. Lipscomb, J. Chem. Phys., 19 (1951) 1432—1433; Acta Crystallogr., 6 (1953) 760—764.
66 R. W. Blue and W. F. Giauque, J. Am. Chem. Soc., 57 (1935) 991—997.
67 J. O. Clayton and W. F. Giauque, J. Am. Chem. Soc., 54 (1932) 2610—2626.
68 F. J. Bockhoff, R. V. Petrella and E. L. Pace, J. Chem. Phys., 32 (1960) 799—804, 36 (1962) 3502—3503; see also D. R. Lide, Jr., D. E. Mann and J. J. Comeford, Spectrochim. Acta, 21 (1965) 497—501.
69 W. F. Giauque and J. W. Stout, J. Am. Chem. Soc., 58 (1936) 1144—1150.
70 E. A. Long and J. D. Kemp, J. Am. Chem. Soc., 58 (1936) 1829—1834.
71 L. Pauling, J. Am. Chem. Soc., 57 (1935) 2680—2684.
72 J. K. Koehler and W. F. Giauque, J. Am. Chem. Soc., 80 (1958) 2659—2662.
73 G. Brodale and W. F. Giauque, J. Am. Chem. Soc., 80 (1958) 2042—2044; K. S. Pitzer and L. V. Coulter, J. Am. Chem. Soc., 60 (1938) 1310—1313.
74 H. W. Ruben, D. H. Templeton, R. D. Rosenstein and I. Olovsson, J. Am. Chem. Soc., 83 (1961) 820—824.
75 J. P. McCullough, H. L. Finke, M. E. Gross, J. F. Messerly and G. Waddington, J. Phys. Chem., 61 (1957) 289—301.
76 H. L. Finke, M. E. Gross, J. F. Messerly and G. Waddington, J. Am. Chem. Soc., 76 (1954) 854—857.
77 J. H. Colwell, E. K. Gill and J. A. Morrison, J. Chem. Phys., 39 (1963) 635—653, 42 (1965) 3144—3155.
78 F. Simon, Z. Phys., 15 (1923) 307—311; F. Simon and F. Lange, Z. Phys., 15 (1923) 312—321.
79 F. Simon, K. Mendelssohn and M. Ruhemann, Naturwissenschaften, 18 (1930) 34—35; K. Mendelssohn, M. Ruhemann and F. Simon, Z. Phys. Chem., Abt. B, 15 (1931) 121—126; R. W. Hill and B. W. A. Ricketson, Philos. Mag., 45 (1954) 277—282.
80 See, for example, R. H. Fowler, Statistical Mechanics, 2nd edn., (1936), University Press, Cambridge, 1966, pp. 82—89.
81 W. Heisenberg, Z. Phys., 41 (1927) 239—267.
82 D. M. Dennison, Proc. R. Soc. London, Ser. A, 115 (1927) 483—486; see also F. Hund, Z. Phys., 42 (1927) 93—120; T. Hori, Z. Phys., 44 (1927) 834—854.
83 W. F. Giauque and H. L. Johnston, J. Am. Chem. Soc., 50 (1928) 3221—3228.
84 W. F. Giauque, J. Am. Chem. Soc., 52 (1930) 4808—4815, 4816—4831.
85 K. F. Bonhoeffer and P. Harteck, Naturwissenschaften, 17 (1929) 182, 321—322; Z. Phys. Chem., Abt. B, 4 (1929) 113—141.
86 C. M. Cunningham, D. S. Chapin and H. L. Johnston, J. Am. Chem. Soc., 80 (1958) 2382—2384.
87 See, for example, A. Farkas, Orthohydrogen, Parahydrogen and Heavy Hydrogen, University Press, Cambridge, 1935; G. Herzberg, Molecular Spectra and Molecular Structure, Vol. I, Spectra of Diatomic Molecules, 2nd. edn., Van Nostrand, New York, 1950, pp. 130—141.
88 For any ortho molecule (nuclear-spin wavefunctions symmetric) $T = 2I$, $2I - 2, \ldots$, 1 or 0; and for any para molecule (nuclear-spin wavefunctions antisymmetric) $T = 2I - 1, 2I - 3, \ldots, 0$ or 1. For each value of T there are $2T + 1$ nuclear-spin states, see ref. 87.
89 See, for example, L. Pauling, Phys. Rev., 36 (1930) 430—443; F. Reif and E. M. Purcell, Phys. Rev., 91 (1953) 631—641.
90 Compare K. Clusius and E. Bartholomé, Z. Elektrochem., 40 (1934) 524—529.
91 K. Clusius and E. Bartholomé, Z. Phys. Chem., Abt. B, 30 (1935) 237—257, 258—264; see also H. L. Johnston and E. A. Long, J. Chem. Phys., 2 (1934) 389—395.
92 D. O. Gonzalez, D. White and H. L. Johnston, J. Phys. Chem., 61 (1957) 773—780; G. Grenier and D. White, J. Chem. Phys., 40 (1964) 3015—3030.

93 W. M. Jones, J. Chem. Phys., 16 (1948) 1077—1081.
94 See C. Kittel, ref. 36, Chapters 9 and 15.
95 See, for example, G. E. Gibson and W. Heitler, Z. Phys., 49 (1928) 465—472; W. F. Giauque, ref. 84.

Chapter 12

CRITERIA OF EQUILIBRIUM AND STABILITY

At the beginning of his treatise "On the Equilibrium of Heterogeneous Substances", Gibbs [1] gave entropy and energy criteria of equilibrium and stability for an isolated system, showed that they are equivalent, and proved that they are necessary and sufficient for equilibrium.

Later he gave work content, enthalpy, and free-energy criteria of equilibrium for such systems when subject to certain restrictions, and showed that they are equivalent to the energy criterion under these conditions.

In his abstract of this paper, Gibbs noted that if a system cannot change its temperature, like the systems treated in theoretical mechanics, its entropy may be assigned the constant value zero. The energy and work content criteria mentioned above for such a system "may therefore both be regarded as extensions of the criterion employed in ordinary statics to the more general case of a thermodynamic system" [1].

We shall consider these results in the present chapter.

12.1. Entropy and energy criteria of equilibrium and stability. Gibbs [2] gave the following entropy and energy criteria of equilibrium.

I. For the equilibrium of any isolated system it is necessary and sufficient that in all possible variations of the state of the system which do not alter its energy, the variation of its entropy shall either vanish or be negative.

II. For the equilibrium of any isolated system it is necessary and sufficient that in all possible variations in the state of the system which do not alter its entropy, the variation of its energy shall either vanish or be positive.

These conditions may be written

$$(\delta S)_E \leqslant 0 \tag{12.1}$$

$$(\delta E)_S \geqslant 0 \tag{12.2}$$

We shall prove below that these criteria are equivalent.

In the application of a criterion of equilibrium to an isolated system we imagine that all "possible variations" in its state (denoted by the symbol δ) are brought about. If the conditions (12.1) and (12.2) hold for all such virtual changes in state, the system is initially in a state of equilibrium. Conversely, if the system is initially in a state of equilibrium, the criteria hold for all such virtual changes in its state.

Gibbs noted that if an isolated system is divided into parts by impermeable, adiabatic, but movable walls, the condition of equilibrium (12.2) becomes [3]

$$(\delta E)_{S'S''\ldots S^\kappa} \geqslant 0 \tag{12.3}$$

12.2. Possible variations. A possible variation in the state of an isolated system is a virtual, infinitesimal change in its state which: (1) is compatible with the imposed constraints; (2) is not prevented by passive forces; and (3) does not violate the general laws of matter.

Constraints are limitations imposed on the variations in the state of a system by the boundary, internal walls (if any), and surroundings. Thus, a rigid boundary prevents any change in the volume of a system, no heat can flow across an adiabatic boundary, and the boundary of a closed system is impermeable to matter. Unless otherwise stated, we shall consider in this chapter only closed systems, and systems whose phases are not separated by internal walls which are rigid, adiabatic, impermeable to any component, or any combination of these constraints.

Passive forces that prevent change are properties of the components of a system (see Section 1.5).

The general laws of matter are given in Section 1.17.

In a possible variation (in theoretical mechanics known as a virtual displacement) an isolated system undergoes an imaginary, infinitesimal change in state which is limited only by the three restrictions listed above. Thus a possible variation does not actually occur. In fact it cannot always be brought about by a real process.

A possible variation in the state of an isolated system may withdraw a virtual, infinitesimal quantity of heat from the surroundings, or produce a virtual, infinitesimal quantity of work in the surroundings, or both.

In order to apply the criterion of equilibrium (12.1) or (12.2), we must in most cases know the changes in the entropy and energy of the system produced by a possible variation in its state. These quantities can be measured or calculated if the variation takes place reversibly, or if the initial and varied states of the system are in equilibrium subject to certain constraints. In some applications we need know only whether a quantity entering into a condition of equilibrium is positive, negative, or zero.

12.3. Equivalence of the entropy and energy criteria of equilibrium. In order to prove the equivalence of the entropy and energy conditions of equilibrium for an isolated system, we must show that, if there is any possible variation in the state of the system which does not satisfy the entropy condition, there is also a possible variation which does not satisfy the energy condition; and conversely, if there is any possible variation in the state of the system which does not satisfy the energy condition, there is also a possible variation which does not satisfy the entropy condition.

Let us apply these tests to the Gibbs conditions of equilibrium (12.1) and (12.2).

If condition (12.1) is not satisfied there must be some possible variation in the state of the system for which

$$\delta S > 0 \text{ and } \delta E = 0 \tag{12.4}$$

Now by following this variation by a virtual flow of heat from the system, we can decrease both the entropy and energy of the system until the total variation from the initial state is

$$\delta S = 0 \text{ and } \delta E < 0 \tag{12.5}$$

This variation does not satisfy condition (12.2).

Conversely, if condition (12.2) is not satisfied, there must be some possible variation in the state of the system for which

$$\delta E < 0 \text{ and } \delta S = 0 \tag{12.6}$$

By following this variation by a virtual flow of heat to the system we can produce a state for which the total variation from the initial state is

$$\delta E = 0 \text{ and } \delta S > 0 \tag{12.7}$$

This variation does not satisfy condition (12.1).

This completes the proof of the equivalence of the entropy and energy criteria of equilibrium. We may use either in a discussion of any aspect of equilibrium in an isolated system.

Clearly, the criteria are equivalent when only the signs of equality or only the signs of inequality are used. Thus

$$(\delta S)_E = 0 \text{ and } (\delta E)_S = 0 \text{ are equivalent} \tag{12.8}$$

and

$$(\delta S)_E < 0 \text{ and } (\delta E)_S > 0 \text{ are equivalent} \tag{12.9}$$

12.4. Kinds of equilibrium states. Let δy be the variation of the dependent variable y of an equation stating a condition of equilibrium in an isolated system. In this expression infinitesimals of higher order than the first are neglected. But to distinguish the kinds of equilibrium states with respect to stability, we must use the exact value Δy of the variation of y in which infinitesimals of higher order are retained. For example, in a simple, closed system

$$E = E(S, V) \tag{12.10}$$

and δE is given by the expression

$$\delta E = \left(\frac{\partial E}{\partial S}\right)_V \delta S + \left(\frac{\partial E}{\partial V}\right)_S \delta V \tag{12.11}$$

The Taylor expansion of ΔE is

$$\Delta E = \delta E + \frac{1}{2!}\left[\left(\frac{\partial^2 E}{\partial S^2}\right)(\delta S)^2 + 2\left(\frac{\partial^2 E}{\partial S \partial V}\right)\delta S \, \delta V + \left(\frac{\partial^2 E}{\partial V^2}\right)(\delta V)^2\right] + \ldots \tag{12.12}$$

In the following discussions [2] we shall give the entropy and energy conditions which are necessary and sufficient for the four kinds of equilibrium states in an isolated system: stable, neutral, unstable, and metastable. Of these four states, stable equilibrium is the most important, dynamic neutral equilibrium is seldom, if ever, observed experimentally, dynamic unstable equilibrium is not met in practice, and metastable equilibrium is maintained by passive forces which prevent a certain change or certain changes in state.

12.4A. Stable equilibrium. An isolated system is in a state of stable equilibrium when, for all possible variations in its state,

$$(\Delta S)_E < 0 \text{ and therefore } (\Delta E)_S > 0 \tag{12.13}$$

while

$$(\delta S)_E \leq 0 \text{ and therefore } (\delta E)_S \geq 0 \tag{12.14}$$

In the following mechanical examples of this kind of equilibrium in a gravitational field, both the ball and the bowl are assumed to be rigid and at rest with respect to the observer.

1. A ball resting on the bottom of a covered hemispherical bowl: see Fig. 12.1a. If a virtual displacement moves the ball an infinitesimal distance up the side of the bowl, the variation in the energy of the system is

$$(\delta E)_S = 0 \text{ and } (\Delta E)_S > 0 \tag{12.15}$$

If the displacement raises the ball an infinitesimal distance vertically upward

$$(\delta E)_S > 0 \text{ and } (\Delta E)_S > 0 \tag{12.16}$$

2. A ball in an inverted covered cone: see Fig. 12.1b. Equations (12.16) hold for every virtual displacement that moves the ball an infinitesimal distance up the cone.

3. An isolated system in which the chemical reaction

$$H_2(g) + I_2(g) = 2HI(g) \tag{12.17}$$

is in a state of dynamic equilibrium. The conditions

$$\left.\begin{array}{l}(\delta S)_E = 0 \text{ and } (\Delta S)_E < 0 \\ (\delta E)_S = 0 \text{ and } (\Delta E)_S > 0\end{array}\right\} \tag{12.18}$$

are satisfied for every possible variation in the state of the system.

Fig. 12.1. Two systems, each in a state of stable equilibrium.

12.4B. Neutral equilibrium. An isolated system is in a state of neutral equilibrium when for some possible variations in its state

$$(\Delta S)_E = 0 \text{ and therefore } (\Delta E)_S = 0 \tag{12.19}$$

while conditions

$$(\Delta S)_E \leq 0 \text{ and therefore } (\Delta E)_S \geq 0 \tag{12.20}$$

and (12.14) are satisfied for all possible variations.

Suppose that a system consists of a ball resting on the bottom of a horizontal, cylindrical pipe closed at each end (Fig. 12.2). If a virtual displacement moves the ball an infinitesimal distance horizontally along the pipe, the variation in the energy of the system is

$$(\delta E)_S = 0 \text{ and } (\Delta E)_S = 0 \tag{12.21}$$

Equations (12.15) and (12.16) hold for other virtual, infinitesimal displacements of the ball.

12.4C. Unstable equilibrium. An isolated system is in a state of unstable equilibrium when, for some possible variations in its state,

$$(\Delta S)_E > 0 \text{ and therefore } (\Delta E)_S < 0 \tag{12.22}$$

while conditions (12.20) are satisfied for all other possible variations, and conditions (12.14) hold for all possible variations.

Imagine that a system consists of a ball resting on the highest point of an inverted, hemispherical bowl with its cover in a horizontal plane (Fig. 12.3). If a virtual displacement moves the ball an infinitesimal distance down the outside of the bowl

$$(\delta E)_S = 0 \text{ and } (\Delta E)_S < 0 \tag{12.23}$$

Equations (12.15), (12.16) and (12.21) hold for other virtual, infinitesimal displacements of the ball.

12.4D. Metastable equilibrium. When an isolated system is in a state of metastable equilibrium, conditions (12.13) and (12.14) are satisfied for all virtual, infinitesimal variations in its state, but there is a finite, virtual variation for which

$$(\Delta S)_E > 0 \text{ and therefore } (\Delta E)_S < 0 \tag{12.24}$$

after which conditions (12.13) and (12.14) are again satisfied for all virtual, infinitesimal variations of the new state.

We shall consider two examples of this kind of equilibrium.

Fig. 12.2. A system in a state of neutral equilibrium.

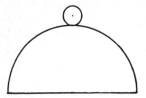

Fig. 12.3. A system in a state of unstable equilibrium.

1. A ball resting on the bottom of the left-hand limb of the apparatus shown in Fig. 12.4. Evidently, eqns. (12.15) and (12.16) hold for every virtual, possible displacement that moves the ball an infinitesimal distance from its initial position. But the equation

$$(\Delta E)_S < 0 \tag{12.25}$$

holds for a virtual, finite displacement that moves the ball to the bottom of the right-hand limb of the apparatus. Thereafter, eqns. (12.15) and (12.16) hold for every virtual, possible displacement that moves the ball an infinitesimal distance from its new equilibrium state.

2. Metastable phases. Frequently a homogeneous system remains in its initial, aggregation state for an indefinite time after it has been heated or cooled through the temperature at which a phase transformation would occur if the system were in a state of dynamic equilibrium with respect to this change.

Pure liquid and crystalline solid metastable phases are considered in Sections 10.11, 11.15, 11.17, and 11.21; metastable glass phases are treated in Section 11.20; and metastable solid solution phases are discussed in Section 11.18.

12.5. Sufficiency of the entropy and energy criteria of equilibrium. In Section 12.3 we proved the equivalence of the criteria of equilibrium (12.1) and (12.2). The sufficiency of these conditions means that if either one of

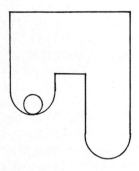

Fig. 12.4. A system in a state of metastable equilibrium.

them is satisfied for all possible variations in the state of an isolated system, the system is in equilibrium.

We shall prove the sufficiency of the above criteria for all four kinds of equilibrium.

12.5A. Stable equilibrium. If an isolated system is in a state of stable equilibrium, it has a larger entropy than in any other state of the same energy [see condition (12.13)]. Hence, any possible change in its state must decrease its entropy at constant energy, which violates the second law of thermodynamics (see Section 8.16).

Therefore, no spontaneous change in state can occur and the system is in a state of equilibrium.

12.5B. Neutral equilibrium. If an isolated system is in a state of neutral equilibrium, it has the largest entropy consistent with its energy and hence the smallest energy consistent with its entropy, but there are other states having the same entropy and energy as the initial state [see conditions (12.19) and (12.20)]. Under these circumstances we find:

1. No flow of heat by conduction or transfer of heat by radiation from one part of the system to another can occur, since such processes increase the entropy of the system at constant energy. This does not satisfy either of the entropy conditions (12.19) or (12.20).

2. Also, no motion of masses can take place because, if any of the energy of the system were converted into kinetic energy of mass motion, a state of the system identical with the given state in all respects except this motion would have less energy but not less entropy. This does not satisfy either of the energy conditions (12.19) or (12.20).

Thus, the above conditions are sufficient for equilibrium in an isolated system in regard to the transfer of heat and the motion of matter. Gibbs gave the following proof that these conditions are also sufficient for all possible variations in the state of the system, including those produced by diffusion and chemical reaction.

Let us suppose that an isolated system has the largest entropy consistent with its energy, but that it is not in equilibrium. Then it must undergo a finite change in its state, subject always to the restrictions that both its entropy and energy remain unchanged.

Consider a short interval of time during which the rate of the supposed process is finite and essentially reversible. In accordance with the above restrictions, no change in state whatever, which starts from the beginning of the above short interval of time, can increase the entropy of the system if its energy remains unchanged.

Hence by some virtual, infinitesimal variation in the circumstances of the case, it will generally be possible to bring about a virtual change in state which differs only infinitesimally from the supposed change defined above, but which produces a decrease in the entropy of the system without changing its energy. This process violates the second law of thermodynamics and therefore cannot occur.

The above variation may be an infinitesimal change in the pressure or temperature of the system, or in the compositions of its phases. It might also be a small change in the properties of the components of the system, subject to the general laws of matter (see Section 1.17).

"If, then, there is any tendency toward change in the system as first supposed, it is a tendency which can be entirely checked by an infinitesimal variation in the circumstances of the case. As this supposition cannot be allowed, we must believe that a system is always in equilibrium when it has the greatest entropy consistent with its energy, or, in other words, when it has the least energy consistent with its entropy" [2].

Gibbs noted that the above procedure may also be applied to any isolated system in a state for which the entropy condition $(\Delta S)_E \leqslant 0$ is satisfied.

12.5C. Unstable equilibrium. If an isolated system is in a state of unstable equilibrium, the condition $(\delta S)_E \leqslant 0$ is satisfied for all possible variations in its state, while the condition $(\Delta S)_E > 0$ holds for some possible variations [see eqns. (12.14) and (12.22)].

We have noted in Section 12.4 that dynamic unstable equilibrium is not met in practice. However, Gibbs gave the following proof of sufficiency of the conditions for equilibrium in the above system [2].

The method used for neutral equilibrium cannot be applied in the present case without modification, since the rate of a possible variation may be infinitesimal at the start of the process, and it is only in the initial state of the system that the condition $(\delta S)_E \leqslant 0$ is satisfied.

Now all orders of the derivatives of the independent state variables of the system with respect to time are functions of these same variables. All must have the value zero for the state for which $(\delta S)_E \leqslant 0$. For example, in the case of the independent variable energy

$$\frac{dE}{dt} = 0, \frac{d^2E}{dt^2} = 0, \text{ etc.} \tag{12.26}$$

for the above state.

Let us suppose that the entropy criterion (12.1) is satisfied for all possible variations in the state of the above system, but that the system is not in equilibrium. Then at least one of the derivatives of eqn. (12.26) must be finite. Now, by some infinitesimal variation in the circumstances of the case, it will generally be possible to bring about a virtual change in state which differs only infinitesimally from the supposed change and which has the value zero for all of the derivatives of eqn. (12.26).

Thus, an infinitesimal modification of the conditions has made a finite change in the value of at least one derivative. But this "is contrary to that continuity which we have reason to expect. Such considerations seem to justify us in regarding such a state as we are discussing as one of the theoretical equilibrium; although as the equilibrium is evidently unstable, it cannot be realized" [2].

12.5D. Metastable equilibrium. In this kind of equilibrium, the conditions

(12.13) and (12.14) are sufficient for equilibrium for all infinitesimal possible variations in the initial state of an isolated system (see Section 12.5A). After the system has undergone a finite change in state, the above conditions are again sufficient for equilibrium for all such variations in the final state of the system.

12.6. Necessity of the entropy and energy criteria of equilibrium. The necessity of the above criteria means that if an isolated system is in equilibrium, the conditions (12.1) and (12.2) are satisfied for all possible variations in its state.

When the equilibrium is dynamic, the generalized force required to produce an infinitesimal variation in the state of the system is at least an infinitesimal of the same order as the conjugate displacement variable.

Now the combined first and second law of thermodynamics may be written in the form

$$dE = TdS - đW_R \qquad (12.27)$$

where $đW_R$ is the reversible work of all kinds produced in the surroundings of the system. Hence, in a possible variation in a state of a system in dynamic equilibrium $đW_R$ is at least an infinitesimal of the second order. Under these circumstances we may omit the signs of inequality in the criteria (12.1) and (12.2), and write them in the form

$$(\delta S)_E = 0 \text{ and therefore } (\delta E)_S = 0 \qquad (12.28)$$

A mechanical example of a system in dynamic equilibrium is a rigid, frictionless pendulum at the lowest point of its swing. Equations (12.15) apply. A chemical example is the reaction (12.17) in an equilibrium state "caused by the balance of the active tendencies of the system" [2]. Equations (12.18) hold for this case. Equations (12.28) are valid in each example.

Suppose that an isolated system remains without change although there is an infinitesimal variation in its state which would decrease its energy by an amount that is not an infinitesimal of higher order than the variations of its independent state variables. Clearly, this change in state is stopped by passive forces that prevent change.

Since the above proposed variation decreases the energy of the system without changing its entropy, it must be theoretically possible to bring about the same change in state by some reversible, adiabatic process that produces work in the surroundings [see eqn. (12.27)]. "Hence we may conclude that the active forces or tendencies of the system favor the variation in question, and that equilibrium cannot subsist unless the variation is prevented by passive forces" [2].

A mechanical example of this kind of equilibrium is a system consisting of a wooden block of mass m resting on a plane inclined at a small angle θ to the horizontal. One infinitesimal variation in its state would move the block a distance δx down the plane. We can produce this variation reversibly by lifting the block off of the plane and replacing it thereon at a distance δx lower down. Under these conditions the entropy of the system has not been altered, but its energy has been decreased by the quantity of work produced in the surroundings, which is $mg \sin\theta \ \delta x$. This work effect is an infinitesimal of the same order as the displacement coordinate δx. Thus, a finite tendency for the block to slide down the plane must have been entirely checked by a passive force that prevented the change.

Therefore the proposed change is not a possible variation in the state of the system, since the equilibrium is maintained by passive forces.

12.7. Uniformity of temperature in an isolated system in equilibrium. The following two theorems are derived from the energy condition (12.2) [4].

12.7A. Thermal equilibrium in a heterogeneous system. Let us suppose that an isolated system in equilibrium consists of two phases. We recall that a phase in a state of equilibrium has a uniform temperature (see Section 1.12).

One reversible, possible variation in the state of the above system increases the entropy of one phase and decreases the entropy of the other without changing the total entropy of the system. In this variation, no change occurs in the volume, mass of any component, or extensive variable of any non-expansion work in either phase.

Application of the energy condition (12.2) gives

$$\delta E = T' \delta S' + T'' \delta S'' \geq 0 \tag{12.29}$$

subject to the restriction

$$\delta S' + \delta S'' = 0 \tag{12.30}$$

Elimination of $\delta S''$ gives

$$(T' - T'') \delta S' \geq 0 \tag{12.31}$$

But $\delta S'$ may be either positive or negative.

$$\left. \begin{array}{l} \text{If } \delta S' > 0, \ T' - T'' \geq 0 \\ \text{If } \delta S' < 0, \ T' - T'' \leq 0 \end{array} \right\} \tag{12.32}$$

The only possibility is

$$T' = T'' \tag{12.33}$$

This method can evidently be extended to prove that temperature is uniform throughout a heterogeneous system of any number of phases in equilibrium.

Therefore, uniformity of temperature is a necessary condition of equilibrium in an isolated system.

12.7B. Effect of non-uniformity of temperature in the varied state. We have just shown that an isolated system in equilibrium necessarily has a uniform temperature. Gibbs proved [5] that condition (12.2) applies without change to such a system if we limit all variations to those ending in states of uniform temperature.

For any variation in the state of the above system which does not satisfy condition (12.2),

$$\delta E < 0 \text{ and } \delta S = 0 \tag{12.34}$$

If a system does not have a uniform temperature in its varied state, we can always increase its entropy without changing its energy by supposing that heat flows from the warmer to the cooler phase. Let such a virtual process continue until the system has the greatest entropy for the energy $E + \delta E$. It would then be in an equilibrium state by condition (12.1), and would necessarily have a uniform temperature by the theorem proved in Section 12.7A. For such a possible variation

$$\delta E = 0 \text{ and } \delta S > 0 \tag{12.35}$$

On combining variations (12.34) and (12.35), we produce a state for which, regarded as a possible variation of the original state of the system,

$$\delta E < 0 \text{ and } \delta S > 0 \tag{12.36}$$

We can always decrease both the energy and entropy of a system by cooling it reversibly. Let us suppose that such a virtual process follows the variation (12.36). The system must eventually reach a state of uniform temperature for which, regarded as a possible variation of the initial state,

$$\delta E < 0 \text{ and } \delta S = 0 \tag{12.37}$$

Neither variation (12.34) nor variation (12.37) satisfies condition (12.2). Thus, if there is a variation to a state with non-uniform temperature which does not obey eqn. (12.2), then there will always be another variation to a state with uniform temperature which does not obey eqn. (12.2). Therefore, in applying the energy criterion of equilibrium, we may limit all possible variations to those which end in states of uniform temperature.

12.8. Uniformity of pressure in a simple, isolated system in equilibrium. Let us suppose that a simple, isolated system in equilibrium is composed of two phases. In Section 1.12 we noted that a simple phase in equilibrium has a uniform pressure at all times.

One reversible, possible variation in the state of the above system increases the volume of one phase and decreases the volume of the other without changing the total volume of the system. In this variation no change occurs in the entropy or mass of any component in either phase.

Application of the energy condition (12.2) gives

$$\delta E = -p' \delta V' - p'' \delta V'' \geqslant 0 \tag{12.38}$$

subject to the restriction

$$\delta V' + \delta V'' = 0 \tag{12.39}$$

Following the method used in Section 12.7A, we find that

$$p' = p'' \tag{12.40}$$

This method can be extended to prove that pressure is uniform throughout a heterogeneous, simple system of any number of phases in equilibrium.

Therefore, uniformity of pressure is a necessary condition of equilibrium in a simple, isolated system.

12.9. Three modifications of the energy criterion of equilibrium. In this section we shall present three versions of the energy condition of equilibrium (12.2) and prove that each is equivalent to this condition, subject to certain constraints. Hence each of these criteria is necessary and sufficient for equilibrium in the given isolated system.

12.9A. A system of uniform temperature. In Section 12.7A we found that uniformity of temperature is always necessary for equilibrium in an isolated system. For such a system, the condition of equilibrium

$$\delta E - T\delta S \geqslant 0 \tag{12.41}$$

and the energy condition

$$\delta E \geqslant 0 \text{ and } \delta S = 0 \tag{12.2}$$

are equivalent. The proof follows the general method used in Section 12.3.

If condition (12.2) is not satisfied, there must be a possible variation in the state of the system for which

$$\delta E < 0 \text{ and } \delta S = 0 \tag{12.42}$$

Let this variation be followed by the virtual, reversible, isometric flow of the quantity of heat $T\delta S$ to the system. For this variation

$$\delta E' - T\delta S = 0 \tag{12.43}$$

The sum of the last two equations is

$$\delta E + \delta E' - T\delta S < 0 \tag{12.44}$$

It may be written in the form

$$\delta E - T\delta S < 0 \tag{12.45}$$

where δE is the change in energy attending the total variation in the state of the system measured from its initial state. This variation does not satisfy condition (12.41).

Conversely, if condition (12.41) is not satisfied, there must be a possible variation in the state of the above system represented by eqn. (12.45). This relation is equivalent to (12.44). Subtraction of (12.43) gives the variation (12.42) which does not satisfy condition (12.2).

Thus we have proved that if condition (12.2) is not satisfied for all possible variations in the state of an isolated system of uniform temperature, condition

(12.41) is not satisfied; and if condition (12.41) is not satisfied, neither is condition (12.2). Therefore, these two criteria of equilibrium are equivalent for the system described above.

12.9B. *A simple system of uniform pressure.* In Section 12.8 we found that uniformity of pressure is necessary for equilibrium in a simple, isolated system. For such a system the condition of equilibrium

$$\delta E + p\delta V \geqslant 0 \text{ and } \delta S = 0 \tag{12.46}$$

and the condition (12.2) are equivalent. The proof of this statement is similar to the method used in Section 12.9A.

If condition (12.2) is not satisfied, there must be a possible variation in the state of the system given by eqn. (12.42).

Let this variation be followed by the production of the quantity of work $p\delta V$ in the surroundings by a virtual, reversible, adiabatic process. For this variation

$$\delta E' + p\delta V = 0 \text{ and } \delta S = 0 \tag{12.47}$$

The sum of eqns. (12.42) and (12.47) is

$$\delta E + \delta E' + p\delta V < 0 \text{ and } \delta S = 0 \tag{12.48}$$

It may be written in the form

$$\delta E + p\delta V < 0 \text{ and } \delta S = 0 \tag{12.49}$$

where δE is the change in energy attending the total variation in the state of the system measured from its initial state. This variation does not satisfy condition (12.46).

Conversely, if condition (12.46) is not satisfied, there must be a possible variation in the state of the above system represented by eqns. (12.49) which is equivalent to (12.48). Subtraction of (12.47) gives the variation (12.42) which does not satisfy condition (12.2).

The above considerations prove the equivalence of the criteria (12.46) and (12.2), for the system defined above.

12.9C. *A simple system of uniform temperature and pressure.* In Sections 12.7A and 12.8 we found that uniformity of temperature and uniformity of pressure are necessary for equilibrium in a simple, isolated system. For such a system, the conditions of equilibrium

$$\delta E - T\delta S + p\delta V \geqslant 0 \tag{12.50}$$

and (12.2) are equivalent (see Sections 12.9A and 12.9B).

If condition (12.2) is not satisfied, there must be a possible variation in the state of the system given by eqn. (12.42).

Let this variation be followed by the virtual, reversible flow of the quantity of heat $T\delta S$ to the system and production of the quantity of work $p\delta V$ in the surroundings. For this variation

$$\delta E' - T\delta S + p\delta V = 0 \tag{12.51}$$

The sum of eqns. (12.42) and (12.51) is

$$\delta E + \delta E' - T\delta S + p\delta V < 0 \qquad (12.52)$$

It may be written in the form

$$\delta E - T\delta S + p\delta V < 0 \qquad (12.53)$$

where δE is the change in energy attending the total variation in the state of the system measured from its initial state. This variation does not satisfy condition (12.50).

Conversely, if condition (12.50) is not satisfied there must be a possible variation in the state of the above system represented by eqn. (12.53) which is equivalent to (12.52). Subtraction of (12.51) gives the variation (12.42) which does not satisfy condition (12.2).

The above considerations prove the equivalence of the criteria (12.50) and (12.2) for the system described above.

12.10. The work-content criteria of equilibrium. In this section we shall derive two work-content conditions of equilibrium for an isolated system of uniform temperature, and prove that each is equivalent to the energy criterion (12.2).

12.10A. A work-content criterion of equilibrium. Let the change in state

$$B(\text{state 1}) = B(\text{state 2}) \qquad (T) \qquad (12.54)$$

of the system B be brought about by a reversible, isothermal process. The corresponding change in the work content of the system is

$$(A_2 - A_1)_T = (\Delta A)_T = -W_{RT} \qquad (12.55)$$

where W_{RT} is the work of all kinds produced in the surroundings during the above process [see eqn. (10.1.5) of Table 10.1].

Now let the change in state

$$B(\text{state 1}) = B(\text{state 2}') \qquad (S) \qquad (12.56)$$

of the above system be brought about by a reversible, adiabatic process. The corresponding change in the energy of the system is

$$(E_{2'} - E_1)_S = (\Delta E)_S = -W_{RS} \qquad (12.57)$$

[see eqn. (8.65)]. Note that the system is in the same initial state in eqns. (12.54) and (12.56).

For an infinitesimal part of each of the above processes

$$(dA)_T = -dW_{RT} \qquad (12.58)$$

$$(dE)_S = -dW_{RS} \qquad (12.59)$$

Gibbs noted that $-A$ is the force function of a system for constant temperature, just as $-E$ is its force function for constant entropy [5].

In modern notation, the quantity $(\Delta A)_T$ (or $-W_{RT}$) of eqn. (12.55) is the potential energy of the system B in state 2, on the basis that its potential energy in state 1 is assigned the value zero. Also, the quantity $(\Delta E)_S$ (or $-W_{RS}$) of eqn. (12.57) is the potential energy of B in the state 2' reduced by its potential energy in the above state 1. Now the potential energy must be a minimum for an isolated system in a state of equilibrium [6]. Hence no possible variation in such a state can decrease $(\Delta A)_T$ (see Section 12.4).

From these considerations we may conclude that, when an isolated system has the same uniform temperature in its initial and varied states, the additional conditions which are necessary and sufficient for equilibrium may be expressed by the criterion

$$(\delta A)_T \geqslant 0 \tag{12.60}$$

We shall prove the equivalence of conditions (12.60) and (12.2) below.

As noted by Gibbs, the validity of eqn. (12.55) does not require the temperature of the system B to remain constant during the reversible process which produces the work and heat effects attending the change in state (12.54), provided that: (1) the proposed process is reversible; (2) the only heat crossing the boundary of the system comes from or goes to a heat reservoir having at all times the same temperature as the system in its initial or final state; and (3) all other external systems used to bring about the above change in state are restored to their initial states at the end of the process.

For example, let the change in state (12.54) be brought about by a reversible process in which the system B is connected by means of a (reversible) Carnot engine to a heat reservoir at the temperature T. The work W_R (including the work of the Carnot engine) produced in the surroundings of the system in this process is

$$W_R = -\Delta A \qquad (T_1 = T_2 = T_0 \equiv T) \tag{12.61}$$

[see eqn. (10.23)]. Here T_1 and T_2 are the respective initial and final temperatures of B, and T_0 is the temperature of the heat reservoir. Note that under the above circumstances W_R depends only on the change in state of the system B, not on the process.

Evidently W_{RT} of eqn. (12.55) is equal to W_R of eqn. (12.61) when each is produced by a reversible process which brings about the change in state (12.54). Therefore, the condition of equilibrium (12.60), which is derived from eqn. (12.55), does not require the temperature of the system to remain constant during the process which brings about the change in state (12.54), provided that the three restrictions listed above are met.

12.10B. A modification of the work-content criterion of equilibrium. We shall now derive from condition (12.41) a modification of the work-content condition of equilibrium in an isolated system of uniform temperature, and show that it is equivalent to criterion (12.2). Then, by application of the method employed in Section 12.9A, we shall prove that condition

(12.60) is equivalent to this new condition. Hence we shall show that (12.60) is equivalent to criterion (12.2) for the above system.

The work content of a system of uniform temperature is defined by the equation

$$A = E - TS \tag{12.62}$$

[see eqn. (10.4)]. Differentiation gives

$$\delta A + S\delta T = \delta E - T\delta S \tag{12.63}$$

In Section 12.9A we proved that the conditions of equilibrium (12.41) and (12.2) are equivalent for an isolated system of uniform temperature. From this result and from eqn. (12.63) we find that the conditions

$$\delta A + S\delta T \geqslant 0 \tag{12.64}$$

and (12.2) are equivalent for this system.

In order to prove the equivalence of the conditions (12.60) and (12.2), we must first prove the equivalence of (12.60) and (12.64) for the above system. This follows the method used in Section 12.9A.

If condition (12.60) is not satisfied, there must be a possible variation in the state of the system for which

$$\delta A < 0 \text{ and } \delta T = 0 \tag{12.65}$$

From eqns. (12.43) and (12.63) we find that there are possible variations in the state of the system for which

$$\delta E' - T\delta S = 0 \text{ and therefore } \delta A' + S\delta T = 0 \tag{12.66}$$

Here neither δS nor δT has the value zero.

Addition of eqn. (12.65) and the right-hand relation of (12.66) gives the expression

$$\delta A + \delta A' + S\delta T < 0 \tag{12.67}$$

which may be written in the form

$$\delta A + S\delta T < 0 \tag{12.68}$$

This variation does not satisfy condition (12.64).

Conversely, starting from eqn. (12.68) which does not satisfy condition (12.64), we can derive (12.65) which does not satisfy condition (12.60).

Hence conditions (12.60) and (12.64) are equivalent.

We have proved the equivalence of conditions (12.60) and (12.64) and also the equivalence of conditions (12.64) and (12.2) for an isolated system of uniform temperature. Therefore, the conditions of equilibrium (12.60) and (12.2) are equivalent for this system.

12.11. *Two enthalpy criteria of equilibrium.* In this section we shall derive two enthalpy conditions of equilibrium for a simple, isolated system of uniform pressure, and prove that each is equivalent to the energy criterion (12.2).

12.11A. An enthalpy criterion of equilibrium. Let the system B undergo the change in state

$$B(\text{state } 1) = B(\text{state } 2) \qquad (p) \qquad (12.69)$$

by a reversible, isopiestic process which produces only the expansion work $p(V_2 - V_1)$ in its surroundings. The corresponding change in the enthalpy of the system is

$$(\Delta H)_p = Q_{Rp} \qquad (12.70)$$

were Q_{Rp} is the heat withdrawn from the surroundings during the above process [see eqn. (6) of Table 4.4].

When the change in state

$$B(\text{state } 1) = B(\text{state } 2') \qquad (V) \qquad (12.71)$$

is brought about by a reversible, isometric process, the corresponding change in energy is

$$(\Delta E)_V = Q_{RV} \qquad (12.72)$$

[see eqn. (5) of Table 4.4].

Gibbs pointed out that H may be called the heat function of a system for constant pressure, just as E may be called its heat function for constant volume [5].

We recall that Q has the value zero and S is not altered in a reversible, adiabatic process.

From the above considerations we may conclude that, when a simple, isolated system has the same entropy and the same uniform pressure in both its initial and varied states, the additional conditions which are necessary and sufficient for equilibrium may be expressed by the condition

$$(\delta H)_{Sp} \geq 0 \qquad (12.73)$$

We shall prove the equivalence of conditions (12.73) and (12.2) below.

12.11B. A modification of the enthalpy criterion of equilibrium. We shall now derive from condition (12.46) a modification of the enthalpy condition of equilibrium in a simple, isolated system of uniform pressure, and show that it is equivalent to criterion (12.2). We shall then prove by application of the method used in Section 12.9B that condition (12.73) is equivalent to this new condition. Hence we shall show that (12.73) is equivalent to criterion (12.2) for the above system.

The enthalpy of a system of uniform pressure is defined by the equation

$$H = E + pV \qquad (12.74)$$

[see eqn. (4.13)]. Differentiation gives

$$\delta H - V\delta p = \delta E + p\delta V \qquad (12.75)$$

In Section 12.9B we proved that the condition of equilibrium (12.46) is equivalent to (12.2) for a simple isolated system of uniform pressure. From this result and from eqn. (12.75) we find that the condition of equilibrium

$$\delta H - V\delta p \geqslant 0 \text{ and } \delta S = 0 \qquad (12.76)$$

and condition (12.2) are equivalent for this system.

The proof of the equivalence of conditions (12.73) and (12.76) for the above system follows the method used in Sections 12.9B and 12.10B.

If condition (12.73) is not satisfied, there must be a possible variation in the state of the above system for which

$$\delta H < 0 \text{ and } \delta S = 0, \delta p = 0 \qquad (12.77)$$

From eqns. (12.47) and (12.75) we find that there are possible variations in the state of the above system for which

$$\delta E' + p\delta V = 0 \text{ and therefore } \delta H' - V\delta p = 0 \qquad (12.78)$$

Here $\delta S = 0$ in each equation, but neither δV nor δp has the value zero.

Addition of eqns. (12.77) and the right-hand relation of (12.78) gives an expression which may be written in the form

$$\delta H - V\delta p < 0 \text{ and } \delta S = 0 \qquad (12.79)$$

This variation does not satisfy condition (12.76).

Conversely, starting from eqn. (12.79) which does not satisfy condition (12.76), we can derive (12.77) which does not satisfy condition (12.73).

Hence conditions (12.73) and (12.76) are equivalent.

We have proved the equivalence of conditions (12.73) and (12.76) and also the equivalence of conditions (12.76) and (12.2) for a simple, isolated system of uniform pressure. Therefore, the conditions of equilibrium (12.73) and (12.2) are equivalent for this system.

12.12. Two free-energy criteria of equilibrium. In this section we shall derive two free-energy conditions of equilibrium for a simple, isolated system of uniform temperature and pressure, and prove that each is equivalent to the energy criterion (12.2).

12.12A. A free-energy criterion of equilibrium. Let the change in state

$$B(\text{state 1}) = B(\text{state 2}) \qquad (T, p) \qquad (12.80)$$

of the simple system B be brought about by a reversible, isothermal, isopiestic process. The corresponding change in the free energy of the system is

$$(\Delta G)_{Tp} = 0 \qquad (12.81)$$

[see eqn. (10.1.12) of Table 10.1].

When the change in state

$$B(\text{state 1}) = B(\text{state 2}') \qquad (S, V) \qquad (12.82)$$

of the above system is brought about by a reversible, adiabatic, isometric process, the corresponding change in the energy of the system is

$(\Delta E)_{SV} = 0$ (12.83)

[see eqn. (11) of Table 4.4].

From the above considerations we may conclude that, when a simple, isolated system has the same uniform temperature and pressure in its initial and varied states, the additional conditions which are necessary and sufficient for equilibrium may be expressed by the criterion

$(\delta G)_{Tp} \geqslant 0$ (12.84)

We shall prove the equivalence of conditions (12.84) and (12.2) below.

From eqn. (10.32) we learn that the restrictions of uniformity of temperature and pressure in condition (12.84) may be relaxed to some extent by applying this relation to a composite system of B and its medium M (see Section 12.10A).

12.12B. A modification of the free-energy criterion of equilibrium. We shall now derive from condition (12.50) a modification of the free-energy condition of equilibrium in a simple, isolated system of uniform temperature and pressure, and show that it is equivalent to criterion (12.2). We shall then prove by application of the method employed in Section 12.9C that condition (12.84) is equivalent to this new condition. Hence we shall show that (12.84) is equivalent to criterion (12.2) for the above system.

The free energy of a system of uniform temperature and pressure is defined by the equation

$G = E - TS + pV$ (12.85)

[see eqn. (10.5)]. Differentiation gives

$\delta G + S\delta T - V\delta p = \delta E - T\delta S + p\delta V$ (12.86)

In Section 12.9C we proved that the condition of equilibrium (12.50) is equivalent to (12.2) for a simple isolated system of uniform temperature and pressure. From this result and from eqn. (12.86) we find that the conditions of equilibrium

$\delta G + S\delta T - V\delta p \geqslant 0$ (12.87)

and (12.2) are equivalent for this system.

Using the methods of Sections 12.9C and 12.10B, we can prove the equivalence of conditions (12.84) and (12.87) for the above system.

If condition (12.84) is not satisfied, there must be a possible variation in its state for which

$\delta G < 0$ and $\delta T = 0$, $\delta p = 0$ (12.88)

From eqns. (12.51) and (12.86) we find that there are possible variations in the state of the system for which

$\delta E' - T\delta S + p\delta V = 0$, and therefore $\delta G' + S\delta T - V\delta p = 0$ (12.89)

Here none of the variables δS, δV, δT and δp has the value zero.

TABLE 12.1

SOME CRITERIA OF EQUILIBRIUM FOR A CLOSED, ISOLATED SYSTEM

Criterion	Property unaltered	Equation number
$\delta S \leqslant 0$	E	(12.1)
$\delta E \geqslant 0$	S	(12.2)
A system of uniform temperature		
$\delta E - T\delta S \geqslant 0$		(12.41)
$\delta A + S\delta T \geqslant 0$		(12.64)
$\delta A \geqslant 0$	T	(12.60)
A simple system of uniform pressure		
$\delta E + p\delta V \geqslant 0$	S	(12.46)
$\delta H - V\delta p \geqslant 0$	S	(12.76)
$\delta H \geqslant 0$	S, p	(12.73)
A simple system of uniform temperature and pressure		
$\delta E - T\delta S + p\delta V \geqslant 0$		(12.50)
$\delta G + S\delta T - V\delta p \geqslant 0$		(12.87)
$\delta G \geqslant 0$	T, p	(12.84)

Addition of eqn. (12.88) and the right-hand relation of (12.89) gives an expression which may be written in the form

$$\delta G + S\delta T - V\delta p < 0 \qquad (12.90)$$

This variation does not satisfy condition (12.87).

Conversely, starting from eqn. (12.90) which does not satisfy (12.87), we can derive (12.88) which does not satisfy condition (12.84).

Hence eqns. (12.84) and (12.87) are equivalent.

We have proved the equivalence of conditions (12.84) and (12.87) and also the equivalence of conditions (12.87) and (12.2) for a simple, isolated system of uniform temperature and pressure. Therefore the conditions of equilibrium (12.84) and (12.2) are equivalent for this system.

12.13. Résumé of criteria of equilibrium. In Table 12.1 we have listed the eleven criteria of equilibria for an isolated system which were defined or suggested by Gibbs [2, 5]. Each is equivalent to criterion (12.2), subject to certain restrictions [7]. The second column of this Table gives the property which is not altered during any possible variation in the state of the system, and the third column lists the ordinal number of the corresponding equation in the present chapter.

REFERENCES

1 J. W. Gibbs, Scientific Papers, Longmans, Green and Co., New York, 1906, pp. 55—62, 89—92, 354—356.
2 J. W. Gibbs, ref. 1, pp. 55—62.
3 Compare G. N. Hatsopoulos and J. H. Keenan, Principles of General Thermodynamics, Wiley, New York, 1965, pp. xxix, 393—396, 427.

4 J. W. Gibbs, ref. 1, pp. 62—65.
5 J. W. Gibbs, ref. 1, pp. 89—92.
6 See A. W. Porter, Surface Tension, Encyclopaedia Britannica, Inc., Chicago, 1969, Vol. 21, p. 443.
7 For a larger number of criteria of equilibrium, see G. N. Hatsopoulos and J. H. Keenan, ref. 3, Chapter 38.

Chapter 13

OPEN SYSTEMS

An open system can exchange matter with its surroundings, whereas no matter can cross the boundary of a closed system (Section 1.2). The laws of thermodynamics are stated for closed systems. However, we have applied them to an open system undergoing a steady-flow process by use of the device of a moving, impermeable boundary which converts the open system to the corresponding closed system (see Sections 3.11F, 3.24, 5.5, 6.3, and 10.5). In the present chapter we shall present a more general treatment of open systems.

13.1. The first- and second-law equation for open, simple phases. Gibbs [1] used the concept of an open system to define the chemical potential of a substance in a homogeneous system when it is in a state of thermodynamic equilibrium. Each phase of a closed, heterogeneous system at equilibrium is an open system at equilibrium. We shall use the term "phase" and the expression "homogeneous system" interchangeably.

Gibbs started his discussion of open, simple, homogeneous systems by writing the first- and second-law differential equation for the energy of a closed phase. He then wrote the corresponding relation for the same phase when open to the exchange of matter with its surroundings. In each case he used the general mathematical principle that the total differential of a function of a set of j independent variables is the sum of its j partial differentials.

Consider a closed, simple, homogeneous system composed of the c substances $B_1, B_2, \ldots, B_c$ whose masses $n_1, n_2, \ldots, n_c$ are fixed. By the first and second laws of thermodynamics, the energy E of the phase when in a state of equilibrium is a function of its entropy S and volume V

$$E = E(S, V) \tag{13.1}$$

The total differential of E is therefore

$$dE = TdS - pdV \tag{13.2}$$

where

$$T = \left(\frac{\partial E}{\partial S}\right)_V, \quad p = -\left(\frac{\partial E}{\partial V}\right)_S \tag{13.3}$$

By the second law, T is the absolute thermodynamic temperature and, by the first law, p is the pressure of the phase.

In the corresponding open, simple, homogeneous system the masses of the c constituents are variable. Hence its energy will be a function of S, V, and the c masses,

$$E = E(S, V, n_1, n_2, \ldots, n_c) \tag{13.4}$$

The total differential of E is

$$dE = TdS - pdV + \Sigma_i \mu_i dn_i \tag{13.5}$$

where the summation extends over the c substances composing the phase, dn_i is positive when B_i enters and negative when it leaves the phase, and

$$T = \left(\frac{\partial E}{\partial S}\right)_{Vn}, \quad p = -\left(\frac{\partial E}{\partial V}\right)_{Sn} \tag{13.6}$$

$$\mu_i = \left(\frac{\partial E}{\partial n_i}\right)_{SVn} \qquad (i = 1, 2, \ldots, c) \tag{13.7}$$

Here the subscript n to a partial derivative denotes that all of the c masses are held constant except the one that appears in the denominator. It is evident that the partial derivatives of eqns. (13.6) are evaluated when the phase is closed. Hence the symbols T and p have the same meaning in eqn. (13.5) as in (13.2). The c quantities μ_i are defined by eqns. (13.7).

Gibbs called μ_i the potential (now termed the chemical potential) of the constituent B_i of a phase. It is completely defined by eqn. (13.7) provided that (1) the differentials $dn_1, dn_2, \ldots, dn_c$ are independent, and (2) they express every possible variation in the composition of the phase, including any produced by the introduction of substances not initially present. It may therefore be necessary to include in eqn. (13.5) $\mu_i dn_i$ terms for substances not initially occurring in the phase if they, or their constituents, are found in some other phase (or phases) of the system.

When the above conditions hold for a substance, it is called a component of the phase. From eqn. (13.7) we learn that the chemical potential of a substance is an intensive state property of a phase.

Definitions expressing the chemical potential of a substance in a phase in terms of various sets of independent variables are given by eqns. (10.68) which are derived from the energy, enthalpy, work-content, and free-energy equations (10.64)–(10.67).

From eqn. (10.57) we find that the chemical potential of a substance in a gaseous, liquid, or solid solution is its partial molar free energy in the phase. Some of the properties of partial molar quantities are derived in Chapter 7.

The introduction of the chemical potential into eqn. (13.5) allowed Gibbs to apply the laws of thermodynamics to open systems, including those in which chemical reactions may occur. This equation is therefore the foundation on which the science of chemical thermodynamics is built.

13.2. Components and species of a phase. A liquid solution of $CuSO_4$ in water has two components, for example, $CuSO_4$ and H_2O, $CuSO_4 \cdot 5H_2O$ and H_2O, or $CuSO_4 \cdot 5H_2O$ and $CuSO_4$ (see Section 1.19).

When the last one of the above three sets of components is chosen, the number of moles of the component CuSO₄ in a given solution will, in general, have a negative value (see Table 1.1). Such a choice of components may be inconvenient for practical purposes, but it is not incorrect thermodynamically.

Other selections of components of the above phase may be made, provided that the conditions given in Section 13.1 are met. A component need not be capable of existing as a physical entity either in the solution or by itself in any aggregation state. For example, we may choose $CuSO_4 \cdot nH_2O$, where n is not an integer, as one of the components of the phase. "In the same homogeneous mass, therefore, we may distinguish the potentials for an indefinite number of substances, each of which has a perfectly determined value" [2].

We shall prove that this statement is correct in the next section, and shall apply it in Section 13.8. If it were not true, chemical thermodynamics as we know it today would not exist.

13.3. Chemical potentials of components and species. In general, the rate of a chemical reaction is affected by pressure, temperature, and the presence of a catalyst or an anticatalyst.

Under conditions where the rate of the homogeneous chemical reaction

$$3H_2(g) + N_2(g) = 2NH_3(g) \tag{13.8}$$

is appreciable, a phase at equilibrium with respect to this reaction has two components: H_2 and N_2, H_2 and NH_3, or N_2 and NH_3.

Under conditions where the rate of this reaction is virtually zero, the above phase at equilibrium has three components: H_2, N_2, and NH_3.

We shall now prove two important theorems on chemical potentials. In the following proofs we shall use the above equation as an example, with H_2, N_2, and NH_3 represented by B_1, B_2, and B_3, respectively.

13.3A. The chemical potential of a component of a phase is independent of the choice of components. The chemical potential μ_1 of the component B_1 of a phase at equilibrium with respect to the chemical reaction (13.8) is defined by either of the two relations

$$\mu_1' = \left(\frac{\partial E'}{\partial n_1'}\right)_{SVn_2}, \text{ or } \mu_1'' = \left(\frac{\partial E''}{\partial n_1''}\right)_{SVn_3} \tag{13.9}$$

Although n_1' does not equal n_1'' (see Table 1.1), we can evidently give dn_1' and dn_1'' the same value in each of the above partial differential coefficients. Then dE' and dE'' will have the same value in each, since each numerator is the change in energy of the given phase attending the same change in state, namely, the addition of dn_1 moles of B_1 to the phase at constant S and V under the condition that the chemical reaction (13.8) is always virtually at equilibrium.

We have shown that $dE' = dE''$ when $dn_1' = dn_1''$ in eqns. (13.9). Hence $\mu_1' = \mu_1''$, and the theorem is proved.

In the same manner, we can show that the chemical potential of H_2O in a solution of $CuSO_4$ in H_2O has the same value whether we choose $CuSO_4$, $CuSO_4 \cdot 5H_2O$, or $CuSO_4 \cdot nH_2O$ as the other component (see Section 13.2).

13.3B. The chemical potential of a constituent of a phase when considered to be a species is equal to its chemical potential when considered to be a component. The chemical potential μ_1 of a component B_1 and the chemical potential μ_{1s} of the corresponding species B_{1s} of the phase considered in Section 13.3A are defined by the expressions

$$\mu_1 = \left(\frac{\partial E'}{\partial n_1}\right)_{SVn_2}, \mu_{1s} = \left(\frac{\partial E''}{\partial n_{1s}}\right)_{SVn_{2s}n_{3s}} \tag{13.10}$$

Although n_1 is not equal to n_{1s} [see eqn. (1.31)] we may give dn_1 and dn_{1s} the same value in the above equations.

Each of the following two processes start from the same equilibrium state of the phase.

1. Let dE' be the change in energy of the above phase attending the addition of dn_1 moles of the component B_1 at constant S, V, and n_2, under the condition that the chemical reaction (13.8) is always essentially at equilibrium. We shall call the final state of the phase produced by this process State'.

2. Let dE'' be the change in energy of the above phase attending the addition of dn_{1s} moles of the species B_{1s} at constant S, V, n_{2s} and n_{3s}, under the condition that no chemical reaction occurs. We shall call the final state of the phase produced by this process State''.

State'' of the phase differs infinitesimally in composition from State'. Both have the same entropy S and volume V. We may consider that State'' can be produced by a possible variation of State' of an isolated system at equilibrium. Application of Gibbs' second criterion of equilibrium, eqn. (12.2), to State' gives

$$(\delta E)_S = dE'' - dE' = 0 \tag{13.11}$$

Here the sign of equality is used in the criterion because "...the active tendencies of the system are so balanced that changes of every kind, except those excluded in the statement of the condition of equilibrium, can take place *reversibly*, (i.e., both in the positive and the negative direction,) in states of the system differing infinitely little from the state in question" [3].

We have found that $dE' = dE''$ when $dn_1 = dn_{1s}$ in eqns. (13.10). Hence

$$\mu_1 = \mu_{1s} \tag{13.12}$$

and the theorem is proved.

13.4. Actual and possible components of a phase. A substance B_a which, in a given problem, has been chosen to be one of the components of a phase of a closed, heterogeneous system at equilibrium was called an *actual component* of the phase by Gibbs if its mass n_a in the phase may be either increased or decreased [1].

A substance B_b was called a *possible component* of this phase by Gibbs if its mass in the phase may be increased but not decreased [1]. Of course, B_b or its constituents must occur in some other phase or phases of the system. Also we must choose the other components of the given phase so that $n_b = 0$.

It is possible to prove that, in an isolated, heterogeneous system at equilibrium, the chemical potential μ_k of a constituent B_k must have the same value M_k in all phases in which it is an actual component, and a value equal to or greater than M_k in all phases in which it is a possible component.

On the other hand, the dilute solution law, applicable alike to gaseous, liquid, and solid solutions, requires the chemical potential of a constituent of a phase to become negatively infinite as its mole fraction tends to zero [4].

Now the dilute solution law requires the chemical potential of a possible component of a phase of a closed, heterogeneous system to have the value $-\infty$ at all times. But thermodynamic theory requires the chemical potential of such a component to have a value in this phase which is equal to or greater than its value in a phase of the system in which it is an actual component. Therefore the substance cannot be a possible component of the phase. It is either an actual component of the phase or not a component at all.

Gibbs introduced the concept of a possible component at the beginning of his application of thermodynamics to the equilibrium of heterogeneous systems because he was unwilling to apply the dilute-solution law until he had shown that it is consistent with eqn. (10.63) now called the Gibbs–Duhem equation, which can be derived from eqn. (13.5).

13.5. Components of a closed, heterogeneous system. The number of components of a closed, heterogeneous system at equilibrium is the sum of the different components of its phases diminished by the number of chemical reactions among them which are at equilibrium and have not already been counted (see Section 1.19).

13.6. Examples of open systems. In this section we shall consider three examples of open systems, namely, physical processes such as those studied by Joule and Thomson, a heterogeneous chemical reaction, and a homogeneous chemical reaction.

13.6A. Physical processes. Joule used a closed system in his free-expansion experiment described in Section 5.4. He also inverted the metal assembly A, CDC, B of Fig. 5.1, immersed bulb A, bulb B, and the connecting unit CDC in separate cans filled with water, and measured the change in temperature of the liquid in each can when the process used in the free-expansion experiment was repeated. Each of these units is an open system. Such systems are considered from the present point of view in Sections 13.11 and 13.16.

In the Joule—Thomson porous-plug experiment described in Section 5.5, the gas contained at any time in the space between the gas inlet I and the gas outlet O of the apparatus (see Fig. 5.2) is an open system which is undergoing a steady-flow process. Other examples of such systems are turbines and reciprocating engines. They are treated in Section 13.13A.

13.6B. A heterogeneous chemical reaction. In the closed, heterogeneous system

$$CuSO_4 \cdot 5H_2O(s), CuSO_4(s), H_2O(g) \tag{13.13}$$

each of the three open phases has one component [5]. These substances take part in the chemical change in state

$$CuSO_4 \cdot 5H_2O(s) = CuSO_4(s) + 5H_2O(g) \qquad (p, T) \tag{13.14}$$

and hence, at equilibrium, the heterogeneous system (13.13) has two components in three phases (see Section 13.5).

Under these conditions, the system is univariant, and all of the intensive properties of each phase depend on a single intensive property, for example, temperature (see Section 10.11).

By slightly decreasing or increasing the pressure applied to the system (13.13), we can cause the change in state (13.14) to take place to the right or to the left in a closed system by a virtually reversible, isopiestic, isothermal process in which $W_x = 0$. Therefore the changes in enthalpy ΔH and entropy ΔS attending this change in state are

$$\Delta H = T\Delta S = Q_p \qquad (p, T = \text{constant}) \tag{13.15}$$

where Q_p is the heat withdrawn from the surroundings [see eqn. (6) of Table 4.4 and eqn. (9.51)].

The corresponding change in free energy ΔG is

$$\Delta G = -\mu_{AW_s(s)} + \mu_{A(s)} + 5\mu_{W(g)} = 0 \qquad (p, T = \text{constant}) \tag{13.16}$$

where A and W denote $CuSO_4$ and H_2O, respectively [see eqn. (10.1.12) of Table 10.1 and eqn. (10.141)].

Thus, for each chemical reaction at equilibrium in a closed heterogeneous system, the same equation will hold for the chemical potentials of the substances taking part in the reaction as for the units of these substances [5].

13.6C. A homogeneous chemical reaction. The following change in state is based on the ammonia synthesis reaction (13.8)

$$3H_2(g, p_1) + N_2(g, p_2) = 2NH_3(g, p_3) \qquad (T) \tag{13.17}$$

Here p_1, p_2, and p_3 are the equilibrium pressures of the respective gases H_2, N_2, and NH_3 in a phase which is at equilibrium with respect to the chemical reaction (13.8) at a total pressure p and a temperature T.

We recall that the equilibrium pressure of a gas (real or perfect) in a gas mixture is defined as the pressure of the pure gas when it is in equilibrium with a gas mixture through a rigid, thermally conducting, semipermeable membrane (see Section 9.17).

In principle, the change in state (13.17) may be brought about by a device known as the van't Hoff equilibrium box [6] which is shown in Fig. 13.1. The vessel A has a constant volume and it contains an equilibrium mixture of the three gases H_2, N_2, and NH_3 at p and T.

The three cylinders 1, 2, and 3 contain $H_2(g, p_1, T)$, $N_2(g, p_2, T)$ and $NH_3(g, p_3, T)$, respectively. Each cylinder is fitted with a weighted, frictionless piston, and it communicates with the vessel A through a rigid head permeable only to the gas contained in it. The system is in thermal equilibrium with a thermostat at a fixed temperature T.

By making slight changes in the external forces applied to the pistons of the three cylinders, we can bring about the change in state (13.17) in the following manner: push 3 moles of H_2 from cylinder 1 and 1 mole of N_2 from cylinder 2 into the vessel A, and draw 2 moles of NH_3 from A into cylinder 3 at such relative rates of flow that each operation is completed in the same time interval. Evidently we can reverse the above procedure at any stage of the process.

The rates of flow of the gases into and out of the phase A are so small that the gas mixture in A is at all times practically at equilibrium with respect to the chemical reaction (13.8), and with respect to the flow of the pure gases into and out of the three cylinders. Hence, throughout the process, the pressures p, p_1, p_2, and p_3, the temperature T, and the composition of the gas in A remain virtually unchanged. During the direct process a quantity of heat $-Q_p$ has flowed from phase A to the thermostat.

Thus the change in state (13.17) has been produced in a closed, heterogeneous system by a virtually reversible, isothermal steady-flow process in which each phase has a constant, separate pressure, and $W_X = 0$. Hence, by eqn. (13.15), each of the quantities ΔH and $T\Delta S$ attending the change in state (13.17) are equal to the heat effect Q_p withdrawn from the surroundings of the system; and, by eqn. (13.16), the corresponding quantity ΔG is zero. Since the phases of the present system have different pressures, we must add eqn. (4.33) to the list of those following eqn. (13.16).

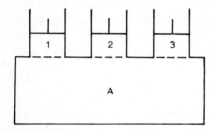

Fig. 13.1. The van't Hoff equilibrium box [6].

Application of eqn. (13.16) to the present process gives

$$3\mu_{H_2} + \mu_{N_2} = 2\mu_{NH_3} \qquad (p, T = \text{constant}) \qquad (13.18)$$

where μ_{H_2}, μ_{N_2}, and μ_{NH_3} are the chemical potentials of H_2, N_2, NH_3 in the equilibrium change in state (13.17), and hence in the cylinders 1, 2, and 3 of Fig. 13.1.

13.7. Chemical potentials in a phase at equilibrium. Let two phases of a heterogeneous system at equilibrium be separated by a rigid, thermally conducting boundary permeable only to a fluid B_1 which is a component of each phase. It is possible to prove that the chemical potential of this substance has the same value in each phase.

If we should stop the process occurring in the system of Fig. 13.1 at any time, phase A would be at equilibrium with respect to the chemical reaction (13.8), and the chemical potentials of H_2, N_2, and NH_3 would have the same values in this phase as in the cylinders 1, 2, and 3, respectively. Hence these quantities satisfy eqn. (13.18).

For each chemical reaction

$$\sum_i \nu_i B_i = 0 \qquad (p, T) \qquad (13.19)$$

which is at equilibrium in a homogeneous system, the relation

$$\sum_i \nu_i \mu_i = 0 \qquad (p, T = \text{constant}) \qquad (13.20)$$

holds for the chemical potentials of the substances taking part in the reaction.

Equations (13.19) and (13.20) show that: "Between the potentials for different substances in the same homogeneous mass the same equations will subsist as between the units of these substances" [2].

13.8. The summation $\sum_i \mu_i dn_i$. We stated in Section 13.1 that the sum $\sum_i \mu_i dn_i$ appearing in the first and second law energy equation (13.5) could be extended over any set of c components of a phase at equilibrium. In this section we shall prove not only that this statement is correct, but that the summation can also be extended over all of the m species of the phase.

Let us use as an example the chemical reaction (13.8) with H_2, N_2, and NH_3 replaced by B_1, B_2, and B_3, respectively, (see Section 13.3).

In order to prove that the summations mentioned above have the same value, we must show that the following equations are correct for a phase at equilibrium with respect to the chemical reaction (13.8)

$$\mu_1 dn_{1s} + \mu_2 dn_{2s} + \mu_3 dn_{3s} = \mu_1 dn_1 + \mu_2 dn_2 \qquad (13.21a)$$
$$= \mu_1 dn_1 + \mu_3 dn_3 \quad (p, T = \text{constant}) \qquad (13.21b)$$
$$= \mu_2 dn_2 + \mu_3 dn_3 \qquad (13.21c)$$

Here μ_i is the chemical potential of $B_i (i = 1, 2, 3)$. Its value for a given substance does not depend on the choice of components, or on whether the substance is considered to be a component or a species of the phase (see Section 13.3). The quantities n_i and n_{is} are the mole numbers of B_i when it is considered to be a component and when it is considered to be a species of the phase, respectively. They are connected by eqn. (1.31), which may be written in the form

$$dn_{is} = dn_i + \frac{\nu_i}{\nu_m} dn_{ms} \qquad (i = 1, 2, 3) \qquad (13.22)$$

where the subscript m denotes the substance which has not been chosen to be a component of the phase [see eqn. (1.27)].

We can now derive the right-hand side of eqn. (13.21a) from the left-hand side by substituting from eqn. (13.22) the expressions for dn_{1s}, dn_{2s}, and dn_{3s}, equating dn_3 to zero in the resulting relation (B_3 is not a component of the phase), and then applying eqn. (13.20). In a similar manner, we can derive the right-hand sides of eqns. (13.21b) and (13.21c) by equating dn_2 and dn_1, respectively, to zero in the above relation derived from the left-hand side, and then applying eqn. (13.20) in each case.

Conversely, by solving eqn. (13.22) for dn_i and substituting from this relation and (13.20) into each of the right-hand sides of eqns. (13.21a, b, and c) we can derive the left-hand side in each case.

We have proved that the sum $\Sigma_i \mu_i dn_i$ in the energy equation (13.5), and hence in the enthalpy, work-content, and free-energy equations (10.65)–(10.67) can be extended over any set of c components of a phase at equilibrium, and also over the m species.

13.9. The first- and second-law equation for open phases producing non-expansion work. When an open, homogeneous system produces reversible expansion work and ω kinds of reversible non-expansion work in its surroundings, eqn. (13.5) becomes

$$dE = TdS - pdV - \Sigma_j y_j dX_j + \Sigma_i \mu_i dn_i \qquad (13.23)$$

[see eqn. (10.177)]. Here

$$T = \left(\frac{\partial E}{\partial S}\right)_{VXn}, \quad p = -\left(\frac{\partial E}{\partial V}\right)_{SXn} \qquad (13.24)$$

$$y_j = -\left(\frac{\partial E}{\partial X_j}\right)_{SVXn} \qquad (j = 1, 2, \ldots, \omega) \qquad (13.25)$$

$$\mu_i = \left(\frac{\partial E}{\partial n_i}\right)_{SVXn} \qquad (i = 1, 2, \ldots, c) \qquad (13.26)$$

where the subscripts X and n to the partial derivatives indicate that all of the ω extensive quantities X_j and all of the c masses n_i, respectively, are held constant except the one appearing in the denominator.

13.10. The first-law equation for open systems. In order to interpret calorimetric measurements made on an open system across whose boundary one or more fluids flow, Gillespie and Coe [7] derived both the first-law and the second-law equations for such systems.

Although the above investigators were primarily interested in the derivation of thermodynamic equations for the flow of gases across the boundary of a system, the resulting relations are valid for the corresponding flow of any fluid, for example, the flow of water into or out of an aqueous solution of sugar through a semipermeable membrane.

An open system may consist of any number of phases, and it may or may not be in a state of thermodynamic equilibrium.

The first-law equation for a closed system is

$$dE = dQ - dW_{EX} - dW_X \tag{13.27}$$

where dE is the change in energy of the system, dQ is the heat withdrawn from the surroundings, and dW_{EX} and dW_X are, respectively, the expansion work and non-expansion work produced in the surroundings in an infinitesimal part of a process which may be either reversible or irreversible.

The corresponding first-law equation for an open system will include terms for the change in energy of the system attending the flow of matter across its boundary when its volume is constant, and both Q and W_X are zero. In the following discussion we shall limit the non-expansion work effect to the shaft work W_S produced in the surroundings of the system by the rotation or reciprocation of a rigid shaft extending through the boundary (see Section 3.3).

Since the same fluid may enter an open system in one state and leave it in another state, we shall identify a stream crossing the boundary and also the pure fluid or the fluid solution composing it by the notation $B_k (k = 1, 2, \ldots, \sigma)$ where σ is the total number of such streams.

Let dn_k moles of a fluid B_k be contained in a cylinder fitted with a weighted, frictionless piston (Fig. 13.2). The cylinder communicates with an open phase A by way of a valve or a rigid, semipermeable membrane which forms a part of the boundary of A. In the following two-step process the fluid B_k flows so slowly across the boundary of the phase that it has a well-defined pressure p_k, temperature T_k, and hence molar volume v_k. For the present, we shall suppose that the kinetic and potential energies of the stream may be neglected.

1. Imagine that the phase A and the dn_k moles of the fluid B_k in the cylinder are enclosed in an impermeable boundary. The energy of the closed, composite system A + dn_k is larger than the energy of A by the amount

$$dE_1 = e_k dn_k \tag{13.28}$$

where e_k is the molar energy of B_k in the state (p_k, T_k).

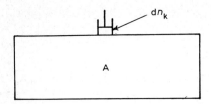

Fig. 13.2. Addition of dn_k moles of a fluid B_k to a phase A.

2. With both Q and W_S equal to zero, push the dn_k moles of B_k across the boundary of A. In this process the fluid in the volume $V + v_k dn_k$ of the system $A + dn_k$ is compressed under the pressure p_k to the volume V of A. Since the composite system is closed, we may apply the usual equations of thermodynamics to the above process. Hence the change in energy dE_2 of A is

$$dE_2 = p_k v_k dn_k \tag{13.29}$$

The total change in energy dE attending the flow of dn_k moles of the fluid B_k across the boundary of the phase A when V is constant and both Q and W_S are zero is

$$dE = dE_1 + dE_2 = (e_k + p_k v_k)\, dn_k = h_k dn_k \tag{13.30}$$

Here h_k is the molar enthalpy of B_k in the state (p_k, T_k) which it has as it enters or leaves A.

If several streams of fluids cross the boundary of an open system, eqn. (13.30) becomes

$$dE = \sum_k h_k\, dn_k \tag{13.31}$$

where the summation extends over all of the σ streams flowing into or out of the system. Here dn_k is positive when B_k enters and negative when it leaves the system.

For the simultaneous flow of heat and matter across the boundary of an open system, and the production of expansion work and shaft work in its surroundings, eqn. (13.27), with W_X replaced by W_S, and eqn. (13.31) give

$$dE = đQ - đW_{EX} - đW_S + \sum_k h_k\, dn_k \tag{13.32}$$

If the volume of the system is held constant throughout a process, eqn. (13.32) becomes

$$dE = đQ - đW_S + \sum_k h_k dn_k \qquad (V = \text{constant}) \qquad (13.33)$$

When the kinetic and potential energies of the stream B_k, as it crosses the boundary of an open system, are not negligible, eqn. (13.33) becomes

$$dE = đQ - đW_S + \sum_k (h_k + \tfrac{1}{2} M_k u_k^2 + M_k g z_k) dn_k \qquad (V = \text{constant}) \qquad (13.34)$$

In eqns. (13.32)–(13.34), E is the energy and V is the volume of the system, n_k is the number of moles of B_k in it at any time, and $đQ$ and $đW_S$ are the heat and shaft work effects, respectively, produced in the surroundings during the time that dn_k moles of B_k cross the boundary. Also, u_k is the velocity of the stream B_k at its place of entry, z_k is the corresponding height of this stream above a horizontal datum plane, and g is the local acceleration of gravity. The quantity h_k is the molar enthalpy of B_k in the state $(p_k, T_k, u_k = 0, z_k = 0)$ where p_k and T_k are the pressure and temperature of the fluid as it passes through the boundary of the system, and M_k is the molecular weight of B_k. The summation extends over all of the σ streams flowing into and out of the system.

If n_k is expressed in the unit of grams in eqns. (13.32)–(13.34), h_k is the specific enthalpy of B_k and M_k is replaced by unity.

Equations (13.32)–(13.34) are various forms of the first-law equation for the flow of matter into or out of a homogeneous or heterogeneous system. The heat and work effects may be either reversible or irreversible. The system may or may not be in a state of equilibrium. At all times the rate of flow of each fluid B_k across the boundary of the system must be so small that B_k has a well-defined pressure, temperature, and hence molar enthalpy.

13.11. Applications of the first-law equation for open systems. Equation (13.33) takes the following form for the flow of a single stream of a fluid, for example a gas B, into or out of a vessel of constant volume

$$dE = đQ - đW_S + h_B dn \qquad (V = \text{constant}) \qquad (13.35)$$

Here h_B is the molar enthalpy of B as it crosses the boundary of the system.

In general, we shall assume that the system has at all times a uniform pressure p and temperature T. The gas B has a pressure p_B and a temperature T_B as it crosses the boundary. The latter quantities may or may not be equal to p and T, respectively.

If the volume of the system varies during the process, or the kinetic or potential energy of B is not negligible, we would base eqn. (13.35) on eqn. (13.32) or (13.34), respectively.

Let an open system composed of a gas B undergo the change in state

$$n_1 B(g, p_1, T_1) = n_2 B(g, p_2, T_2) \qquad (V) \qquad (13.36)$$

under the conditions mentioned above. Integration of eqn. (13.35) from n_1 to n_2 moles of B in the system gives

$$E_2 - E_1 = Q - W_S + \int_{n_1}^{n_2} h_B \, dn \qquad (V = \text{constant}) \qquad (13.37)$$

where h_B is a function of p_B and T_B, and n_1 and n_2 may be evaluated for the conditions (p_1, V, T_1) and (p_2, V, T_2), respectively, from tables of thermodynamic properties of the gas or from its equation of state.

We shall now consider several applications of eqn. (13.35) or eqn. (13.37).

13.11A. *Relation between ΔE and h_B.* When both Q and W_S are zero, the change in energy $E_2 - E_1$ of an open system of constant volume, whether or not in internal equilibrium, is the difference between the total enthalpy entering and the total enthalpy leaving the system. From eqn. (13.37) we find

$$E_2 - E_1 = \int_{n_1}^{n_2} h_B \, dn \qquad (V = \text{constant}) \qquad (13.38)$$

13.11B. *Flow of a gas into a system.* Let the change in state (13.36) be brought about by the flow of a gas B into a vessel of constant volume. The rate of flow of B is such that at all times the pressure p and temperature T of the system are practically uniform; and the pressure p_B, temperature T_B, and hence molar enthalpy h_B of the gas entering the system have well-defined and virtually constant values.

Let us now impose the following conditions on the above process:

1. When both Q and W_S are zero, eqn. (13.37) becomes

$$n_2 e_2 - n_1 e_1 = h_B (n_2 - n_1) \qquad (V = \text{constant}) \qquad (13.39)$$

where e_1 and e_2 are the molar energies of B in the states (p_1, T_1) or (v_1, T_1) and (p_2, T_2) or (v_2, T_2), respectively, and h_B, which is a function of p_B and T_B, is constant. Here v, which is a function of p and T, is the molar volume of B in the system.

2. When W_S is zero and the system is in thermal equilibrium with a thermostat at a fixed temperature T, we must replace T_1 and T_2 by T in the change in state (13.36). Then eqn. (13.37) becomes

$$Q = n_2 e_2 - n_1 e_1 - h_B (n_2 - n_1) \qquad (V, T = \text{constant}) \qquad (13.40)$$

In eqns. (13.39) and (13.40)

$$n_2 e_2 - n_1 e_1 = n_2(e_2 - e_1) + e_1(n_2 - n_1) \qquad (13.41)$$

The quantity $e_2 - e_1$ can be evaluated from the line integral derived by integrating the expression

$$de = c_v \, dT + \left(\frac{\partial e}{\partial v}\right)_T dv \qquad (13.42)$$

from (v_1, T_1) to (v_2, T_2) along any curve. Here c_v is the molar constant volume heat capacity of the gas B in the state (v, T).

We can express $(\partial e/\partial v)_T$ for a real gas as a function of the variables (v, T) by substitution from its equation of state in the form $p = p(v, T)$ into the second-law relation (9.38). From this equation and thermal data, or by an equivalent method, the values of e have been computed at various pressures and temperatures for some real gases, and are listed in tables of thermodynamic properties.

For a perfect gas, the quantity $(\partial e/\partial v)_T$ is zero, and hence both e and h are functions of temperature alone [see eqns. (5.50)].

13.11C. Flow of a gas out of a system. Let the change in state (13.36) be brought about by the flow of a gas B out of a vessel of constant volume [7]. The rate of flow of B is such that at all times the pressure p and temperature T of the system are virtually uniform; and p, T, and all molar properties of the gas crossing the boundary are equal to the corresponding properties of the system.

Under these conditions we find that, at any stage of the process, the relations

$$E = ne, \quad dE = n\,de + e\,dn \qquad (13.43)$$

$$h = e + pv \qquad (13.44)$$

hold for the gas B in the system.

Substitutions from the above equations into (13.35) give the following expression

$$đQ - đW_S = n\,de - pv\,dn \qquad (V = \text{constant}) \qquad (13.45)$$

Substituting from eqn. (13.42) and the relations

$$v = \frac{V}{n}, \quad dv = -\frac{V}{n^2}\,dn \qquad (V = \text{constant}) \qquad (13.46)$$

into eqn. (13.45), we find that

$$đQ - đW_S = nc_v\,dT - \left[\frac{V}{n}\left(\frac{\partial e}{\partial v}\right)_T + \frac{pV}{n}\right]dn \qquad (V = \text{constant}) \qquad (13.47)$$

Let us now impose the following conditions on the above process:

1. When both Q and W_S are zero, eqn. (13.47) becomes

$$nc_V\,dT = \left[\frac{V}{n}\left(\frac{\partial e}{\partial v}\right)_T + \frac{pV}{n}\right]dn \qquad (V = \text{constant}) \qquad (13.48)$$

Equation (13.48) takes the following form for a perfect gas

$$\frac{dT}{T} = \frac{R}{c_V}\frac{dn}{n} \qquad (V = \text{constant}) \qquad (13.49)$$

When a perfect gas undergoes the change in state (13.36) by the process described above, the following relations hold for the variables p, T, and n

$$\frac{T_2}{T_1} = \left(\frac{n_2}{n_1}\right)^{R/c_V}, \frac{p_2}{p_1} = \left(\frac{n_2}{n_1}\right)^{c_p/c_V}, \frac{T_2}{T_1} = \left(\frac{p_2}{p_1}\right)^{R/c_p} \qquad (V = \text{constant}) \qquad (13.50)$$

The molar heat capacities c_V and c_p are assumed to be constant in the temperature range $T_1 - T_2$.

2. When W_S is zero and the system is in thermal equilibrium with a thermostat at a fixed temperature T, eqn. (13.47) becomes

$$đQ = -\left[\frac{V}{n}\left(\frac{\partial e}{\partial v}\right)_T + \frac{pV}{n}\right]dn \qquad (V, T = \text{constant}) \qquad (13.51)$$

For a perfect gas, eqn. (13.51) takes the form

$$đQ = -RT\, dn \qquad (V, T = \text{constant}) \qquad (13.52)$$

Integration from n_1 to n_2 moles of gas in the system gives

$$Q = RT(n_1 - n_2) \qquad (V, T = \text{constant}) \qquad (13.53)$$

13.12. The first-law steady-flow equation. At every point in an open system undergoing a steady-flow process, all intensive properties of the fluid and its direction and rate of flow are independent of time. Hence, at all times the rate of flow of mass into the system is equal to its rate of flow out. This rate, as well as the rates of flow of heat across the boundary and production of shaft work in the surroundings, are constant. Such a system is said to be in a steady state. It may or may not be in a state of internal equilibrium (see Section 3.11F). Examples of steady-state systems are the phase A of Fig. 13.1 and the steam turbine.

The first-law equation derived below for a steady-flow process has also been successfully applied to any number of complete cycles of a process in which all of the intensive properties of the system are cyclic, that is, all return simultaneously to their respective values existing at an earlier time. The reciprocating steam engine operates in such a cycle.

When a system is in a steady state, its energy E and volume V are constant. All molar properties of the fluid B_k as it enters or leaves the system in the stream k ($k = 1, 2, \ldots, \sigma$) are independent of time. Hence, integration of eqn. (13.34) over the time interval during which n_k moles of B_k flow across the boundary gives

$$Q - W_S = -\sum_k (h_k + \tfrac{1}{2} M_k u_k^2 + M_k g z_k) n_k \qquad (13.54)$$

where the summation extends over the k streams flowing into or out of the system.

In a steady-flow process, the change in state

$$\sum_i \nu_i B_i \text{ (state } i) = 0 \qquad (13.55)$$

is produced in the surroundings of an open system (see Section 13.6C). Here the notation B_i identifies a stream crossing the boundary of the system and also the pure substance or the fluid solution composing it (see Section 13.10). The coefficient ν_i is negative for a stream entering the open system (a reactant in the above change in state), and it is positive for a stream leaving the system (a product in this change in state). Hence

$$\nu_i = -n_k \qquad (i \equiv k = 1, 2, \ldots, m) \tag{13.56}$$

[see eqns. (1.10) and (13.31)].

We can now write eqn. (13.54) in the form

$$Q - W_S = \Sigma_i(h_i + \tfrac{1}{2} M_i u_i^2 + M_i g z_i)\nu_i \tag{13.57}$$

or in the more familiar form

$$Q - W_S = \Delta H + \tfrac{1}{2}\Delta(Mu^2) + g\,\Delta(Mz) \tag{13.58}$$

where, for example, ΔH is the change in enthalpy attending the change in state (13.55).

13.13. Applications of the first-law steady-flow equation to physical processes. When one mole of a fluid B in state 1 enters a steady-state system in one stream and 1 mole of the same fluid in state 2 leaves it in another stream, $B_1 = B_2 = B$ and $-\nu_1 = \nu_2 = 1$ in eqn. (13.55). Thus, the physical change in state produced in the surroundings of the system may be written in the form

$$B(\text{state 1}) = B(\text{state 2}) \tag{13.59}$$

For the change in state (13.59), eqn. (13.58) becomes

$$Q - W_S = \Delta h + \tfrac{1}{2} M \Delta u^2 + Mg\Delta z \tag{13.60}$$

We shall now apply this equation to several steady-flow processes [8].

13.13A. Adiabatic steady-flow processes. When the change in state (13.59) is brought about by an adiabatic process, $Q = 0$ in eqn. (13.60), which becomes

$$W_S = -(\Delta h + \tfrac{1}{2} M \Delta u^2 + Mg\Delta z) \tag{13.61}$$

This is the first-law equation for an adiabatic steady-flow process. We shall consider the following examples:

1. When $Q = 0$ and potential-energy effects are negligible, eqn. (13.61) becomes

$$W_S = -(\Delta h + \tfrac{1}{2} M\Delta u^2) \tag{13.62}$$

This relation holds for the process occurring in an adiabatic steam turbine.

2. When $Q = 0$, and kinetic- and potential-energy effects are negligible, eqn. (13.61) gives

$$W_S = -\Delta h \tag{13.63}$$

This relation holds for the process occurring in an adiabatic reciprocating steam engine.

3. When $Q = 0$, $W_S = 0$, and kinetic- and potential-energy effects are negligible, eqn. (13.61) becomes

$$\Delta h = 0 \tag{13.64}$$

This equation applies to the process taking place in the adiabatic porous-plug experiment in which the Joule—Thomson coefficient $(\partial T/\partial p)_H$ of a gas has been measured (see Section 5.5).

13.13B. Steady-flow heating and cooling processes. When $W_S = 0$, and kinetic- and potential-energy effects are negligible, eqn. (13.60) gives

$$Q = \Delta h \tag{13.65}$$

Besides its industrial use in the heating and cooling of fluids, this relation is used in the calculation of the thermal properties of fluids from measurements made in steady-flow calorimeters (see Section 3.24).

13.13C. The isothermal porous-plug experiment. In the pressure—temperature region in which the (adiabatic) Joule—Thomson coefficient $(\partial T/\partial p)_H$ of a gas is positive, we can measure electrically the quantity of heat Q required to maintain the temperature of a known amount of the gas constant in a steady-flow expansion through a porous plug (see Sections 5.5 and 5.7).

Since in this process $W_S = 0$, and potential- and kinetic-energy effects are negligible, eqn. (13.65) applies. From a series of such measurements, all starting from the same initial pressure, we can calculate the isothermal Joule—Thomson coefficient $(\partial H/\partial p)_T$ of a gas without a knowledge of its constant-pressure heat capacity.

13.14. Applications of the first-law steady-flow equation to chemical processes. In Section 13.6C we found that the equilibrium change in state (13.17) could, in principle, be produced in the surroundings of the open phase A of Fig. 13.1 by a steady-flow process. Application of eqn. (13.58) to this phase, subject to the conditions that $W_S = 0$ and kinetic- and potential-energy effects are negligible, shows that the change in enthalpy ΔH attending the above change in state is equal to the quantity of heat Q withdrawn from the surroundings during the process. In Section 13.6C we found the same result by applying the ordinary laws of thermodynamics to the closed, heterogeneous system of Fig. 13.1 composed of four phases at equilibrium.

The change in enthalpy attending a change in state such as (13.19) has been calculated from measurements made in a steady-flow calorimeter (see Section 6.3). In this method, each pure reactant enters an open system in a separate stream, and the mixture of the products and unused reactant, if any, leaves the system in a single stream. In general, the reactants and the mixture of the products have the same pressure, but different temperatures.

From eqn. (13.57) we find that ΔH for the change in state (13.55) is equal to the heat Q withdrawn from the surroundings during the process.

In order to find ΔH for the desired change in state (13.19) we must apply corrections to the above value for: (1) The difference between the temperatures of the reactants and the products; and (2) the change in enthalpy attending the mixing of the products.

13.15. The second-law equation for open systems. The second-law equation for a *reversible process* in a closed, homogeneous system is

$$dS = \frac{dQ_R}{T} \tag{13.66}$$

where S and T are the entropy and temperature of the system, respectively, and Q_R is the heat withdrawn from the surroundings during the process.

Gillespie and Coe [7] derived the second-law equation for the corresponding open system by a process similar to that used to derive the first-law expression (see Section 13.10), except that in the present case the process must be reversible. This will require the use of one or more semipermeable membranes for some processes.

We shall use the notation B_k ($k = 1, 2, \ldots, \sigma$) to identify both a stream crossing the boundary of an open system and the substance composing it.

Let a piston and cylinder device containing dn_k moles of a fluid B_k in a state (p_k, T) communicate with an open phase A by way of a valve or a semipermeable membrane (Fig. 13.2). The pressure p and temperature T of the phase A are uniform at all times. We shall now carry out the following two-step process in which B_k flows reversibly across the boundary of A.

1. Imagine that the phase A and the dn_k moles of the fluid B_k in the cylinder are enclosed in an impermeable boundary. The entropy of the closed, composite system A + dn_k is larger than the entropy of A by the amount

$$dS_1 = s_k \, dn_k \tag{13.67}$$

where s_k is the molar entropy of B_k in the state (p_k, T).

2. With both Q and W_S equal to zero, push the dn_k moles of B_k reversibly across the boundary of A. The corresponding change in entropy dS_2 of A in this step of the process is

$$dS_2 = 0 \tag{13.68}$$

The total change in entropy dS of the phase A attending the reversible addition of dn_k moles of B_k is

$$dS = dS_1 + dS_2 = s_k \, dn_k \tag{13.69}$$

When a substance B_k flows reversibly across the boundary of a phase composed entirely of B_k, it will have the same pressure and temperature as the phase. Hence its molar entropy s_k is equal to the molar entropy of the fluid composing the phase.

When an open phase is a gaseous or liquid solution, the condition of reversibility requires a pure fluid constituent B_k to enter or leave it through a semipermeable membrane (see Fig. 13.1). Under these conditions, the pressure p_k, temperature T, and hence the molar entropy s_k of the fluid B_k crossing the boundary are fixed by the pressure, temperature, and composition of the fluid in the open phase.

When several substances flow reversibly across the boundary of an open system, eqn. (13.69) gives

$$dS = \sum_k s_k \, dn_k \tag{13.70}$$

where the summation extends over the m substances flowing into or out of the system. Here dn_k is positive when B_k enters the system and negative when it leaves the system.

From eqns. (13.66) and (13.70), we find that the change in entropy dS attending the reversible flow of both heat and matter across the boundary of a system is given by the relation

$$dS = \frac{đQ_R}{T} + \sum_k s_k \, dn_k \tag{13.71}$$

This is the second-law equation for an open system. It may be written in the form

$$đQ_R = T\,dS - T\sum_k s_k \, dn_k \tag{13.72}$$

If the process is irreversible, the above equation becomes

$$\frac{đQ}{T} < dS - \sum_k s_k \, dn_k \tag{13.73}$$

where T is the temperature of the part of the system into which the quantity of heat $đQ$ flows from the surroundings (see Sections 8.13 and 8.15).

13.16. An application of the second-law equation for open systems. The heat effect Q attending the isothermal flow of a gas (CO_2) from an open homogeneous system which is in thermal equilibrium with an ice calorimeter (Fig. 3.14) has been measured by Gillespie and Coe [7]. The apparatus consisted of a vessel of constant volume connected by a capillary to a reducing valve which controlled the rate of flow of the gas to the surroundings. The system at any time was the gas currently in the volume V composed of the volumes of the vessel and a short length of the contiguous capillary, which were immersed in the ice calorimeter.

The change in state of the system produced by the above process is

$$n_1 CO_2(g, p_1) = n_2 CO_2(g, p_2) \qquad (273.15 \text{ K}) \tag{13.74}$$

Here n_1 and n_2 are the numbers of moles of CO_2 calculated from an equation of state of CO_2 for the conditions (p_1, V, 273.15 K) and (p_2, V, 273.15 K).

In a series of seven "slow expansions", in which the times of escape of CO_2 from the system for the same values of p_1 and p_2 varied ninefold, the quantities of heat Q withdrawn from the ice calorimeter were found to agree within experimental error. Also, in these runs the ratios of the measured values of Q to those calculated from an equation derived from (13.72) varied randomly from 0.975 to 1.021. These results indicate that for slow rates of flow of CO_2 out of the system the above expansions "simulate a reversible process" [7]. Hence in these runs both the pressure p and temperature T (273.15 K) of the system were virtually uniform throughout the process; and the gas leaving the system at any instant was virtually in the state $(p, 273.15\,\text{K})$.

At higher rates of expansion, the ratio of the observed to the calculated value of Q decreased steadily with increase in the rate of flow, which indicates that these processes were irreversible.

We shall now derive the second-law equation for the above reversible process. Since the pressure of the system is virtually uniform throughout the experiment, the equations

$$S = ns, \quad \mathrm{d}S = n\mathrm{d}s + s\mathrm{d}n \tag{13.75}$$

hold at all times. Here S is the total and s is the molar entropy of the gas in the system. Substitution into eqn. (13.72) gives

$$\mathrm{d}Q_R = nT\mathrm{d}s \qquad (V, T = \text{constant}) \tag{13.76}$$

The molar entropy of a gas may be expressed in terms of the independent variables (p, T) or (v, T) where v is its molar volume. Hence, at constant temperature eqn. (13.76) may be written in the following forms

$$\mathrm{d}Q_R = nT\left(\frac{\partial s}{\partial p}\right)_T \mathrm{d}p = nT\left(\frac{\partial s}{\partial v}\right)_T \mathrm{d}v \qquad (V, T = \text{constant}) \tag{13.77}$$

In view of the Maxwell relations (9.22) and (9.23), these equations become

$$\mathrm{d}Q_R = -nT\left(\frac{\partial v}{\partial T}\right)_p \mathrm{d}p \qquad (V, T = \text{constant}) \tag{13.78}$$

$$\mathrm{d}Q_R = nT\left(\frac{\partial p}{\partial T}\right)_v \mathrm{d}v \qquad (V, T = \text{constant}) \tag{13.79}$$

The latter is the more useful relation, since equations of state of gases are generally in the form $p = p(v, T) = p(V/n, T)$ where n is the number of moles of gas in the volume V.

Substitution from eqn. (13.46) into (13.79) gives

$$\mathrm{d}Q_R = -T\frac{V}{n}\left(\frac{\partial p}{\partial T}\right)_v \mathrm{d}n \qquad (V, T = \text{constant}) \tag{13.80}$$

We note that this same equation is obtained by replacing the partial derivative $(\partial e/\partial v)_T$ in the first-law equation (13.51) by its value $T(\partial p/\partial T)_V - p$ given by the second-law equation (9.38).

Integration of eqn. (13.80) from n_1 to n_2 gives the following relation for the heat Q_R withdrawn from the surroundings when the change in state (13.74) is produced by the reversible, isothermal flow of a gas out of a vessel of constant volume

$$Q_R = -T \int_{n_1}^{n_2} \frac{V}{n} \left(\frac{\partial p}{\partial T}\right)_v dn \qquad (V, T = \text{constant}) \qquad (13.81)$$

When a perfect gas undergoes the above process, eqn. (13.81) becomes

$$Q_R = RT(n_1 - n_2) \qquad (V, T = \text{constant}) \qquad (13.82)$$

[see eqn. (13.53)].

When the above process is carried out with a real gas, we can calculate Q_R by substituting from its equation of state into eqn. (13.81). Values of Q_R so obtained were designated "calculated" in the above discussion of the experimental results.

13.17. The second-law steady-flow equation. The entropy S of an open system undergoing a reversible, steady-flow process is constant (compare Section 13.12). Hence, integration of eqn. (13.71) over the time interval during which n_k moles of B_k flow across the boundary of the system gives the following relation for the quantity of heat Q_R withdrawn from the surroundings

$$Q_R = -T\Sigma_k s_k n_k \qquad (13.83)$$

When the change in state (13.19) is brought about in the surroundings of an open system at equilibrium, eqn. (13.83) becomes

$$Q_R = T\Sigma_i \nu_i s_i = T\Delta S \qquad (13.84)$$

[see eqn. (13.56)]. This is the same relation we found in Section 13.6C.

13.18. The combined first- and second-law equation for open, simple phases. The first-law equation (13.32) for the change in energy attending a reversible process in an open, simple phase, may be written in the form

$$dE = đQ_R - pdV + \Sigma_i(e_i + p_i v_i)\, dn_i \qquad (13.85)$$

Substitution from the corresponding second-law equation (13.72) gives

$$dE = TdS - pdV + \Sigma(e_i - Ts_i + p_i v_i)\, dn_i \qquad (13.86)$$

In the above equations, the summations may be extended over the c components or the m species of the phase (see Section 13.8).

Comparison of eqn. (13.86) with (13.5) shows that for each species of an open phase at equilibrium

$$\mu_i = e_i - Ts_i + p_i v_i \qquad (13.87)$$

where μ_i is the chemical potential of B_i in the phase. Here e_i, s_i, and v_i are the molar energy, entropy, and volume, respectively, of the substance B_i as it flows across the boundary of the phase in a reversible, adiabatic process in which W_X is zero.

13.19. Fundamental equations. In Section 13.1 we noted that the energy E of an open, simple phase composed of c components may be expressed as a function of the extensive quantities S, V, $n_1, n_2, \ldots, n_c$. The partial derivatives of E with respect to these variables are the intensive quantities T, $-p$, μ_1, $\mu_2, \ldots, \mu_c$, respectively. Among these $2c + 5$ quantities there are $c + 3$ independent relations (13.5)–(13.7). Hence $c + 2$ of the variables are independent.

An equation from which we may deduce for any phase all of the relations subsisting among the above $2c + 5$ quantities was called by Gibbs a "fundamental equation" [9].

The total differential of the fundamental energy equation

$$E = E(S, V, n_1, n_2, \ldots, n_c) \tag{13.4}$$

is

$$dE = TdS - pdV + \sum_i \mu_i dn_i \tag{13.5}$$

This relation is homogeneous of the first degree in mass. Integration by Euler's theorem on homogeneous functions gives

$$E = TS - pV + \sum_i \mu_i n_i \tag{13.88}$$

Differentiation of this equation, and substitution from (13.5) gives

$$SdT - Vdp + \sum_i n_i d\mu_i = 0 \tag{13.89}$$

which is the Gibbs–Duhem relation.

In thermodynamics we have need of fundamental equations which have different sets of independent variables from that of the energy equation (13.4). The Legendre transforms [10] of this relation give a group of such functions. In each, the independent variable is replaced by the conjugate dependent variable in one or more terms of the right-hand side of eqn. (13.5).

We can carry out $2^{c+2} - 1$ Legendre transforms of eqn. (13.4) by subtracting from E all possible combinations of the $c + 2$ products

$$(TS), (-pV), (\mu_1 n_1), (\mu_2 n_2), \ldots, (\mu_c n_c) \tag{13.90}$$

taken 1 at a time, 2 at a time, $\ldots, c + 2$ at a time. Each of these relations is entirely equivalent to the energy expression (13.4) and hence each is a fundamental equation. When the energy relation is included, there are 2^{c+2} functions related by Legendre transformations.

Inversely, we can recover eqn. (13.4) by adding the proper product or products of the set (13.90) to any of the above Legendre transformations.

The quantities H, A, and G (Sections 4.8 and 10.1) are partial Legendre transforms of E that replace the independent variables V by p, S by T, and both S by T and V by p, respectively. The corresponding complete transform of E, denoted here by ϕ, is zero by eqn. (13.88). These four relations are

$$H = E + pV \tag{13.91}$$
$$A = E - TS \tag{13.92}$$
$$G = E - TS + pV \tag{13.93}$$
$$\phi = E - TS + pV - \sum_i \mu_i n_i = 0 \tag{13.94}$$

Substituting from eqn. (13.88), into eqns. (13.91)–(13.94), we find

$$H = TS + \sum_i \mu_i n_i \tag{13.95}$$
$$A = -pV + \sum_i \mu_i n_i \tag{13.96}$$
$$G = \sum_i \mu_i n_i \tag{13.97}$$
$$\phi = 0 \tag{13.98}$$

(see Section 10.7).

Differentiation of eqns. (13.91)–(13.94) and substitution from (13.5) give

$$dH = TdS + Vdp + \sum_i \mu_i dn_i \tag{13.99}$$
$$dA = -SdT - pdV + \sum_i \mu_i dn_i \tag{13.100}$$
$$dG = -SdT + Vdp + \sum_i \mu_i dn_i \tag{13.101}$$
$$d\phi = -SdT + Vdp - \sum_i n_i d\mu_i = 0 \tag{13.102}$$

These relations may also be derived by differentiating eqns. (13.95)–(13.98) and subtracting (13.89) from each result.

All of the following relations are fundamental equations of an open, simple phase

$$E = E(S, V, n_1, n_2, \ldots, n_c) \tag{13.4}$$
$$H = H(S, p, n_1, n_2, \ldots, n_c) \tag{13.103}$$
$$A = A(T, V, n_1, n_2, \ldots, n_c) \tag{13.104}$$
$$G = G(T, p, n_1, n_2, \ldots, n_c) \tag{13.105}$$
$$\phi = \phi(T, p, \mu_1, \mu_2, \ldots, \mu_c) = 0 \tag{13.106}$$

We note that all of the independent variables of the energy equation are extensive properties of a phase, and all of the independent variables of the function ϕ are intensive properties. We may write the latter expression in the form

$$dp = \frac{S}{V} dT + \sum_i \frac{n_i}{V} d\mu_i \tag{13.107}$$

The $c + 1$ quantities derived from this equation by differentiation and eqns. (13.106) and (13.88), give $c + 3$ relations among the $2c + 5$ variables of the energy equation (13.4). Hence $c + 2$ of the latter are independent.

Now a relation among the quantities

$$E, S, V, n_1, n_2, \ldots, n_c \tag{13.108}$$

is a fundamental equation, but one among

$$E, T, V, n_1, n_2, \ldots, n_c \tag{13.109}$$

is not. Since $(\partial E/\partial S)_{Vn} = T$, we can derive the second equation from the first, but we cannot derive the first from the second.

An equation among the quantities (13.109) is entirely equivalent to one among

$$\left(\frac{\partial S}{\partial E}\right)_{Vn}, E, V, n_1, n_2, \ldots, n_c \tag{13.110}$$

We cannot express S completely in terms of the other variables of this equation since the necessary integration introduces an arbitrary function of $V, n_1, n_2, \ldots, n_c$. Hence this relation and (13.109) are not fundamental equations.

The fundamental energy equation of an open phase which produces ω different kinds of reversible, non-expansion work in its surroundings is

$$E = E(S, V, X_1, X_2, \ldots, X_\omega, n_1, n_2, \ldots, n_c) \tag{13.111}$$

Its total differential is eqn. (13.23).

We can carry out $2^{c+\omega+2} - 1$ Legendre transforms of E by subtracting from eqn. (13.111) all possible combinations of the products

$$(TS), (-pV), (-y_1 X_1), (-y_2 X_2), \ldots, (-y_\omega X_\omega), (\mu_1 n_1), (\mu_2 n_2), \ldots, (\mu_c n_c) \tag{13.112}$$

taken 1 at a time, 2 at a time, ..., $c + \omega + 2$ at a time [see eqns. (10.178) and (13.90)]. All of these transformations are fundamental equations.

Massieu has shown that all of the thermodynamic properties of a closed, simple fluid may be derived from a single "characteristic function" (see Section 10.1).

From eqn. (13.4) we find that the fundamental equation for the entropy of an open, simple phase is

$$S = S(E, V, n_1, n_2, \ldots, n_c) \tag{13.113}$$

and, by (13.5), its total differential is

$$dS = \frac{1}{T}dE + \frac{p}{T}dV - \sum_i \frac{\mu_i}{T} dn_i \tag{13.114}$$

Since this equation is homogeneous of the first degree in mass, its integral, by Euler's theorem, is

$$S = \frac{1}{T}E + \frac{p}{T}V - \sum_i \frac{\mu_i}{T} n_i \tag{13.115}$$

[see eqn. (13.88)].

The $2^{c+2}-1$ Legendre transforms of S are carried out by subtracting from eqn. (13.113) all possible combinations of the products

$$\left(\frac{1}{T}E\right), \left(\frac{p}{T}V\right), \left(-\frac{\mu_1}{T}n_1\right), \left(-\frac{\mu_2}{T}n_2\right), \ldots, \left(-\frac{\mu_c}{T}n_c\right) \tag{13.116}$$

taken 1 at a time, 2 at a time, ..., $c+2$ at a time. All are fundamental equations.

Three of the partial Legendre transforms of S, denoted here by S', S'', and S''', are

$$S' = S - \frac{1}{T}E = -\frac{A}{T} \tag{13.117}$$

$$S'' = S - \frac{p}{T}V = \frac{1}{T}\left(E - \sum_i \mu_i n_i\right) \tag{13.118}$$

$$S''' = S - \frac{1}{T}E - \frac{p}{T}V = -\frac{G}{T} \tag{13.119}$$

where A and G are given [10] by eqns. (13.96) and (13.97).

REFERENCES

1 J. W. Gibbs, Scientific Papers, Longmans, Green and Co., New York, 1906, Vol. 1, Thermodynamics, pp. 62—64.
2 J. W. Gibbs, ref. 1, pp. 92—96: see especially p. 93.
3 J. W. Gibbs, ref. 1, pp. 61, 222.
4 J. W. Gibbs, ref. 1, pp. 64—67, 135—138.
5 Compare J. W. Gibbs, ref. 1, pp. 67—70.
6 J. H. van't Hoff, Z. Phys. Chem., 1 (1887) 481—508: see especially pp. 498—500.
7 L. J. Gillespie and J. R. Coe, Jr., J. Chem. Phys., 1 (1933) 103—113.
8 See, for example, H. C. Weber and H. P. Meissner, Thermodynamics for Chemical Engineers, Wiley, New York, 1957, p. 72.
9 J. W. Gibbs, ref. 1, pp. 85—89.
10 See H. B. Callen, Thermodynamics, Wiley, New York, 1960, pp. 90—102.

INDEX OF NAMES

Abraham, B. M., 245, 271
Ahlberg, J. E., 259, 260, 272
Alcock, C. B., 79, 97
Anderson, R. L., 34, 46
Arnett, R. L., 137, 142

Bailey, S. M., 137, 142
Barth, J. A., 233, 270
Bartholomé, E., 267, 273
Baum, J. L., 245, 271
Beattie, J. A., 34, 37, 39, 40, 46, 47, 185, 196, 248, 258, 272
Beenakker, J. J. M., 259, 272
Bekkedahl, N., 87, 98
Benedict, M., 34, 37, 39, 46, 47, 185, 196
Bentz, D. R., 247, 250, 271, 272
Berry, R. J., 4, 23
Berthelot, D., 194, 196, 233, 270
Bever, M. B., 83, 98
Black, J., 99, 104, 114
Blaisdell, E. B., 34, 39, 46
Blanchard, E. R., 259, 272
Blue, R. W., 262, 273
Bockhoff, F. J., 263, 273
Boltzmann, L., 236, 271
Bonhoeffer, K. F., 265, 273
Borchers, H., 134, 142
Braun, R. M., 137, 142
Brewer, D. F., 245, 259, 271, 272
Brewer, L., 147, 156, 252, 255, 272
Brickwedde, F. G., 35, 47, 83, 87, 98
Bridgman, P. T., 211, 231
Brodale, G., 263, 273
Brönsted, J. N., 254, 272
Browne, W. R., 158, 177
Bryan, G. H., 167, 177
Buchdahl, H. A., 158, 177
Buckingham, E., 185, 196
Burgess, F. K., 31, 46

Callen, H. B., 228, 231, 317, 320
Carathéodory, C., 158, 177
Carnot, S., 101, 103, 114, 157, 169, 176
Chandler, D., 239, 271
Chapin, D. S., 265, 273
Clapeyron, E., 101, 114, 157, 177
Clausius, R., 102, 107, 114, 158, 177

Clayton, J. O., 251, 262, 272, 273
Clement, J. R., 35, 47
Clusius, K., 267, 273
Coe, J. R., 63, 97, 305, 309, 313, 314, 315, 320
Cohen, E. G. D., 259, 272
Cohen, E. R., 43, 47
Collins, S. C., 127, 130
Colwell, J. H., 263, 268, 273
Comeford, J. J., 263, 273
Coops, J., 83, 97
Corak, W. S., 250, 272
Coughlin, J. P., 137, 142
Coulter, L. V., 263, 273
Cunningham, C. M., 265, 273

Das, P., 244, 271
Daunt, J. G., 245, 271
Davidson, N., 252, 255, 272
Davy, H., 100, 114
Debye, P., 247, 272
de Donder, Th., 18, 23
Defay, R., 18, 21, 23
Dennison, D. M., 264, 273
Dickinson, H. C., 80, 97
van Dijk, H., 35, 47
Douglas, T. B., 85, 86, 98
Dugdale, J. S., 246, 271
Dulmage, W. J., 262, 273
Dulong, P. L., 236, 271
Du Mond, W. M., 43, 47
Durieux, M., 35, 47

Eastman, E. D., 234, 238, 239, 254, 257, 271, 272
Edsinger, R. E., 34, 46, 47
Edwards, D. O., 245, 259, 271, 272
Egan, C. J., 83, 98
Einstein, A., 246, 271
Epstein, P. S., 235, 271
Eucken, A., 79, 97
Evans, E. Ll., 79, 97
Evans, W. H., 137, 142

Fairbank, H. A., 246, 271.
Fairbank, W. M., 259, 272
Farkas, A., 265, 273
Finke, H. L., 263, 273

Flannagan, G. N., 184, 196
Flood, H., 211, 231
Fowler, R. H., 239, 264, 271, 273
Franck, J. P., 246, 271
Friaf, J. B., 142
Furukawa, G. T., 85, 86, 98

Gaines, J. M., 39, 47
Garfunkel, M. P., 250, 272
Gerry, H. T., 217, 231
Giauque, W. F., 83, 98, 251, 254, 259, 262, 263, 265, 267, 270, 272, 273, 274
Gibbs, J. W., 2, 3, 5, 10, 12, 13, 15, 18, 23, 49, 97, 190, 192, 196, 197, 200, 201, 203, 205, 213, 231, 256, 272, 275, 277, 282, 283, 284, 285, 288, 291, 294, 296, 298, 299, 300, 301, 303, 317, 320
Gibson, G. E., 259, 261, 270, 272, 273
Gill, E. K., 263, 268, 273
Gillespie, L. J., 63, 97, 189, 196, 217, 231, 305, 309, 313, 314, 315, 320
Ginnings, D. C., 85, 86, 87, 98
Gonzalez, D. O., 268, 273
Gopal, E. S. R., 74, 97, 246, 247, 248, 271
Gorter, C. J., 35, 47
Graham, G. M., 245, 271
Green, C. B., 80, 81, 97
Grenier, G., 268, 273
Grilly, E. R., 244, 245, 246, 271
Gross, M. E., 263, 273
Guggenheim, E. A., 239, 271
Guildner, L. A., 34, 46, 47

Haber, F., 234, 270
Hadley, W. B., 83, 98
Hatcher, J. B., 83, 98
Hallet, A. C. Hollis, 245, 271
Halow, I., 137, 142
Harteck, P., 265, 273
Hatsopoulis, G. N., 48, 97, 172, 177, 275, 294, 295
Hausen, H., 137, 142
Heisenberg, W., 264, 273
Heitler, W., 270, 274
Hellwege, K. H., 137, 142
Helmholtz, H., 101, 114
Henning, F., 261, 272
Herzberg, G., 265, 273
Heuse, W., 87, 98
Hill, R. W., 264, 267, 273
Hilsenrath, J., 247, 272
Hinkley, R. B., 217, 231
van't Hoff, 302, 320
Hori, T., 264, 273

Howlett, B. W., 83, 98
Hubbard, W. N., 83, 97, 135, 142
Huffman, H. M., 83, 87, 98
Hund, F., 264, 273

Jacobus, D. D., 39, 47
Jaffee, I., 137, 142
Jessup, R. S., 80, 81, 97
Johnston, H. L., 262, 265, 267, 268, 273
Jones, G. O., 259, 260, 261, 272
Jones, G. W., 142
Jones, W. M., 268, 274
Jost, W., 242, 244, 271
Joule, J. P., 89, 98, 102, 114, 119, 122, 130
Justice, J. L., 184, 196

Karo, A. M., 247, 271
Kaye, J., 34, 37, 47, 185, 196
Keenan, J. H., 48, 66, 97, 172, 177, 203, 231, 275, 294, 295
Keesom, W. H., 11, 12, 23, 40, 47, 244, 245, 271
Kelley, K. K., 137, 142, 254, 272
Kemp, J. D., 255, 263, 272, 273
Kelvin (Lord Kelvin) — see W. Thomson
Keyes, F. G., 35, 47, 120, 121, 125, 127, 130, 185, 196, 213, 231
King, E. G., 137, 142
Kirkwood, J. G., 159, 177
Kistemaker, J., 40, 47
Kistiakowsky, G. B., 183, 196
Kittel, C., 249, 268, 272, 274
Kivelson, D., 59, 97
Kleppa, O. J., 211, 231
deKlerk, D., 35, 47
Kobe, K. A., 211, 231
Koehler, J. K., 263, 273
Koenigsberger, L., 114
Kohnstamm, Ph., 11, 23
Kopp, H., 236, 271
Kostkowski, H. J., 35, 47
Kubaschewski, O., 79, 97

Lamb, A. B., 234, 270
Lange, E., 83, 98
Lange, F., 254, 259, 261, 264, 266, 272, 273
Latimer, W. M., 137, 142, 261, 272
Lax, E., 247, 248, 271
Leach, J. S. Ll., 83, 98
Le Chatelier, H. M., 233, 270
Lee, R. D., 35, 47
van Leeuwen, J. M. J., 259, 272

LePair, C., 244, 271
Levine, S., 137, 142
Lewis, B., 142
Lewis, G. N., 147, 156, 218, 231, 238, 252, 255, 271, 272
Lide, D. R., Jr., 263, 273
Lindemann, F. A., 249, 272
Lipscomb, W. N., 262, 273
Logan, J. K., 35, 47
Long, E. A., 263, 267, 273
Lundberg, W. O., 259, 272
Lurie, E., 189, 196, 217, 231
Lynn, R. E., Jr., 211, 231

McCoskey, R. E., 85, 86, 98
McCullough, J. P., 263, 273
MacDougall, F. H., 247, 248, 271, 272
McGavock, W. C., 254, 272
McWilliams, A., 245, 271
Magie, W. F., 158, 177
Mann, D. E., 263, 273
Maple, T.L.G., 247, 249, 271
Massieu, M. F., 198, 231
Mayer, J. E., 252, 255, 257, 272
Mayer, J. R., 101, 114
Mayer, M. G., 252, 255, 257, 272
Meissner, H. P., 311, 320
Mendelssohn, K., 264, 267, 273
Menzies, A. W. C., 211, 231
Messerly, J. F., 263, 273
Meyers, C. H., 87, 98
Meyers, E. A., 262, 273
Miller, E. E., 123, 130
Mills, R. L., 244, 245, 246, 271
Milner, R. T., 257, 272
Morrison, J. A., 263, 268, 273
Morrow, J. C., 247, 249, 271
Moser, H., 40, 47
Moussa, M. R., 35, 47
Muijlwijk, R., 35, 47
Murrell, T. A., 123, 130, 185, 196

Nernst, W., 234, 236, 237, 238, 239, 241, 249, 270, 271, 272
van Nes, K., 83, 97
Neumann, F. E., 236, 271
Newman, E. S., 83, 98
Newton, R. F., 259, 260, 272

Oblad, A. G., 259, 260, 272
Ogg, A., 158, 177, 237, 271
Olovsson, I., 262, 263, 273
Onnes, H. K., 11, 12, 23
Oppenheim, I., 59, 97, 159, 177, 239, 271

Orehotsky, R. S., 75, 97
Osborne, D. W., 83, 87, 98, 245, 271
Osterberg, H., 124, 130
Otto, J., 40, 47
Ouboter, R. DeBruyn, 244, 259, 271, 272

Pace, E. L., 263, 273
Parker, V. B., 137, 142
Parks, G. S., 261, 272
Partington, J. R., 79, 97
Pauling, L., 263, 266, 273
Pernott, G. St. J., 142
Petit, A. T., 236, 271
Petrella, R. V., 263, 273
Phillips, H. B., 170, 177
Pielemier, W. H., 183, 196
Pimentel, G. C., 137, 142
Pitzer, K. S., 137, 142, 147, 156, 252, 255, 263, 272, 273
Planck, M., 158, 177, 237, 271
Porter, A. W., 289, 295
Potter, R. L., 247, 249, 271
Prigogine, I., 18, 21, 23
Purcell, E. M., 266, 273

Quinn, T. J., 34, 47

Randall, M., 147, 156, 218, 231, 238, 252, 255, 271, 272
Rands, R. D., 87, 98
Regnault, H. V., 102, 114, 158, 177
Reid, R. C., 217, 231
Reif, F., 266, 273
Rhuemann, M., 264, 267, 273
Richards, T. W., 234, 236, 270
Ricketson, B. W. A., 264, 267, 273
Robie, R. A., 137, 142
Robinson, A. L., 83, 98
Roebuck, J. R., 121, 123, 124, 130, 185, 196
Rosenstein, R. D., 262, 263, 273
Rossini, F. D., 135, 137, 142
Rowlinson, J. S., 12, 23
Ruben, H. W., 262, 263, 273
Rumford, 100, 114

Sackur, O., 252, 272
Satterthwaite, C. B., 250, 272
Schafer, K., 137, 142, 247, 248, 271
Scheel, K., 87, 98
Schmidt, E., 137, 142
Schumm, R. H., 137, 142
Scott, D. W., 83, 97, 135, 142
Scott, R. B., 87, 98

Sears, F. W., 120, 130
Seligman, P., 259, 272
Shilling, W. G., 79, 97
Simon, F., 234, 239, 241, 244, 256, 259, 260, 261, 264, 266, 267, 271, 272, 273
Skertic, M., 259, 272
Skinner, H. A., 135, 142
Sligh, T. S., 87, 98
Smith, D., 75, 97
Smith, H. S., 114
Southard, J. S., 83, 85, 98
Spencer, H. M., 184, 196
Steenland, M. J., 35, 47
Stephenson, C. C., 75, 97, 247, 249, 250, 254, 271, 272
Stevenson, D. A., 247, 249, 250, 271, 272
Stimson, H. F., 4, 23, 30, 31, 34, 46, 70, 87, 97, 98
Stout, J. W., 83, 98, 263, 273
Stull, D. R., 137, 142, 220, 231
Swenson, C. A., 244, 271

Taconis, K. W., 244, 271
Tammann, G., 211, 231
Telfair, D., 183, 196
Teller, E., 255, 272
Templeton, D. H., 262, 263, 273
Tetrode, H., 252, 272
Thomas, W., 34, 40, 47, 157, 177
Thomsen, J., 233, 270
Thomson, B. — see Rumford
Thomson, W., (Lord Kelvin), 102, 103, 114, 122, 130, 169, 177

Ticknor, L. B., 83, 98
Tisza, L., 10, 23
Todd, S. S., 87, 98
Topley, B., 255, 272

Vignos, J. H., 246, 271
van der Waals, J. D., 11, 23

Waddington, G., 83, 87, 97, 98, 135, 142, 263, 273
Wagman, D. D., 137, 142
Waldbaum, D. R., 137, 142
Walters, G. K., 259, 272
Washburn, E. M., 83, 97, 135, 137, 142, 211, 231
Weber, H. C., 311, 320
Weinstock, B., 245, 271
Wells, L. S., 83, 98
West, E. D., 254, 272
Westrum, E. F., 83, 98
Wexler, A., 250, 272
White, D., 268, 273
Wilks, J., 242, 244, 245, 246, 259, 260, 271, 272
Wilson, A. H., 231, 258, 272
Wilson, E. B., 255, 272
Witt, R. K., 255, 272

Yaqub, M., 259, 272

Ziegler, G. G., 247, 272

SUBJECT INDEX

Absolute temperature, 29, 30
Adiabatic boundary, 27
Adiabatic processes, 53, 54
   reversible in perfect gases, 188
Aggregation states, 10—12
Apparent molar properties,
   enthalpy, 152
   volume, 147
Available energy, 203

Berthelot's equation of state, 194
Boundaries,
   definition, 1
   location of, 50, 61, 62, 89
Boyle's law, 41

Caloric theory, 99
Calorie, 69
Calorimeter, 79
   bomb, 80—83, 135
   flow, 87—89
   high-temperature, 85—87
   low-temperature, 83—85
   open, 83, 135
Calorimetric coefficients, 94, 95, 181
Calorimetry, 79
Carathéodory's principle, 159
Carathéodory's theorem, 158
Carnot cycle, 162
   efficiency of, 164
Carnot function, 169
Celsius temperature scale, 31
Change in state, 6—8
   path of, 8
Chemical equilibrium, criterion for, 224, 303
Chemical potentials, 205
   definition, 206
   for chemical reactions in equilibrium, 303
   of components and species, 298
Chemical reactions, 6—8, 17, 131, 301—303
   criterion for equilibrium, 224, 303
Chemical units of mass, 14
Clapeyron equation, 212
   fusion and transition curves, 215
   sublimation and vaporization curves, 214

Clausius—Clapeyron equation, 215
Clausius inequality, 170
Clausius' statement of the second law, 158
Closed system, 2
Coefficient,
   calorimetric, 94, 95, 181
   of adiabatic compressibility, 97
   of isothermal compressibility, 97
   of thermal expansion, 96
   of thermal pressure increase, 96
Coefficient of performance of a Carnot engine, 164
Components,
   actual, 299
   composition in terms of, 17—23
   definition, 14
   number of, 15—17
   of a phase, 297
   possible, 299
Compounds, 13
Conservation of energy, 107
Criteria of equilibrium,
   energy criterion for an isolated system, 274
   enthalpy criteria, 290—292
   entropy criterion for an isolated system, 275
   equivalence of entropy and energy criteria, 276
   for chemical changes in state, 224
   for system with uniform temperature, 200, 201
   for system with uniform temperature and pressure, 203
   free-energy criteria, 292—294
   heterogeneous equilibrium
      uniformity of pressure, 285
      uniformity of temperature, 284
   metastable equilibrium, 279
   necessity of entropy and energy criteria, 283, 284
   neutral equilibrium, 279
   stable equilibrium, 278
   sufficiency of entropy and energy criteria, 280—283
   unstable equilibrium, 279
   work-content criteria, 288—290

Cycle,
  definition, 8
  reversible, 9

Debye's heat-capacity equation, 247
Diathermal boundary, 26
Differential enthalpy of dilution, 150
Differential enthalpy of transfer, 151
Dissipative effects, 50
Dissolved state, 134

Efficiency of a Carnot engine, 164
Einstein's heat-capacity equation, 246
Elastic coefficients, 96, 97
  adiabatic compressibility, 97
  isothermal compressibility, 97
Elements, 13
Energy,
  conservation of, 101, 107
  definition, 105
  effect of volume, 115
  of reaction, 131
  properties of, 108
  units of, 108
Enthalpy,
  definition, 109
  effect of pressure, 115
  of reaction, 131
    calculation of, 137
    effects of pressure and temperature, 138—141
  of solution, 149—152
  relative molar, 155
  standard molar, 153, 219, 225
Entropy,
  calorimetric, 250, 255
  changes attending chemical changes in state, 194
  changes for finite changes in state, 181
  consequences of third law, 240
  definition in a closed system, 172
  isentropic processes, 180
  of glasses at 0 K, 258
  of isotope mixing, 269
  of liquid solutions of ^{4}He and ^{3}He below 1 K, 258
  of mixing, 190, 239
  of nuclear spins, 270
  of paramagnetic substances, 268
  of phase transformations, 184
  of polymorphic forms at 0 K, 253
  of solid solutions at 0 K, 257
  of solids with configurational disorder at 0 K, 261

  principle of increase in, 174, 175
  properties, 175, 176
  relation to heat effects, 174
  spectroscopic, 252, 254
  standard molar, 193, 225
  variations with temperature, 179
  variations with volume and pressure, 180
Equations of state, 53
  perfect gas, 42
  virial equations, 44, 45
Equilibrium,
  chemical — see Chemical equilibrium
  thermodynamic — see Thermodynamic equilibrium
Equilibrium pressure, 189, 216
Euler's theorem on homogeneous functions, 144
Expansion work, 52
  equations for, 63—67
Explosion temperatures, 141
Extensive properties,
  definition, 5
Extent of reaction, 18
Extra-thermodynamic laws, 13

First law of thermodynamics, 105
  applications of, 115
  for closed systems, 110—114
  for open systems, 296, 305—307
  for systems with flow, 305—312
Flame temperatures, 141
Free energy (Gibbs function, thermodynamic potential), 197
  changes attending chemical changes in state, 222
  changes attending finite changes in state, 207—210
  definition, 197
  for perfect gases, 218
  partial molar, 205
  relation to work of certain processes, 201
  standard molar free energy of a pure substance, 219, 225
  variation with pressure and temperature, 204
Fundamental equations, 317

Gas,
  equations of state, 42, 44, 45
  Joule coefficient, 120
  Joule—Thomson coefficient, 124
  perfect — see Perfect gas
  standard molar enthalpy, 219
  standard molar entropy, 193

standard molar free energy, 219
standard state, 133
work of expansion, 64
Gas constant, 43
Gas mixtures,
  equation of state, 53
  perfect, 189
Gas thermometer, 36, 37, 40, 185
Gay-Lussac's law, 41
Gibbs, function — see Free energy
Gibbs—Duhem relation, 317
Gibbs—Helmholtz equations, 205, 230
Gibbs paradox, 192

Heat, 48, 67, 89, 92, 93
  of reaction, 131
  units of, 68—70
Heat capacity,
  at low temperatures, 247, 249
  constant-pressure, 72
  constant-volume, 72
  Debye's equation, 247
  definition, 70
  Einstein's equation, 246
  electronic, 249
  isothermal variations, 182
  lamda anomaly, 74
  law of Dulong and Petit, 236
  measurement of, 183
  relations among, 117, 182
  rotational of hydrogen, 264
  saturation, 75
  Schottky anomaly, 74
  variation with temperature, 184
Heat engine, 159
Heat reservoir, 159
Helmholtz free energy — see Work content
Hess' law, 133
Heterogeneous equilibrium,
  uniformity of pressure, 285
  uniformity of temperature, 284
Homogeneous functions, 143

Integral enthalpy of dilution, 150
Integral enthalpy of solution, 149
Intensive properties,
  definition, 5
Internal energy — see Energy
International practical temperature scale, 32
Inversion temperature,
  Joule, 121
  Joule—Thomson, 124

Jouguet's criterion, 21

Joule, 70
Joule coefficient,
  definition, 120
  expression for, 184
  relation to Joule—Thomson coefficient, 126
  values of, 121
Joule free-expansion experiment, 63, 119, 300
Joule paddle-wheel experiment, 89—91
Joule's law, 122
Joule—Thomson coefficient,
  definition, 124
  properties of, 124—126, 184
  relation to Joule coefficient, 126
Joule—Thomson experiment, 67, 122, 203, 301, 312
  isothermal, 127, 312

Kelvin temperature scale, 30
Kelvin's principle, 158
Kirchhoff's formula, 139

Latent heat,
  of dilation, 76, 77, 118
  of phase change, 75
  of pressure increase, 94, 118
Legendre transforms, 228

Macroscopic properties, 1
Maxwell relations, 180
Mean molar property, 144
Mechanical equivalent of heat, 102—104
Medium,
  definition, 2
Mixtures, 12
Mole, 44

Open systems, 296

Partial molar enthalpies,
  apparent, 152—153
  calculation of, 152
  effect of pressure and temperature on, 152
Partial molar free energy — see Chemical potential
  apparent, 147, 152
  definition of, 143
  determination of, 147—149
  effect of pressure and temperature on, 145
  for binary solutions, 145—147
  relations among, 144
  relative, 155
Path of a process, 53

Perfect gas, 41, 178, 186
  Boyle's law, 41, 178
  equation of state, 42
  Gay-Lussac's law, 41
  Joule's law, 122, 178
  mixtures, 189
  properties of, 128—130, 186—192
Perfect gas temperature scale, 41
Phase,
  definition, 9
Phase diagrams, 210—212
Polytropic expansion, 65
  in perfect gases, 189
Possible variations, 276
Pressure, 24
  conversion factors for units, 25
Principle of maximum work (Principle of Thomsen and Berthelot), 233
Principle of the unattainability of the absolute zero, 239
Processes,
  adiabatic, 53
  cyclic, 8, 57, 103
  definition, 8
  irreversible, 9, 57
  isothermal, 54
  path of, 53
  quasistatic, 8, 59
  reversible, 9, 54
Pure substances, 12

Quasistatic processes, 8, 59

Relative molar enthalpy, 155
Reversible process, 9, 54

Second law of thermodynamics, 159
  and entropy, 172
  applications, 178
  Carathéodory's statement, 159
  Carnot's statement, 157
  Clausius inequality, 170
  Clausius' statements, 158
  corollaries of, 164—167, 170
  for open systems, 313, 314
  mathematical statement, 160, 161
  Planck's statement, 158
  Thomson's statements, 158
Species,
  composition in terms of, 17—23
  definition, 14
Standard enthalpy of formation, 136
Standard states, 133, 153
  dissolved state, 134

  of pure substances, 133
  of substances in solution, 134
State of thermodynamic equilibrium, 2, 5
State variables, 5
Steady-flow process, 66, 203
  first-law description, 310—312
  second-law equation, 316
Stoichiometric coefficients, 7
Surroundings,
  definition, 1
Systems,
  closed, 2
  definition, 1
  heterogeneous, 10
  homogeneous, 10
  isolated, 2
  open, 2, 296

Temperature, 26
  absolute scale, 29, 30
  definition, 28
  scales, 30
    Celsius, 31
    International practical, 31
    Kelvin, 30, 169
    perfect gas, 41, 178
    thermodynamic, 167
Thermal conductor, 26
Thermal contact, 26
Thermal equilibrium, 26
Thermal insulator, 26
Thermochemistry, 131
Thermodynamic equilibrium, 2—5
  metastable, 279
  neutral, 279
  stable, 278
  unstable, 279
Thermodynamic postulate, 6
Thermodynamic properties, 2
Thermodynamic temperature scales, 167
  Kelvin's scale, 169
Thermometers, 28, 35
  gas, 36
    constant bulb-temperature, 40
    constant-pressure, 37
    constant-volume, 36
    correction of gas thermometers to Kelvin scale, 185
Thermometric coefficients, 95, 96
  thermal expansion, 96
  thermal pressure increase, 96
Thermometric property, 28
Thermometry, 28
Thermostat, 46

Third law of thermodynamics, 240
　apparent deviations, 256
　consequences of, 241—246
　mathematical statement, 235
　Planck's statement, 237
　principle of unattainability of the absolute zero, 239
　Simon's statements, 238

van't Hoff equilibrium box, 302
Virial coefficients, 44, 45
Virial equations of state, 44, 45
Vapor pressure,
　effect of pressure on, 216
　effect of temperature on, 217

Work, 48, 89, 92, 93
　definition, 49
　expansion, 52
　kinds of, 51
　units of, 50
Work content (Helmholtz free energy), 197
　changes attending chemical changes in state, 223
　changes attending finite changes in state, 207—210
　definition, 197
　relation to work of certain processes, 198
　variation with volume and temperature, 204

Zeroth law of thermodynamics, 27

OCT 29

CHEMISTRY LIBRARY
100 Hildebrand
RETURN CHEMISTRY LIBRARY
TO 100